Defense Against Biological Attacks

Sunit K. Singh • Jens H. Kuhn
Editors

Defense Against Biological Attacks

Volume I

Springer

Editors
Sunit K. Singh
Molecular Biology Unit
Institute of Medical Sciences, Banaras
Hindu University
Varanasi, India

Jens H. Kuhn
Integrated Research Facility at Fort Detrick
NIH/NIAID, Division of Clinical Research
Frederick, MD, USA

ISBN 978-3-030-03052-0 ISBN 978-3-030-03053-7 (eBook)
https://doi.org/10.1007/978-3-030-03053-7

Library of Congress Control Number: 2019933210

This Springer imprint is published by the registered company Springer Nature Switzerland AG
The registered company address is: Gewerbestrasse 11, 6330 Cham, Switzerland

Preface 1

Biological attacks comprise a multitude of highly diverse forms of aggression using biological agents such as bacteria, toxins, and viruses. These forms range from lone-actor biocriminal activities (e.g., contamination of salad bars with bacteria, deliberate infections of others with HIV-1, murder using insulin injections) at one end of the spectrum to research and development programs that are well-funded, long-term, and clandestine nation-controlled on the other end. These latter programs may be aimed at developing deployable tactic or strategic weaponry (e.g., US, UK, Soviet biological weapons programs).

"Biodefense" has become the catch-all phrase for measures mitigating the effect of an attack with a biological agent. These measures include public and specialist education, national legislation and pathogen surveillance, international arms-control treaties and confidence-building measures, and intelligence gathering aimed at preventing the construction and deployment of biological weapons and/or their use. In addition, biodefense measures also include all responses to a biological attack, such as rapid biological agent diagnostics, emergency patient management, application of efficacious and safe medical countermeasures, and remediation of attack sites. Biodefense is therefore a highly interdisciplinary nexus for multiple subspecialties of the life sciences, humanities, and political sciences. Biodefense is a topic that has become increasingly complex; we argue that biodefense is in fact too complex for any single individual to comprehend in all of its facets. At the same time, biodefense activities and general public health measures overlap considerably. Hence, we would also argue that most public health experts can contribute significantly to biodefense and vice versa.

This two-volume book attempts to provide an overview of various priorities in biodefense in a format that is aimed to engage both laypersons and specialists. The book deliberately joins experts from various subspecialties with the hope of furthering communication between them and the readership. Volume I begins with an overview of the historical development and use of biological weapons to set the scene for past accomplishments and failures in offense research and development. Since all known nation-sponsored biological weapons programs were ultimately terminated many years ago, would such programs be organized differently and have different successes if they were undertaken today with current scientific methodologies in a different political climate? Subsequent chapters discuss whether novel technologies, such as synthetic biology, big data analysis, and CRISPR/Cas9, could be used for nefarious purposes and whether offensive activities involving these technologies are even covered by nonproliferation treaties. Volume II provides high-level overviews of the biological agents that are most commonly associated with biodefense activities. Additional chapters present the status quo of antibacterial and antiviral therapy and diagnostic development. The book concludes with a chapter that serves as a reminder that biodefense also includes protection from attacks against nonhuman targets, a fact that is often drowned out in discussions about weapons targeting humans directly.

Of course, even a two-volume book cannot cover all aspects of biodefense, and each chapter represents only the at-times subjective assessments of individual authors rather than the consensus views of entire fields (if such views exist). However, we are proud of having recruited such a diverse set of highly renowned authors and hope that the reader shares our enthusiasm for the resulting mix of well-articulated viewpoints that demonstrate that biodefense is indeed a field of great importance.

Varanasi, India Sunit K. Singh
Frederick, MD, USA Jens H. Kuhn

Preface 2

Biodefense is a nexus for multiple subspecialties of the life sciences, humanities, and political sciences. Achieving biosafe and biosecure environments for large populations requires increased interdisciplinary communication and collaboration. This book was written for policy and life science professionals, faculty, students, journalists, and laypersons to provide an overview of the multiple and often complicated facets of biodefense.

We acknowledge the very patient and professional support of Rakesh Kumar Jotheeswaran, Project Coordinator (Books) at Springer Nature, who guided us (editors) and ultimately helped us to bring this book to completion.

The content of this book does not necessarily reflect the views or policies of the US Department of Health and Human Services or of the institutions and companies affiliated with the authors or the editors. This work was supported in part through Battelle Memorial Institute's prime contract with the US National Institute of Allergy and Infectious Diseases (NIAID) under Contract No. HHSN272200700016I (editor J. H.K.).

Varanasi, India — Sunit K. Singh
Frederick, MD, USA — Jens H. Kuhn

Contents

About the Editors

Sunit K. Singh, PhD, is a Professor and Head in the Molecular Biology Unit (MBU) and the Professor Incharge of Centre for Experimental Medicine and Surgery (CEMS) at the Institute of Medical Sciences, Banaras Hindu University (BHU), Varanasi. Prof. Singh completed his PhD degree from the University of Wuerzburg, Wuerzburg, Germany, in the area of molecular infection biology and postdoctoral training from the Department of Internal Medicine, Yale University, School of Medicine, New Haven, Connecticut, USA, and the Department of Neurology, University of California Davis Medical Centre, Sacramento, California (USA). Prof. Singh also worked as a visiting faculty at the Department of Pathology, Albert Einstein College of Medicine, New York (USA); the Department of Arbovirology, Institute of Parasitology, Ceske Budejovice, Czech Republic; the Department of Pathology and Microbiology, Chonbuk National University, South Korea; and the Department of Immunology, University of Geneva, Geneva, Switzerland. Prof. Singh has been a faculty member at CSIR-Centre for Cellular and Molecular Biology (CCMB), Hyderabad, India, and led a research group in the area of neurovirology and inflammation biology. Prof. Singh is member of various international professional organizations such as the National School of Tropical Medicine, Texas, USA, and the European Virus Bioinformatics Center, Jena, Germany. Prof. Sunit K. Singh has done seminal work in the area of neurovirology and inflammation biology. Prof. Singh's work has a very broad impact in terms of understanding the molecular mechanisms of the pathogenesis of infectious diseases in general and neurotropic viruses in particular. In addition to his original research publications in high impact factor peer-reviewed international journals, Prof. Singh has published many books in the area of infectious diseases such as

Neuroviral Infections-RNA Viruses and Retroviruses, *Neuroviral Infections-General Principles and DNA Viruses*, *Viral Hemorrhagic Fevers*, *Human Respiratory Viral Infections* from CRC Press/Taylor & Francis group, USA, Viral Infections and Global Change, *Human Emerging and Re-emerging Infectious Diseases— Vol-I and Vol-II* by Wiley Blackwell Publications, USA, and *Neglected Tropical Diseases-South Asia* by Springer, USA. Prof. Singh has been associated with many reputed peer-reviewed international journals as Associate Editor and Editorial Board Member.

Jens H. Kuhn, MD, PhD, PhD, MS, is a Research Leader at Battelle Memorial Institute, Columbus, OH, USA, tasked as the Virology Lead (Contractor) at NIH/NIAID/DCR's Biosafety Level 4 facility, the Integrated Research Facility at Fort Detrick (IRF-Frederick) in Frederick, MD, USA. Dr. Kuhn specializes in highly virulent viral human and animal pathogens. He is the author of *Filoviruses: A Compendium of 40 Years of Epidemiological, Clinical, and Laboratory Studies* (Vienna: Springer, 2008) and coauthor of *The Soviet Biological Weapons Program—A History* (Cambridge: Harvard University Press, 2012) and has studied and worked in Germany, Italy, Malta, Russia, South Africa, and South Korea. In the USA, he rotated through or worked at Harvard Medical School, Boston, MA; the Arthropod-borne Infectious Disease Laboratory (AIDL) in Fort Collins, CO; the Centers for Disease Control and Prevention (CDC) in Atlanta, GA; and the US Army Medical Research Institute of Infectious Diseases (USAMRIID) in Frederick, MD. Dr. Kuhn was the first Western scientist with permission to work in a former Soviet biological warfare facility, SRCVB "Vektor" in Siberia, Russia, within the US Department of Defense's Cooperative Threat Reduction (CTR) Program. Dr. Kuhn was a contributor to the Center for International and Security Studies at Maryland's Controlling Dangerous Pathogens Project and a member of the Center for Arms Control and Nonproliferation's CBW Scientist Working Group. He is currently a Subcommittee Chair and Member of the Executive Committee of the International Committee on Taxonomy of Viruses (ICTV), Chair of the *ICTV Bunyavirales, Filoviridae*, and *Mononegavirales* Study Groups, and a member of the ICTV *Arenaviridae*, *Bornaviridae*, *Mononegavirales*, *Nairoviridae*, and *Nyamiviridae* Study Groups. He furthermore serves as a Subject Matter Expert for NCBI for all mononegaviruses as a member of the NCBI Genome Annotation Virus Working Group and the database RefSeq. He served or serves on the editorial boards of 10 journals and has been a peer reviewer for more than 60 journals, including *Cell, Cell Host and Microbe*, *Emerging Infectious Diseases*, *JAMA*, *The Lancet. Infectious Diseases*, *Nature*, *Nature Microbiology*, *Nature Protocols*, *PLoS Pathogens*, *Science*, and *Journal of Virology*. Dr. Kuhn was a member of the 2009–2011 US National Academy of Sciences' Committee on "Animal Models for Assessing Countermeasures to Bioterrorism Agents" and was continuously involved with AAAS's and the US State Department's bioengagement efforts in the BMENA Region, Turkey, and the NIS countries. Dr. Kuhn can be found on LinkedIn at www.linkedin.com/pub/jens-h-kuhn/1b/817/72 and on ResearchGate at https://www.researchgate.net/profile/Jens_Kuhn.

Historical Aspects of Biological Weapons Development and Use

1

Gregory D. Koblentz

1.1 Introduction

Biological weapons (BW) are the least well understood member of the so-called weapons of mass destruction that also include nuclear and chemical weapons. Despite the growing awareness of the threat posed by biological weapons, the history of the development and use of biological weapons is mired in myth and cloaked in legend. Biological weapons have never been used openly on the battlefield and their development has always been shrouded by intense secrecy. After biological weapons became the first class of weapons to be completely outlawed by an international treaty, the 1972 Biological Weapons Convention (BWC), the study of biological weapons, never high on the list of priorities of scholars, slipped even lower. As a result, the vast majority of the public, scientists, and policy-makers remain unaware of the history of biological weapons. As Thomas Schelling, the Nobel Prize-winning economist and strategist, has observed, "the tendency in our planning is to confuse the unfamiliar with the improbable. The contingency we have not considered seriously looks strange; what looks strange is thought improbable; what is improbable need not be considered seriously" [1]. By familiarizing readers with the history of biological weapons, this chapter is an effort to encourage the serious consideration of the threat posed by biological weapons and the development of strategies to prevent their proliferation, deter their use, and defend against their effects.

G. D. Koblentz (✉)
Schar School of Policy and Government, George Mason University, Arlington, VA, USA
e-mail: gkoblent@gmu.edu

S. K. Singh, J. H. Kuhn (eds.), *Defense Against Biological Attacks*,
https://doi.org/10.1007/978-3-030-03053-7_1

1.2 Biological Warfare

Biological warfare is the use of microorganisms, toxins derived from living organisms, or bioregulators to deliberately cause the death or illness of humans, plants, or animals.[1] Biological weapons are unique among the instruments of warfare because they are composed of, or derived from, living organisms. This feature of biological weapons has several important implications for their use as weapons by states and non-state actors [2].

Disease-causing microorganisms such as bacteria, viruses, and fungi are called pathogens. Pathogens require a human, plant, or animal host to multiply and cause disease. Because these organisms are self-reproducing, a small dose can initiate an infection. Once a pathogen infects a host, its effects are determined by a complex interaction between the microorganism and the host's immune system. The time between infection and when the signs and symptoms of the disease appears is called the incubation period, and it can last for days or weeks. Pathogens also vary in their virulence, the severity of the disease they cause. Some pathogens kill only a low percentage of those infected and instead cause temporary incapacitation or long-term illness. Other pathogens have high levels of lethality. If the disease is transmissible from person to person, a small number of infections could spark an epidemic.

Since toxins and bioregulators are nonliving molecules that do not replicate in the body, the initial exposure dose is what causes the illness. This means that toxins tend to be faster acting than pathogens, causing effects within hours or at most a day or two. Their effects are still slower than some chemical weapons, such as nerve agents, which can kill victims within minutes. Toxins can be derived from a variety of sources such as plants (ricin from the castor bean), animals (saxitoxin from shellfish), fungi (aflatoxin from *Aspergillus flavus*), or bacteria (botulinum toxin from *Clostridium botulinum*). The number of toxins that are highly lethal and easily obtainable in large quantities, however, is far more limited than the number of pathogens with these attributes.

Bioregulators are a recent addition to the traditional definition of biological weapons. Bioregulators are chemicals normally produced in the human body that control communication between cells and play a crucial role in governing the nervous, endocrine, and immune systems. Small imbalances in the level of bioregulators can have dramatic effects on cognition, emotion, and physiological processes [3].

Biological weapons present both attackers and defenders with some unique challenges and opportunities. Biological weapons, whether pathogens, toxins, or bioregulators, are selective in their targets. Traditional biological warfare agents affect only living things and do not damage or destroy vehicles, buildings, or machinery. Bacteria that have been genetically engineered to produce enzymes that can consume and degrade substances such as fuel, lubricants, and cement

[1]For the purpose of this chapter, viruses are included in the category of microorganisms despite the debate about whether or not viruses are truly alive.

have the potential to be used as anti-materiel weapons. Most biological agents are also fragile creatures that require special measures to maintain their viability and virulence during production, storage, delivery, and dissemination. Bacterial, viral, and toxin agents can be produced in two versions for use as weapons: as a liquid slurry or as a dry powder. Although the slurry from is easier and safer to produce, dry agents are more concentrated, easier to store, and easier to disseminate.

There are several ways to disseminate biological agents. The most primitive tactics use fomites (a physical object that serves to transmit an infectious agent) or vectors (a living organism such as a human or insect that transmits the agent) as crude munitions. The most effective means of disseminating a biological agent to infect a large population is by dispersing the agent as an aerosol or cloud of microscopic droplets with diameters between 1 and 5 μm. Aerosols composed of particles in this size range have several advantages: they can stay airborne longer, require a smaller dose to cause an infection, the severity of the resulting disease is typically more severe, and they are invisible to the human senses. Creating an aerosol containing viable organisms and particles of the correct size is one of the most difficult steps in the development of modern biological weapons. In addition, most biological warfare agents will die if exposed to sunlight or oxygen or extremes of temperature or humidity and aerosol clouds of biological agents are subject to the vagaries of meteorological conditions. Thus, the use of aerosolized biological agents as weapons is fraught with uncertainties for the attacker.

There are two principal categories of munitions designed to disseminate aerosols of biological agents: point sources and line sources. Point source munitions use explosive or gaseous energy to disseminate their payload of biological agent from a stationary position. The most efficient type of point source munition divides the biological agent payload into a large number of bomblets to achieve a wider distribution of the agent. Large numbers of bomblets dispersed from a bomb or missile warhead will saturate the center of the impact area with the biological agent regardless of wind direction and speed. Line sources are created by dispersing a biological agent from a moving vehicle, ship, aircraft, or cruise missile in a line perpendicular to the direction of the prevailing wind. This type of dissemination is the most efficient means for delivering biological agents since it can cover a large area, is well-suited to generating aerosols in the optimal size range, and it is difficult to detect such aerosols at stand-off ranges.

Defending against biological weapons is complicated by the range of available agents, the agent-specific nature of most defenses, and the time lag required to develop new vaccines and treatments. Biological weapons, however, are in some ways more susceptible to countermeasures than high explosives, chemical weapons, or nuclear weapons. They are unique among weapon systems in that vaccines can protect war-fighters and civilians before an actual attack. The incubation period following infection with a pathogen provides a window of opportunity for a well-prepared defender to detect an attack and launch a public health and medical intervention to mitigate the consequences of an attack. For these reasons, a strong public health system is the best defense against a biological attack.

Another key aspect of biological weapons is the multiuse dilemma: the skills, materials, and technology needed to produce biological weapons are also necessary to develop defenses against them and to conduct civilian activities such as biomedical research and pharmaceutical production. Many of the raw materials and equipment required for the research, development, production, and weaponization of biological weapons are also used in civilian industries or in biodefense programs. The multiuse property of biotechnology allows a nation developing biological weapons to hide its activities in civilian institutes that appear to be, or actually are, conducting legitimate pharmaceutical or medical research. This exacerbates the difficulty in determining the true purpose behind suspicious activities or facilities through intelligence and verification. At the same time, the multiuse nature of biotechnology also increases the likelihood that civilian facilities or peaceful activities may be misinterpreted as being part of a military program [2].

1.3 Early History of Biological Warfare

Biological warfare has been practiced since ancient times, though the number of confirmed attacks is quite small. Until the twentieth century, disease was particularly common and deadly during war. Throughout this pre-modern era, naturally occurring diseases commonly killed more soldiers than the enemy, and thus they had a significant impact on military conflicts. The natural impact of disease on military campaigns probably motivated early attempts to harness disease as a weapon. Although military forces did not know what caused disease until the development of the germ theory in the late 1800s, they could easily identify sources of disease such as dead bodies and they had a basic understanding of the contagion effect.

Due to the lack of documentation and the poor understanding of the causes of disease outbreaks in ancient times, there are few clear cases where biological weapons were used deliberately in war. More common have been allegations that an outbreak was deliberately caused by an enemy. Given the difficulty in determining the origins of disease outbreaks, especially before germ theory was understood, some of those making such allegations may have been sincere in their belief that an outbreak they were suffering from was artificially created by an adversary. In many cases, however, allegations of biological warfare were motivated by a desire to deflect blame for a natural event onto an adversary for propaganda purposes. The earliest known example of such allegations were the scapegoating ritual practiced by the Hittites in which a goat was sacrificed to appease the angry deity thought to be the source of the scourge with a request that the disease be afflicted on the group's enemies instead [4]. In other cases, the judgment that an outbreak was caused deliberately was the result of a retrospective analysis, usually based on circumstantial evidence.

The earliest uses of biological weapons relied on the use of fomites or vectors as crude munitions. Primitive tribes used toxins derived from plants and animals to poison their arrows, the Mongols catapulted corpses of individuals who had died from plague into the besieged city of Kaffa in 1346, and British soldiers at Fort Pitt gave blankets contaminated with variola virus (the virus that causes smallpox) to hostile Native American tribes in 1763 [4, 5]. Given the prevalence of these diseases at the time and the lack of authoritative records, however, it is difficult to determine if any of these attacks was successful.

1.4 Modern History of Biological Warfare

The modern era of biological warfare was ushered in by the development of germ theory, which identified microorganisms as the causative agent of disease, and the emergence of the discipline of microbiology which allowed scientists to grow bacteria and viruses artificially and study them. During the 1940s, advances in the large-scale production of microorganisms and the development of the field of aerobiology, the scientific study of the dispersion and effects of airborne biological materials such as microorganisms, greatly magnified the potential destructive power of biological weapons.

Defenses against disease and biological weapons also improved markedly during the modern era. Once germ theory was widely accepted, it became possible to block disease transmission by improving sanitation and hygiene. Applied microbiology also led to the creation of vaccines to prevent a number of common diseases. The discovery of penicillin provided physicians for the first time with the means to cure a range of bacterial diseases. This section will first discuss the history of BW program from World War I through World War II and then discuss major BW programs during the Cold War and post-Cold War eras.

1.4.1 Biological Warfare Programs from World War I to World War II

During World War I, Germany became the first country to apply the new field of microbiology to warfare. Germany employed an international network of agents to cultivate and deliver animal pathogens, particularly *Bacillus anthracis* (the bacterium that causes anthrax) and *Burkholderia mallei* (the bacterium that causes glanders), against cavalry and draft animals being shipped from neutral countries to the Allies. While German biological sabotage operations were extensive, they were ultimately ineffective [6]. In response to the horrors of chemical warfare during World War I, the use of chemical and biological weapons was banned in 1925 under the Geneva Protocol. The Geneva Protocol did not prohibit the development of chemical and biological weapons and most of the signatories reserved the right to retaliate with these weapons if they were attacked first. As a result, this treaty did not prevent countries from developing biological weapons during the interwar period.

By the beginning of World War II, most of the great powers, including Britain, Canada, France, Japan, the Soviet Union, and the United States, had offensive and defensive biological warfare programs [7]. Hungary, Italy, and Poland also initiated small BW programs during the 1930s, but they do not appear to have advanced beyond the research stage [5]. The Polish program, however, contributed to a widespread sabotage and assassination campaign using chemical and biological agents that was carried out by the Polish resistance against German occupying forces [4]. Surprisingly, Nazi Germany, which developed a robust chemical weapon capability, did not engage in offensive BW activities. For unknown reasons, Adolf Hitler prohibited the development of biological weapons and ordered that only defensive measures were permissible [8].

France and the Soviet Union were the first countries to launch BW programs during peacetime. France began exploring the potential for using pathogens as a method of warfare in 1921 and for almost 20 years conducted laboratory research and field trials. While the French were innovators, in that they were the first to recognize the potential advantages of aerosol dissemination of biological agents, the lack of political interest and funding kept the program limited. The program was accelerated in the late 1930s in response to the threat from Germany, but the program did not produce any biological weapons before the German invasion and occupation in 1940 [9].

The Soviet Union launched a biological weapons program in the 1925. Although the Soviet program initially engaged in vigorous research and testing activities, it was undermined by periodic purges of microbiologists during Stalin's Great Terror. As a result, the Soviet Union was not prepared to engage in biological warfare following Germany's invasion in 1941. Claims that the Soviet Union used *Francisella tularensis* (the bacterium that causes tularemia) against German forces during the Battle of Stalingrad are not supported by the available evidence [10, 11].

Japan's aggressive BW program, launched in 1931 under the leadership of the military scientist Ishii Shiro, was the largest of its kind during this era. Japanese scientists in Unit 731 and related organizations conducted gruesome experiments on thousands of prisoners and used biological weapons against Chinese civilians and soldiers on multiple occasions. Despite the scope of their research and the amount of resources invested in it, the Japanese could not overcome important scientific and technical hurdles. They were able to produce hundreds of kilograms of BW agents, but their production methods were crude and inefficient. Japan also failed to develop an effective munition to disseminate BW agents. Instead, they were forced to rely on vectors such as fleas infected with *Yersinia pestis* (the bacterium that causes plague) and the contamination of food and water supplies with *Vibrio cholerae* (the bacterium that causes cholera) and *Salmonella typhi* (the bacterium that causes typhoid fever) to spread disease. Japan first used biological weapons on a limited scale against Soviet forces in 1939. Between 1939 and 1942, the Japanese also conducted a number of biological attacks against Chinese civilians and soldiers with *B. anthracis*, *B. mallei*, *V. cholerae*, *S. typhi*, and *Y. pestis*. Although the Japanese succeeded in causing widespread epidemics, their techniques proved unreliable, caused Japanese casualties as well, and did not provide Japan with a significant advantage over the Chinese opposition forces. These attacks are the only confirmed large-scale use of biological weapons in the twentieth century. The Japanese BW program ended in 1945 with the country's surrender and occupation by the United States. Ishii Shiro and his colleagues avoided war crime trials by providing information on their BW research to the United States [12].

During World War II, the United States, Britain, and Canada collaborated on the development of biological weapons. The only operational capability developed by the Allies were cattle cakes contaminated with *B. anthracis* that the British intended to use against German cattle herds in the event that Germany initiated biological warfare [13]. The most advanced BW project was the development of the Mark 1 bomblet to disseminate a liquid slurry of *B. anthracis* in an aerosol cloud. The munition, based on a British design and field tested in Canada, was to be mass produced by the United States. The United States was also responsible for producing *B. anthracis* on an industrial scale for filling into the bomblet. The war ended, however, before the industrial facilities for the production of the bomblet and *B. anthracis* were completed [14].

1.4.2 Biological Warfare Programs During the Cold War and Post-Cold War Period

During the Cold War, the United States and Soviet Union developed sophisticated biological weapons programs. In addition to the superpowers, biological weapons were also developed and used by Rhodesia and South Africa, and were deployed by Iraq on the eve of the 1991 Persian Gulf War. Most recently, potential Syrian and North Korean BW programs have become major international concerns.

The U.S. BW program was downgraded at the end of World War II but was revitalized in 1950 with the outbreak of the Korean War. The United States conducted research and development at Camp (later Fort) Detrick in Maryland, produced biological agents at Pine Bluff Arsenal in Arkansas, and tested biological weapons at Dugway Proving Ground in Utah and at overseas locations. During the course of its program, the United States selected *B. anthracis*, *F. tularensis*, yellow fever virus, botulinum toxin, and saxitoxin for use as lethal antipersonnel agents; *Brucella suis* (a bacterium that causes brucellosis), Venezuelan equine encephalitis virus (VEEV), and staphylococcal enterotoxin type B for use as incapacitating biological weapons; and *Puccinia graminis* (the fungus that causes stem rust of wheat) and *Magnaporthe oryzae* (the fungus that causes rice blast disease) for use as anti-crop agents. The United States abandoned research on anti-animal agents in 1954. The United States stockpiled a small number of spray tanks to disseminate these agents as a strategic weapon and a large quantity of specialized tactical munitions for use by special operations forces operating behind enemy lines. In 1969, after a comprehensive review of U.S. chemical and biological weapons policy, President Richard M. Nixon decided to unilaterally renounce the use of biological weapons, terminate the offensive BW program, and destroy the existing stockpiles of biological agents and munitions. Since that time, the United States military has conducted only defensive research and development [15].

The United Kingdom gave its BW program a high priority immediately after World War II, but once Britain acquired a nuclear deterrent, its offensive research and testing was gradually phased out until the program had a strictly defensive orientation [16]. Canada continued to participate in the tripartite cooperation on biological weapons that began during World War II, but eschewed the development of its own BW capability in favor of relying on those developed by its partners. Canadian BW research and testing activities, conducted in cooperation with and the support of its American and British partners, were halted in 1969 following the end of the U.S. BW program [17]. The French BW program followed a similar trajectory to that of the British program although the French offensive program had a brief resurgence in the early 1960s based on the potential for using toxins and incapacitating biological weapons on the battlefield. From the mid-1960s, however, the French program shifted away from offensive research to a focus on defense. The *de facto* abandonment of an offensive BW program was transformed into a *de jure* renunciation when France signed the BWC in 1972 [18].

During the Cold War, the Soviet Union created the world's largest biological warfare program. The Soviet military ran BW facilities at Zagorsk, Sverdlovsk, and Kirov as well as a testing facility on Vozrozhdenie Island in the Aral Sea. These military sites maintained stockpiles of BW agents such as *B. anthracis*, *Y. pestis*, and variola virus. The facility at Sverdlovsk was the site of an accidental release of *B. anthracis* in 1979 that caused an anthrax outbreak in the city which killed at least 68 people. In 1974, the Soviet Union created a new quasi-civilian organization, Biopreparat, whose purpose was to apply advances in biotechnology to create new and improved biological weapons. The Scientific-Research Institute of Applied Microbiology in Obolensk and the Scientific-Research Institute of Molecular Biology (also called Vector) in Koltsovo were tasked with engineering pathogens to be more lethal. The Institute of Ultra-Pure Biological Preparations in Leningrad was dedicated to developing improved methods for stabilizing, drying, milling, and disseminating BW agents. Biopreparat also maintained a network of six stand-by production plants that could be mobilized during wartime to produce thousands of tons of BW agents a year. By the time the Soviet Union collapsed, Biopreparat had become a massive complex of 50 research and production facilities with over 30,000 employees. Over the course of its program, the Soviet Union selected the following agents for use as antipersonnel biological weapons: *B. anthracis*, *F. tularensis*, *Y. pestis*, brucellae (bacteria that cause brucellosis), *B. mallei*, *Burkholderia pseudomallei* (the bacterium that causes melioidosis), *Coxiella burnetii* (the bacterium that causes Q fever), *Rickettsia prowazekii* (the bacterium that causes typhus), variola virus, VEEV, Marburg virus, and botulinum toxin. The Soviet military developed bomblets and spray tanks to deliver these agents on the battlefield and against cities. The Soviet BW program achieved some progress in developing genetically engineered weapons in the laboratory, but the Soviet Union collapsed before Biopreparat was able to demonstrate these new weapons in field trials. The Soviet Ministry of Agriculture also conducted research on anti-animal and anti-crop pathogens for use as weapons. Although President Boris Yeltsin officially terminated the offensive BW program that Russia inherited from the Soviet Union, Russia maintains three military microbiological sites that were once part of the offensive program and these facilities remain closed to outsiders [11]. The United States remains concerned that Russia is not in full compliance with the BWC [19].

In the 1970s, Rhodesia used crude chemical and biological agents as part of a brutal counterinsurgency campaign against rebel groups seeking to overthrow the white minority government. While the primary tactic was to contaminate clothing, food, and drinks used by guerilla groups and their supporters with various poisons, Rhodesians also used *V. cholerae* to contaminate water supplies that the guerillas relied on. A massive anthrax outbreak that struck humans and cattle in rural Rhodesia in 1978–1980 has been attributed by some to the Rhodesian security services, but the most recent research indicates that this was a natural outbreak exacerbated by the breakdown of veterinary and public health services during the country's civil war [20]. During the 1980s, South Africa's apartheid-era chemical and biological weapons program, Project Coast, adopted similar techniques. The program supplied members of South African security services with small quantities of poisons, toxins, and pathogens such as *B. anthracis*, *V. cholerae*, and *Salmonella typhimurium* (a bacterium that causes salmonellosis), to contaminate food and beverages and to assassinate anti-apartheid activists [21].

Iraq launched a biological warfare program in 1985 at the height of the Iran-Iraq War and made significant progress over the next 6 years. By the 1991 Persian Gulf War, Iraq had mass-produced *B. anthracis*, botulinum toxin, and aflatoxin and filled those agents into Scud missile warheads and aerial bombs. Iraq secretly destroyed its biological agents and weapons during the summer of 1991 out of fear that they would be discovered by the United Nations Special Commission (UNSCOM), which was charged with disarming Iraq of its nuclear, biological, and chemical weapons. By 1995, UNSCOM's detective work and the defection of a high-ranking Iraqi official forced Iraq to reveal its past production, testing, and weaponization activities. Neither UNSCOM nor its successor, the United Nations Monitoring and Verification Commission (UNMOVIC), however, were able to fully verify Iraq's account of its past BW work. In 2003, the United States accused Iraq of possessing mobile biological agent production vehicles and a stockpile of biological weapons. Following the U.S. invasion of Iraq, the intelligence supporting these assessments was found to be deeply flawed and the Iraq Survey Group determined that Iraq had indeed abandoned its BW program in the 1990s and had not subsequently resumed its work on those weapons [2].

Western intelligence sources have long believed that Syria has a biological warfare program. In 2014, the Defense Intelligence Agency (DIA) reported to Congress that, "Syria may be capable of limited agent production, however, we do not believe Syria has achieved a capability to use biological agents as effective mass-casualty weapons" [22]. At the same time, senior Syrian officials have alluded to the possession of biological weapons, or at least not denied it [23]. Little evidence has been presented publicly, however, to support the existence or status of a Syrian BW program [24]. In 2014, Syria declared a ricin production facility as part of its chemical weapons program to the Organization for the Prohibition of Chemical Weapons (OPCW), the international organization that implements the Chemical Weapons Convention (CWC). Syria claims to have destroyed all of the ricin it had produced prior to its joining the CWC in September 2013, but the OPCW has been unable to verify its account of the production and disposal of the toxin [25, 26]. As a toxin, ricin is considered both a chemical weapon and a biological weapon. It is unclear if the ricin program was the sole basis for Western intelligence assessments of Syrian interest in biological weapons or if there is classified information on Syrian development of bacterial or viral agents for use as weapons.

North Korea is another state that has long been suspected of harboring a BW program, but details supporting this assessment are sparse. According to the U.S. State Department, "North Korea has a longstanding BW capability and biotechnology infrastructure that could support a BW program" [27]. In 2012, the South Korean Ministry of Defense reported that North Korea's BW program included *B. anthracis*, variola virus, *Y. pestis*, *F. tularensis*, and an unspecified hemorrhagic fever virus [28]. Much of this intelligence appears to be based on reports from defectors who can be unreliable [29]. The Pyongyang Bio-technical Institute, which produces *Bacillus thuringiensis* for use as a biopesticide, has been identified as a potential *B. anthracis* production facility based on an analysis of pictures of the equipment installed at the institute [30]. The existence of dual-use equipment at this institute, however, is not sufficient evidence that the facility has the capability to safely produce *B. anthracis* or is even part of a BW program [31].

1.5 Explaining the Non-Use of Biological Weapons

Despite the major advances in biological warfare since the 1940s, there is no evidence that modern biological weapons based on aerosol dissemination technology have been successfully employed. On the rare occasions when states have resorted to the use of biological weapons, as in the cases of Rhodesia and South Africa, they have used less sophisticated means of dissemination. Accounting for the rare use of these weapons in modern times is important to determine the likelihood that the conditions that led to this restraint will continue to remain strong. There are three likely reasons why biological weapons have been rarely used in modern times.

First, there is a strong normative barrier to the use of these weapons. The use of pathogens as weapons has been considered taboo by multiple civilizations and religions since antiquity. This sense of revulsion at using poison or disease as an instrument of war has been codified in national legal prohibitions for centuries and in international law in the twentieth century. The 1925 Geneva Protocol prohibits the use of biological weapons and the 1972 Biological Weapons Convention prohibits the development, production, or acquisition of these weapons. The BWC, the first treaty to outlaw an entire class of weapons, reinforced the long-standing norm against these weapons by declaring the use of biological weapons as being "repugnant to the conscience of mankind."

Second, military organizations have had practical reasons for not assimilating biological weapons into their arsenals and war plans. Safely storing and handling these weapons presents logistical difficulties, and employing them in combat can pose significant operational problems. Early generations of biological weapons had limited effectiveness and uncertain results due to their reliance on vectors, such as insects or rats, or contamination of food or water to infect the enemy. Modern biological weapons based on aerosol dissemination are susceptible to vagaries in atmospheric conditions. This creates uncertainty for military planners and poses the risk of accidentally infecting friendly troops due to a change in wind direction.

A third reason for the rare use of biological weapons is political and strategic. The use of these weapons may have been limited by the fear of retaliation or escalating a conflict. Japan, Rhodesia, and South Africa only engaged in biological warfare against opponents that were unable to retaliate in kind. The domestic or international reaction to the use of these illegitimate weapons, rather than the prohibition itself, may have also had a restraining influence.

Unfortunately, there is cause for concern that all three of these constraints on the use of biological weapons may be eroding. The BWC remains the weakest of the major WMD nonproliferation treaties in terms of institutional capacity and resources. Syria's continued use of chemical weapons and the inability of the international community to hold the Syrian regime accountable for its blatant violations of the CWC demonstrates the limits of normative constraints. The second constraint, posed by operational and logistical problems, may be less daunting now due to technological advances that may make these weapons easier and safer to produce and store and more effective when employed. Finally, the dominance of the United States and its allies in conventional military technology may provide dissatisfied actors with a strong incentive to employ biological weapons as part of an asymmetric strategy that outweighs the political and strategic hazards of using these weapons. Revisionist states may calculate that they can use their biological weapons as force multipliers to bring about a fait accompli, tailor their use of these weapons to avoid provoking regime-threatening retaliation, or conduct anonymous attacks and avoid retaliation entirely. As with nuclear weapons, the lack of large-scale use of biological weapons since World War II is a cause for celebration, but it is not grounds for complacency.

1.6 Biological Terrorism

The prospect of terrorists acquiring and using biological weapons emerged as a security threat during the mid-1990s. The terrorist attacks on September 11, 2001 and the Amerithrax letter attacks that fall elevated biological terrorism to the top of the international security agenda [32]. This section reviews the four most important cases of bioterrorism and provides an assessment of the current threat of biological terrorism.

Biological terrorism has been exceedingly rare. During the twentieth century, only eight terrorist groups succeeded in acquiring a biological agent [33]. Prior to the anthrax letter attacks in 2001, only one group, the disciples of guru Bhagwan Shree Rajneesh in Oregon, managed to cause any casualties with a biological agent. Although terrorists have fewer resources to develop biological weapons compared to states, their needs are more limited. Unlike states, terrorists can achieve their objectives without developing BW agents that can be produced in large quantities, stored for lengthy periods of time, disseminated by highly efficient and reliable devices, and delivered by systems designed for use under battlefield conditions. Nevertheless, terrorists who seek to inflict mass casualties still face significant hurdles in acquiring and producing virulent agents and designing effective dissemination devices. Terrorists whose interest is limited to causing a small number of casualties or mass disruption face fewer obstacles. As the proliferation of anthrax hoax letters since 2001 has demonstrated, individuals interested in causing terror and disruption at a local level can achieve their objectives with nothing more than a powdery substance and a threatening note.

1.6.1 Rajneeshees

In 1984, members of the Rajneeshee cult poisoned salad bars in The Dalles, Oregon with *Salmonella typhimurium*, sickening 751 townspeople. From the establishment of their ranch in The Dalles in 1981, the Rajneeshee found themselves in a series of disputes with state and local authorities. As part of a strategy to influence a local election, the cult contaminated ten salad bars in the town with *S. typhimurium* that they had produced in their medical clinic. Public health officials believed that the cause of the mass food poisoning was unsanitary practices by food handlers at the restaurants. The identities of the perpetrators weren't revealed until the mastermind of the attack had a falling out with other members of the cult and the group's leader publicly accused her of poisoning local officials and the townspeople [34].

1.6.2 Aum Shinrikyo

The Japanese cult Aum Shinrikyo is the only group that is known to have tried to create an aerosolized biological weapon to cause mass casualties. Aum's efforts, however, were unsuccessful due to scientific, technical, operational, and organizational deficiencies. Aum, led by its guru Shoko Asahara, was characterized by an apocalyptic ideology that justified the murder of nonbelievers. Although Aum was motivated by a mix of religious beliefs, it also had an extremely ambitious political objective: the overthrow of the Japanese government. Despite its significant financial resources, the scientific backgrounds of many of its members, and its ability to operate unmolested by Japanese authorities, none of Aum's ten attempted BW attacks conducted between 1990 and 1995 resulted in any casualties. Aum's inability to develop an effective aerosolized biological weapon led the cult to turn to chemical weapons. The group released the nerve gas sarin in Matsumoto in June 1994 and on the Tokyo subway system in March 1995 killing a total of 19 and injuring over 1000.

Aum's experience sheds some light on the difficulties that terrorists face in developing biological weapons. Although Aum was well funded and well equipped, their BW effort suffered several handicaps. At the scientific level, Aum's program was not run by a microbiologist who knew how to work with bacteria. As a result, Aum was unable to cultivate a lethal strain of botulinum toxin from the wild, and the only strain of *B. anthracis* it could acquire was a vaccine strain used for animals. At the technical level, the slurry of *B. anthracis* that was produced was very low quality. Aum also lacked the engineering capability to disseminate a liquid slurry of *B. anthracis*. Their rooftop sprayer was prone to breaking down, leaking, clogging, and was either incapable or highly inefficient at producing particles in the 1–5 micron size. At the operational level, Aum did not demonstrate an understanding of the proper environmental conditions conducive to a BW attack. Aum attempted to disseminate its biological agents during the day, which exposed the agents to UV radiation and thermal updrafts, reducing the viability of the agents and the area covered by the aerosol [35].

Aum's failure indicates that biological terrorism capable of causing mass casualties through an aerosolized agent is not as easy as commonly portrayed. Developing biological weapons requires the right strain of a pathogen, the ability to produce the organism in a form suitable for dissemination, and a means of effectively disseminating the agent at the desired location. Aum failed at all of these tasks. Aum's experience demonstrates that money, equipment, and educated personnel alone are not sufficient to produce biological weapons; scientific and organizational skills are needed as well.

1.6.3 Bruce Ivins and the Anthrax Letters

In September and October 2001, envelopes containing a dry powder of *B. anthracis* spores were mailed to Senators Thomas Daschle (D-SD) and Patrick Leahy (D-VT) and five media outlets in Florida and New York City. The letters caused 22 cases of anthrax, including 11 cases of cutaneous anthrax and 11 cases of inhalation anthrax. Five of the inhalation anthrax cases were fatal. The anthrax letter attacks also had pervasive ripple effects, forcing thousands of people to take antibiotics as a precaution, disrupting the U.S. Postal Service, temporarily shutting down the U.S. Senate, causing nationwide anxiety about the safety of the mail, and triggering a flood of false alarms and hoaxes involving white powders. All told, the cost of the incident was estimated at $6 billion. The anthrax letter attacks were dubbed Amerithrax by the FBI.

In August 2008, the FBI announced that its sole suspect in the Amerithrax case was Bruce E. Ivins, a microbiologist and anthrax vaccine researcher with the United States Army Medical Research Institute for Infectious Diseases (USAMRIID) at Fort Detrick, Maryland. Only days earlier, Ivins, who had been under investigation for over a year and knew he was about to be indicted for the anthrax letter attacks, killed himself with an intentional drug overdose. The strongest evidence presented by the FBI links the *B. anthracis* used in the attacks to a flask of *B. anthracis* in Ivins's lab at USAMRIID. The FBI established this link using the new discipline of microbial forensics which employs powerful analytical techniques to determine the genetic, chemical, and physical properties of a pathogen or toxin agent used as a weapon. Ivins also spent extended periods of unsupervised time in his lab on nights and weekends immediately before the anthrax letters were mailed to the senators and media outlets [36].

Ivins's motive for sending the anthrax letters was probably his frustration with the slow pace of anthrax vaccine development. By the fall of 2001, this work had virtually ground to a halt due to technical, bureaucratic, political, and financial problems. According to the FBI, Ivins was also suffering at this time from serious mental health problems. In the aftermath of September 11, Ivins may have feared that the next attack could be with biological weapons that could cause even more harm than 9/11 had. Ivins may have intended the anthrax letters as a warning to the nation about the dangers posed by biological weapons and the need for stronger defenses against these weapons.

Notwithstanding the important unanswered questions regarding whether, how, and why Ivins conducted the anthrax letter attacks, it is possible to make several observations about the implications of this case for assessing the threat posed by bioterrorism. Ivins possessed a level of experience, set of skills, and extensive tacit knowledge that could only be found in an individual affiliated with a state-run biodefense program. Ivins was a PhD microbiologist with over 20 years of experience working with *B. anthracis*, was considered an expert in the growth, sporulation, and purification of the bacteria, and had extensive experience preparing liquid anthrax spore preparations for animal aerosol challenges. Ivins's employment at USAMRIID also afforded him advantages such as access to a highly virulent strain of *B. anthracis*, a well-equipped biocontainment laboratory, experience working in such a lab, immunization against anthrax, and knowledge of decontamination procedures. These are resources that a terrorist group would find difficult to acquire on its own.

The high concentration and very good aerosolization properties of the *B. anthracis* spores sent to Senators Daschle and Leahy led many to assume that the powder was produced using sophisticated equipment and/or the use of special additives or coatings. The FBI's contention that Ivins was able to produce such high-quality powder of *B. anthracis* spores with standard laboratory equipment and without the use of any special additives has raised concern that the technical threshold for sophisticated biological weapons is lower than commonly assumed. This inference, however, ignores the high level of tacit knowledge that Ivins possessed about *B. anthracis*. Even if Ivins employed a low-tech method to produce the powder in the anthrax letters, it does not mean that this method did not require a high level of skill to apply successfully [2].

1.6.4 Jihadist Groups

Jihadist groups, most notably Al Qaeda and the Islamic State, have expressed an interest in developing biological weapons. In December 1998, Osama bin Laden declared that it was a "religious duty" to acquire nuclear, biological, and chemical weapons. In May 2003, a Saudi cleric issued a fatwa legitimating the use of nuclear, biological, and chemical weapons against infidels.

Al Qaeda's chemical and biological weapon program, code-named Project al-Zabadi (Arabic for yogurt), was created in 1999 and headed by Ayman al-Zawahiri, the second-ranking official in al Qaeda. Zawahiri was attracted to biological weapons because he believed that these weapons were as lethal as nuclear weapons, that they could be produced simply, that the delayed effects of a biological attack would increase the number of casualties, and that defending against these weapons was very difficult. By 2001, the group had established two laboratories in Afghanistan, obtained scientific literature on several bacterial pathogens, procured dual-use production equipment, recruited microbiologists, and had a small cell dedicated to producing *B. anthracis*. Al Qaeda's progress was stymied by its inability to obtain a virulent strain of *B. anthracis* or to master the techniques necessary to aerosolize a biological agent. According to captured al Qaeda operatives, the group's BW efforts were in the early "conceptual stage" when it was disrupted by the U.S. invasion of Afghanistan. Al Qaeda's BW ambitions were set back further by the death or arrest of most of the key participants in the program [2]. Although al Qaeda's aspirations in this area have outstripped its capabilities, the fact that this group has been as interested in these weapons for as long as it has sets them apart from other terrorist organizations.

The Islamic State exhibits many of the risk factors that Jonathan Tucker has identified in other groups that acquired and used chemical and biological weapons, including a charismatic leadership with grandiose goals driven by an apocalyptic ideology that has commitment to extreme levels of violence and a demonstrated ability to innovate and experiment [37]. So far, however, the Islamic State has only been able to develop and use crude chemical weapons based on chlorine and mustard agents [38, 39]. The primary piece of evidence cited to demonstrate the Islamic State's interest in BW is the so-called "laptop of doom" recovered from a member of the Islamic State in 2014 which contained a document on how to develop biological weapons, including *Y. pestis* [40, 41]. The BW "manual" was a copy of a jihadi document that has been available online for years and includes only basic information on how to grow bacteria without providing any instructions on how to turn the bacteria into a viable biological weapon [38, 42]. This manual appears to be similar to other jihadi BW "manuals" which tend to be rudimentary, lack important details or include incorrect information, are unsuited for producing pathogens or toxins of sufficient quantity or quality to cause mass casualties, and do not describe techniques for disseminating biological agents [42]. Individuals or small groups inspired by or directed by the Islamic State may be capable of producing a small amount of a crude bacterial agent or toxin suitable for poisoning food or beverages, but, without extensive expert assistance, will likely remain unable to aerosolize a virulent biological agent capable of causing mass casualties.

1.7 Conclusion

The development of biological weapons by states and terrorists has lagged far behind the number of countries and groups capable of producing such weapons. The magnitude of the future threat posed by biological weapons will be determined by two poorly understood and difficult to influence variables. The first variable is the net impact of the biotechnology revolution on the balance between offense and defense in biological warfare. While advances in the life sciences strengthen the defender and provide new capabilities to verify biological arms control agreements, they also reduce the barriers for attackers to develop more sophisticated weapons and the means of concealing them. This type of assessment is complicated by the accelerating pace of innovation in the life sciences, the inevitable global diffusion of these technologies, and the multiuse nature of biotechnology. In addition, social factors such as tacit knowledge and communities of practice can mediate the capability of a state or terrorist group to transform scientific or technical breakthroughs into weapons [43, 44]. An unfortunate by-product of the multiuse dilemma is that investing in biodefense research also generates new knowledge that could be applied to the development of biological weapons.

The second key variable is the level of interest of states and non-state actors in developing biological weapons. Since the capabilities to develop these weapons are already widespread, the main factor driving the pursuit of these weapons will be intent. Will the norm against BW continue to limit the appeal of these weapons or will security concerns or the bureaucratic ambitions of scientific and military leaders overwhelm this inhibition? What is the likelihood that non-state actors will emerge that combine technical acumen, a desire to cause mass casualties, and an interest in biological weapons? Thus far, no terrorist group has combined both the capability and motivation to use biological weapons to cause mass death. Terrorists still overwhelmingly prefer to use guns and bombs to wreak havoc. The events of September 11, 2001 and the surprising rise of the Islamic State are vivid reminders that past experience is not always a reliable predictor of future threats. Given the difficulty in tracking terrorist groups and detecting activities to develop biological weapons, it is possible that such a group will arise with little or no warning. Preventing the emergence of such groups and the misuse of the biotechnology revolution will be major security challenges for the twenty-first century.

References

1. Schelling T. Foreword. In: Wohlstetter R, editor. Pearl Harbor: warning and decision. Palo Alto: Stanford University Press; 1962.
2. Koblentz GD. Living weapons: biological warfare and international security. Ithaca: Cornell University Press; 2009.
3. Tucker JB. The body's own bioweapons. Bull At Sci. 2008;64:16–22–56–57.
4. Carus WS. The history of biological weapons use: what we know and what we don't. Health Secur. 2015;13(4):1–37.

5. Carus WS. A short history of biological warfare: from pre-history to the 21st century. Washington, DC: National Defense University; 2017.
6. Wheelis ML. Biological sabotage in World War I. In: Geissler E, Moon J, editors. Biological and toxin weapons: research, development and use from the middle ages to 1945. Oxford: Oxford University Press; 1999. p. 35–62.
7. Geissler E, Moon J, editors. Biological and toxin weapons: research, development and use from the middle ages to 1945. Oxford: Oxford University Press; 1999.
8. Geissler E. Biological warfare activities in Germany, 1923-1945. In: Geissler E, Moon J, editors. (1999) Biological and toxin weapons: research, development and use from the Middle Ages to, vol. 1945. Oxford: Oxford University Press; 1999. p. 91–126.
9. Lepick O. French activities related to biological warfare, 1919-45. In: Geissler E, Moon J, editors. Biological and toxin weapons: research, development and use from the middle ages to 1945. Oxford: Oxford University Press; 1999. p. 70–89.
10. Bojtzov V, Geissler E. Military biology in the USSR, 1920–45. In: Geissler E, Moon J, editors. Biological and toxin weapons: research, development and use from the middle ages to 1945. Oxford: Oxford University Press; 1999. p. 153–67.
11. Leitenberg M, Zilinskas RA. The Soviet Biological Weapons Program: a history. Cambridge: Harvard University Press; 2012.
12. Harris SH. Factories of death: Japanese biological warfare, 1932–45, and the American cover-up. 2nd ed. London: Routledge; 2002.
13. Carter GB, Pearson GS. British biological warfare and biological defence, 1925-45. In: Geissler E, Moon J, editors. Biological and toxin weapons: research, development and use from the Middle Ages to 1945. Oxford: Oxford University Press, vol. 1999; 1999. p. 168–89.
14. Moon J. US biological warfare planning and preparedness: the dilemmas of policy. In: Geissler E, Moon J, editors. biological and toxin weapons: Research, development and use from the Middle Ages to 1945. Oxford: Oxford University Press; 1999. p. 215–54.
15. Moon J. The US Biological Weapons Program. In: Wheelis ML, Lajos R, Dando M, editors. Deadly cultures: biological weapons since 1945. Cambridge: Harvard University Press; 2006. p. 9–46.
16. Balmer B. The UK Biological Weapons Program. In: Wheelis ML, Lajos R, Dando M, editors. Deadly cultures: biological weapons since 1945. Cambridge: Harvard University Press; 2006. p. 47–83.
17. Avery D. The Canadian Biolgoical Weapons Program and the Tripartite Alliance. In: Wheelis ML, Lajos R, Dando M, editors. Deadly cultures: biological weapons since 1945. Cambridge: Harvard University Press; 2006. p. 84–107.
18. Lepick O. The French Biological Weapons Program. In: Wheelis ML, Lajos R, Dando M, editors. Deadly cultures: biological weapons since 1945. Cambridge: Harvard University Press; 2006. p. 108–31.
19. Department of State. Adherence to and compliance with arms control, nonproliferation and disarmament agreements and commitments. Washington, DC: Department of State; 2017.
20. Glenn C. Dirty war: Rhodesia and chemical biological warfare 1975–1980. West Midlands: Helion; 2017.
21. Gould C, Folb P. Project Coast: Apartheid's Chemical and Biological Warfare Programme. Geneva: United Nations Publications; 2002.
22. Flynn M. Statement before the senate armed service committee: annual threat assessment. Washington, DC: Defense Intelligence Agency; 2014.
23. Lentzos F. Syria and bioweapons: the need for transparency. Bull At Sci. 2013. http://thebulletin.org/syria-and-bioweapons-need-transparency. Accessed 22 Sept 2017.
24. Ben Ouagrham-Gormley S. Bioweapons alarmism in Syria. Bull At Sci. 2013. http://thebulletin.org/bioweapons-alarmism-syria. Accessed 22 Sept 2017.
25. Deutsch A. Exclusive: Syria reveals more chemical weapons facilities to watchdog – sources. Reuters. 2014. http://www.reuters.com/article/2014/09/17/us-syria-crisis-chemicalweapons-exclusiv-idUSKBN0HC1GA20140917?utm_source=twitter. Accessed 22 Sept 2017.

26. Lynch C. The world may never know if Syria really destroyed all its chemical weapons. Foreign Policy. 2016. http://foreignpolicy.com/2016/08/31/the-world-may-never-know-if-syria-really-destroyed-all-its-chemical-weapons/. Accessed 22 Sept 2017.
27. Department of State. Adherence to and compliance with arms control, nonproliferation and disarmament agreements and commitments. Washington, DC: Department of State; 2015.
28. Ministry of National Defense. 2012 Defense White Paper. Ministry of National Defense, Seoul. 2012. http://www.mnd.go.kr/user/mnd_eng/upload/pblictn/PBLICTNEBOOK_201308130553561260.pdf. Accessed 22 Sept 2017.
29. International Institute for Strategic Studies. North Korean security challenges: a net assessment. London: International Institute for Strategic Studies; 2011.
30. Hanham M. Kim Jong un tours pesticide facility capable of producing biological weapons. 38 North. (2015). http://www.38north.org/2015/07/mhanham070915/. Accessed 22 Sept 2017.
31. Ben Ouagrham-Gormley S. Potemkin or real? North Korea's Biological Weapons Program. Bull At Sci. 2017. http://thebulletin.org/potemkin-or-real-north-korea%E2%80%99s-biological-weapons-program10957. Accessed 22 Sept 2017.
32. Koblentz GD. Predicting peril or the peril of prediction? Assessing the risk of CBRN terrorism. Terrorism Polit Viol. 2011;23:501–20.
33. Carus WS. Bioterrorism and biocrimes: the illicit use of biological agents in the 20th century. Washington, DC: National Defense University; 2001.
34. Carus WS. The Rajneeshees (1984). In: Tucker JB, editor. Toxic terror: assessing terrorist use of chemical and biological weapons. Cambridge: MIT Press; 2000. p. 115–37.
35. Rosenau W. Aum Shinrikyo's Biological Weapons Program: why did it fail? Stud Conflict Terrorism. 2001;24:289–301.
36. Koblentz GD, Tucker JB. Tracing an attack: the promise and pitfalls of microbial forensics. Survival. 2010;52:159–86.
37. Tucker JB, editor. Toxic terror: assessing terrorist use of chemical and biological weapons. Cambridge: MIT Press; 2000.
38. Stenersen A. Islamic State dabbles with chemical weapons. Jane's Intell Rev. 2015.
39. Quillen C. The Islamic State's evolving chemical arsenal. Stud Conflict Terror. 2016;39 (11):1019–30.
40. Doornbos H, Moosa J. Found: the Islamic State's terror laptop of doom. Foreign Policy. 2014. http://foreignpolicy.com/2014/08/28/found-the-islamic-states-terror-laptop-of-doom/. Accessed 22 Sept 2017.
41. Doornbos H, Moosa J. Recipes from the Islamic State's laptop of doom. Foreign Policy. 2014a. http://foreignpolicy.com/2014/09/09/recipes-from-the-islamic-states-laptop-of-doom/. Accessed 22 Sept 2017.
42. Stenersen A. Chem-bio cyber-class: assessing jihadist chemical and biological manuals. Jane's Intell Rev. 2007:8–13.
43. Ben Ouagrham-Gormley S. Barriers to bioweapons: the challenges of expertise and organization for weapons development. Ithaca: Cornell University Press; 2014.
44. Vogel KM. Phantom menace or looming danger? A new framework for assessing bioweapons threats. Baltimore: Johns Hopkins University Press; 2013.

Overview of Global Coordination on Bio-Nonproliferation and Its Limits

2

Maria Hecht, Stefan Kloth, and Min-Hi Lee

Due to ongoing globalization, multilayered interactions between individuals, governments and companies and increasing mobility, pathogens are able to spread more rapidly than ever before and are an issue of international importance. In this context, not only geographical but also thematic demarcations are blurred since preventive measures, like nonproliferation of agents that pose a risk as potential biological weapons, require concerted action of responsible actors comprising sectors such as public health, biosafety/security and governmental and non-governmental organizations. Basic knowledge on biological agents is easily available nowadays since research is going further, technology is developing rapidly and respective information is accessible at almost any point in time and place. Therefore, research results coming from biology, microbiology or related field of work might be classified as "dual-use". That means that scientific findings can be used for peaceful purposes but also for hostile objectives. Scientists should be aware of the risks this knowledge bears [1]. An important factor is the fast-evolving technology. Scientists are well educated and are therefore able to utilize the knowledge and technical devices [2].

To cope with this challenge, many approaches focusing on different aspects of bio-nonproliferation and related topics have been initiated. In this chapter, relevant programmes, regulations and binding international agreements are introduced to give an overview of the activities countries, institutions and clusters are dealing with in terms of bio-nonproliferation (Table 2.1).

M. Hecht · S. Kloth (✉)
Strengthening Global Health and Biosecurity, Robert Koch Institute, Berlin, Germany
e-mail: KlothS@rki.de

M.-H. Lee
Strengthening Global Health and Biosecurity, Robert Koch Institute, Berlin, Germany

Federal Information Centre for Biological Threats and Special Pathogens (IBBS), Robert Koch Institute, Berlin, Germany

S. K. Singh, J. H. Kuhn (eds.), *Defense Against Biological Attacks*,
https://doi.org/10.1007/978-3-030-03053-7_2

Table 2.1 Instruments discussed within this chapter according to international law and their main focus[a] (the time of entry into force, respectively, the year of founding in parentheses)

2.1 Binding in terms of international law	2.2 Not binding in terms of international law
2.1.1 Geneva Protocol (1928)[b]	2.2.1 Australia Group (1985)[b]
2.1.2 Biological and Toxin Weapons Convention (1975)[b]	2.2.2 Global Health Security Initiative (2001)[c,d]
2.1.3 Convention on Biological Diversity (1993)[c,d] 2.1.5 WHO Advisory Committee on Variola Virus Research (1999)[c,d] 2.1.6 Security Council Resolution 1540 (2004)[b]	2.2.3 Global partnership against the Spread of Weapons and Materials of Mass Destruction (2002)[b,c] 2.2.4 EU CBRN Centres of Excellence Initiative (2010)[b,c]
2.1.4 Nagoya Protocol (2014)[c,d]	2.2.5 Global Health Security Agenda (2014)[c,d]

[a]Content mainly aligned to [b]bio-nonproliferation, [c]biosafety/biosecurity and [d]health situation

2.1 Binding in Terms of International Law

2.1.1 Geneva Protocol

The Convention on the Prohibition of the Development, Production and Stockpiling of Bacteriological (Biological) and Toxin Weapons and on their Destruction (BTWC) is—along with the Convention on the Prohibition of the Development, Production, Stockpiling and Use of Chemical Weapons and on their Destruction (CWC)—a development of the 1925 signed Geneva Protocol for the Prohibition of the Use in War of Asphyxiating, Poisonous or other Gases, and of Bacteriological Methods of Warfare [3]. The Geneva Protocol, agreed in the aftermath of the First World War, during which chemical weapons were heavily used, is an outcome of the rapid development of new weapon systems. The Protocol prohibits the use of biological or chemical weapons. Nevertheless, it is only referring to the actual use of biological or chemical weapons but does not mention the development, production and stockpiling of these weapons.

2.1.2 Biological and Toxin Weapons Convention

The BTWC [4] was a special regime when it was opened for signature in 1972 and set into force in 1975. It was the first multilateral disarmament treaty which prohibited an entire category of weapons. By now, it has 182 state parties and five signatory states. The BWC is neither ratified nor signed by Chad, Comoros, Djibouti, Eritrea, Israel, Kiribati, Federated States of Micronesia, Namibia, South Sudan and Tuvalu.

Today, there are increasing tendencies that the BTWC is morphing from a regime dealing with disarmament to a regime working on health security aspects [5]. Having hard and controversial discussions about verification in the past and strong interests in the three regional groups, Western Group (WG), Eastern European Group (EG) and Non-Aligned Movement (NAM) [6], the focus of the originally disarmament regime was shifted to what might be seen as lowest common denominator without arguing over contentious topics. One reason might be that there was no extensive use of biological weapons since the treaty was put into force. Nevertheless, there is proof that there were several programmes on biological warfare agents: one was run by the Soviet Union [7], another by South Africa [8] and a third by Iraq [9]. While Iraq was a signatory state to the BTWC at that time, the former Soviet Union and South Africa were already state parties.

After the convention was set into force in 1975, confidence-building measures (CBMs) were introduced in 1986 [10]. State parties are requested to submit, for instance, information regarding their research on biological agent-related topics and all sites under national authority and their related industry. The submission of this information is political binding for all state parties. But since there are no sanctions for those who do not submit their CBMs, the documentation is somewhat sporadic [5]. State parties initiated a discussion about a verification system, to prove some kind of compliance to the BTWC. Negotiations about how this verification system could look like started in 1992, were pushed forward as from 1995 and lasted until 2001. After years of debates, some state parties took the view that the BTWC is not verifiable for different reasons [11], mentioning that technology is developing too fast and research and industrial or economic interests are too much overlapping. Especially industry espionage and violation of patent protection were seen as potential consequences of verification. In 1993, the CWC was signed. In contrast to the BTWC, the CWC has a verification mechanism and procedures on how to use it.

After the negotiations about a verification system failed, the discussions on the BTWC stagnated. In consequence, the focus during the next years was to reassemble the key interests of the BTWC and to keep all state parties together [12]. For that, annual Meetings of States Parties to discuss varying articles of the convention were initiated. The Seventh Review Conference in 2011 was expected to become a new milestone. But the high expectations were—mainly—unfulfilled [13]. Although state parties were able to agree on a final document to summarize the debates during the past 3 weeks, the language was neither new nor path-breaking for future discussions.

Nevertheless, the time after the Seventh Review Conference was used for some minor adjustments and new thematic dialogues, especially regarding a reform of the confidence-building measures and procedures in case of a bioweapon attack. The intersessional process—the time between Review Conferences—was used to have a Meeting of Experts and a Meeting of States Parties each year. The Meeting of Experts was supposed to bring discussions on specific science and technology topics forward, whereas the Meeting of States Parties was seen as place for discussions on future steps. However, especially the different positions with regard to strengthening the BTWC with legally binding instruments or measures concerning compliance are still challenging [14]. Neither the Meeting of Experts nor the Meeting of States

Parties has the mandate to agree on substantial changes. Such changes can only be implemented as a result of the Review Conference.

In December 2016 the Eighth Review Conference took place in Geneva. After 3 weeks of discussion, the results were frustrating: Although a final document on agreed topics and further steps was available [15], no content was added to the document agreed on after the Seventh Review Conference in 2011 [16].

2.1.3 The Convention on Biological Diversity

The development and the entry into force of the Convention on Biological Diversity (CBD) was one of the first approaches to systemically facilitate the conservation of biological diversity within an international legal context not only focusing on specific ecosystems or species but also including socio-economic and ethical aspects. Furthermore, it considers the importance and promotion of equity and fairness between providers and users of respective resources. In May 1992, the CBD was adopted as a global treaty, opened for signature from 1992 to 1993 and entered into force in December 1993. The CBD comprises three main objectives:

1. The conservation of biological diversity
2. The sustainable use of its components
3. The fair and equitable sharing of benefits arising out of the utilization of genetic resources [. . .] (Art. 1, CBD)

Further development and advancement of the CBD is governed at the Conference of the Parties (COP) [17], which is administratively supported by the Secretariat of the CBD. The COP takes place on a regular basis. During the COPs, decisions promoting the implementation of the CBD are taken, and its progress is reviewed. Two protocols have been developed up to 2017: the Cartagena Protocol on Biosafety to the CBD and the Nagoya Protocol on Access to Genetic Resources and the Fair and Equitable Sharing of Benefits Arising from their Utilization to the CBD [18].

The CBD Secretariat is based in Montreal, Canada, and linked to the United Nations Environment Programme, assisting not only the COP but also other bodies of the CBD such as the Subsidiary Body for Scientific, Technical and Technological Advice (SBSTTA).

The SBSTTA covers multidisciplinary competencies provided by experts of participating parties. Main tasks of the SBSTTA are situation assessment concerning biological diversity and advice for the COP, e.g. in the form of scientific and technical recommendations [18]. In 2017, the CBD had 196 contracting parties. These parties are obliged to implement the provisions of the CBD on national level according to their "particular conditions and capabilities [. . .]" (Art. 6, CBD).

2.1.4 The Nagoya Protocol on Access to Genetic Resources and the Fair and Equitable Sharing of Benefits Arising from Their Utilization

The Nagoya Protocol on Access to Genetic Resources and the Fair and Equitable Sharing of Benefits Arising from their Utilization was developed as a supplementary, legally binding agreement on the CBD, "contributing to the conservation of biological diversity and the sustainable use of its components" by "the fair and equitable sharing of the benefits arising from the utilization of genetic resources, including by appropriate access to genetic resources and by appropriate transfer of relevant technologies, taking into account all rights over those resources and to technologies [. . .]" (Article 1, Nagoya Protocol). The protocol has been adopted in October 2010 and entered into force in October 2014. Ninety-seven parties signed the Nagoya Protocol until 2017. The main goal of the Nagoya Protocol is the further development of the legal framework concerning access and benefit sharing (ABS) and suitable ways of its accomplishment in practice and promotion of compliance and commitment of users and providers. Applying the Nagoya Protocol's principles, a Prior Informed Consent (PIC) of the providing country has to be obtained by a potential user, followed by the establishment of mutually agreed terms (MAT) according to conditions of access and use of the genetic resource of interest but also with regard to dealing with subsequent benefits. Nevertheless, the Nagoya Protocol does not include specific lists with genetic resources, either positive or negative, but refers to access to genetic resources within the scope of Article 15 of the CBD and to the benefits arising from their utilization. The implementation into national legislation in compliance with the Nagoya Protocol by its parties is subject to subsequent uncertainties—such as meeting the requirements of all stakeholders since benefits and costs are to be shared between providers and users and of numerous application areas like research, commercial use and education. In addition, there are many technical challenges, e.g. with regard to digital data such as sequence information or synthetic biology [19] and issues concerning the temporal scope of measures to be taken if genetic material had been acquired prior the effective date of the Nagoya Protocol [20].

For facilitation of the implementation of the Nagoya Protocol, the information exchange platform Access and Benefit-sharing Clearing-House (ABSCH) has been established (Art. 14, Nagoya). It represents a component of the clearing-house mechanism according to Article 18.3 (CBD) and refers to other articles of the CBD such as Article 15, addressing both users and providers in terms of sharing related information on implementation of the Nagoya Protocol and on procedures for access and monitoring of genetic resources and their utilization, thus enhancing (legal) certainty and supporting transparency [18, 21].

2.1.5 WHO Advisory Committee for Variola Virus Research

In 1980, following a global campaign of surveillance and vaccination, the WHO declared smallpox eradicated [22]. Shortly afterwards, the World Health Assembly decided to ultimately destroy the remaining stocks of live variola virus (the causative agent of smallpox), with interim use of the virus permitted only for defined WHO-approved research projects. Variola virus stocks were then voluntarily consolidated in the early 1980s to the WHO Collaborating Centre laboratories, in Russia (Russian Scientific Research Centre of Virology and Biotechnology "Vektor" in Koltsovo) and in the United States (US Centers for Disease Control and Prevention in Atlanta), which thereby remain the only two WHO-approved sites for research with live variola virus [23]. In 1999, a decision by WHO was made to increase the amount of research utilizing live variola virus. The WHO Advisory Committee on Variola Virus Research (ACVVR) was subsequently composed of members from all WHO regions and advised by scientific academic experts from areas such as public health, fundamental applied research and regulatory agencies. The committee's responsibility is to oversee [22]:

- The research using live variola virus
- Biosafety and biosecurity inspections of the two repository sites
- Sequencing the viral genome from variola virus isolates
- Distribution of live variola virus DNA to other researchers under specific rules

The members of the ACVVR were asked to consider whether live variola virus was needed for further essential research for public health benefits. The committee decided that live variola virus was needed for further development of antiviral agents against smallpox. In the report of the 18th Meeting (November 2016), the Advisory Committee noted that the advent of synthetic biology allows individuals to create viruses such as variola, as relevant information is available in public domain. Thus, the threat that the virus poses to public health will not be eliminated by simply destroying the virus stocks housed in the two global repositories [24].

2.1.6 UN Security Council Resolution 1540

The Security Council Resolution 1540 is "[A]ffirming that proliferation of nuclear, chemical and biological weapons, and their means of delivery, constitutes a threat to international peace and security" [25]. The resolution was adopted unanimously by the Security Council at its 4956th meeting 2004 in New York [26]. In line with many other nonproliferation instruments, Security Council Resolution 1540 focuses on the prevention of weapons of mass destruction, its trafficking and transport without prohibiting peaceful purposes such as research, energy production or mutual technical assistance and collaboration. All state members to the United Nations are obligated to abide by the Security Council Resolution 1540 because they are bound to the Security Council's agreements. However, there is the ongoing threat of terrorism and of non-state actors laying hands on nuclear, biological or chemical materials to threat peace, security and stability. All states are requested that, by any means, terroristic groups may not get access to any materials and/or sensitive production facilities to minimize the risk of the transport, trading or use of weapons of mass destruction. Furthermore, all states are obliged to undertake binding national legal obligations—not just referring to Security Council Resolution 1540 but also in line with other relevant disarmament treaties. Therefore, this resolution and its related committee watching the implementation are seen as a forum for coordination, consultation and mutual assistance. In 2006 and 2011, the Security Council confirmed the Security Council Resolution 1540 and extended the mandate of the so-called 1540 Committee, a group of experts supporting the full implementation of the resolution. The committee offers (1) help to participating states mainly when it comes to (technical) assistance requests regarding the resolution and (2) liaison and cooperation with International Organizations and (3) is regularly reviewing its implementation. For now, the mandate for the committee is set until 2021.

2.2 Not Binding in Terms of International Law

2.2.1 Australia Group

The Australia Group (AG) is an informal forum of countries which, through the harmonization of export controls, seeks to ensure that exports do not contribute to the development of chemical or biological weapons [27].

In 1984, investigation under the United Nations Secretary-General's Mechanism provided the evidence for the use of chemical weapons in the Iran-Iraq war in violation of the 1925 Geneva Protocol [28]. It was assumed that at least some of the chemicals used had been obtained from places outside Iraq. In response, several countries introduced export controls on certain chemicals potentially relevant for manufacturing chemical weapons. To increase efficiency of control measures, Australia proposed a meeting of the countries with export controls with the aim of harmonizing their national licensing measures and enhancing cooperation. Since then, annual meetings were held.

Since its foundation in 1985 the forum has grown from 15 to 41 countries plus the European Union. All states participating in the Australia Group are parties to the Chemical Weapons Convention (CWC) and the Biological Weapons Convention (BWC) and therefore support all efforts under those conventions. Thus, this forum assists its participants' coordination of national export control measures to fulfil the obligation under CWC and BWC.

The scope of the export controls discussed by the group has evolved to address emerging threats and challenges. Evidence of the diversion of dual-use materials to biological weapon programmes in the early 1990s led to participants' adoption of export controls on specific biological agents [27]. The control lists developed by the group have expanded to include technologies and equipment that can be used in the manufacturing or dispersal of chemical and biological weapons.

2.2.2 Global Health Security Initiative

The Global Health Security Initiative (GHSI) is an informal, international partnership of Ministers/Secretaries/Commissioner of Health from the countries Canada, France, Germany, Italy, Japan, Mexico, the United Kingdom and the United States and the European Commission and the World Health Organization as technical advisors to globally strengthen health preparedness and response to threats of biological, chemical and radio-nuclear terrorism (CBRN) and pandemic influenza [29, 30]. The GHSI was set up in 2001, following September 11 terrorist attacks, addressing health issues like global health security, focusing on the following areas:

1. To explore joint cooperation in procuring vaccines and antibiotics
2. To engage in a constructive dialogue regarding the development of rapid testing, research in variations of vaccines and our respective regulatory frameworks for the development of vaccines, in particular smallpox vaccines
3. To further support the World Health Organization's disease surveillance network and WHO's efforts to develop a coordinated strategy for disease outbreak containment
4. To share emergency preparedness and response plans, including contact lists, and consider joint training and planning
5. To agree on a process for international collaboration on risk assessment and management and a common language for risk communication
6. To improve linkages among laboratories, including level four laboratories
7. To undertake close cooperation on preparedness and response to radio-nuclear and chemical events
8. To share surveillance data from national public health laboratories and information on real or threatened contamination of food and water supplies along with information on risk mitigation strategies to ensure safe food supplies

To achieve the set objectives, there are three different levels generating together an initiative, that is, although having an informal status, appreciated for its function as a trusted network to exchange views on health issues: (1) The Ministerial Forum is meeting every year to discuss current questions regarding health security on the highest level. The Ministerial Forum established (2) the Global Health Security Action Group (GHSAG) to develop and implement concrete actions to improve global health security. This group of senior officials on vice-minister or assistant-secretary level is the core element within a network of rapid communication and reaction in the event of a crisis. Furthermore, there is the level of (3) GHSI Working Groups carrying out actions laid out in the initiative's strategic work plan. An additional element of delegation liaisons and chairs of the working groups is charged with the integration of GHSI priorities and linking the level of senior officials (GHSAG) with the Working Groups [31]. Current GHSI Working Groups are the:

- Risk Management and Communications Working Group
- Global Laboratory Network Working Group

- Pandemic Influenza Working Group
- Chemical Events Working Group
- Radio-Nuclear Threats Working Group

In 2002, the scope of the GHSI mandate was broadened to include the public health threat posed by pandemic influenza. In 2017, as a result of the increased frequency of terrorism attacks and experiences gathered during outbreaks of infectious diseases, the mandate of the GHSI was underlined which calls for concerted global action to strengthen public health preparedness and response to the threat of international chemical, biological or radio-nuclear (CBRN) terrorism.

Global Health and Global Health Security

The core issues of global health are the trends and impacts of health in an increasingly globalized and interdependent world including dissemination of infectious and noninfectious public health risks [32]. The term refers to transnational factors biased by globalization and related to health situations, determinating factors and risks not being under control of single countries. For Lee and Collin, these factors are "determinants circumvent, undermine or are oblivious to the territorial boundaries of states and, thus, beyond the capacity of individual countries to address through domestic institutions" [33].

During the last decades, the term global health with its scope has replaced the term international health with its historic emphasis on tropical diseases and cross-border infection control [34]. The term could also be described as being related to health issues in developing countries[1] and the flow of resources and knowledge from the developed to the developing world, whereas global health has three primary concerns: (1) the global distribution of health and disease and their determinants, (2) the impact of globalization on health and (3) the changing nature of global health governance [35]. Global health thus reflects a deepening globalization with global health-related challenges.

This conceptuality also includes socio-economic and socioecological impact on markets and changing lifestyles. Development therefore might be defined as one overlapping theme determining global health action (despite health security and public health) and underlines the correlation between countries' gross domestic product and their relevant health indicators as development raises living standards and improves basic services and drivers of health [32]. Throughout the transition process from low- to middle-income countries, a shift of health risks can be observed: as a country develops, the types of diseases that affect a population shift from primarily infectious diseases, such as diarrhoea and pneumonia, to primarily noncommunicable diseases, such as cardiovascular disease and cancers [36]. This shift is caused by improvements in medical care, ageing of the population and public health interventions.

The central approach of global health strategies is composed of interventions based on global partnerships as it emphasizes transnational health issues, involves disciplines within and beyond the health science and promotes interdisciplinary collaboration [34]. These partnerships involve national and international cooperation between companies, civil societies, private foundations, non-government organizations, governments and supranational and international institutions. Global health is therefore closely linked to the achievement of the Millennium Development Goals (MDGs) of the United Nations. Three of the eight MDGs are focused on health (reduce child mortality; improve maternal health; combat

[1] Like the spread of infectious diseases agents, access to clean water and sanitary facilities, reproductive health and mother-child health.

HIV/AIDS, malaria and other diseases), whereas health is also a component of several other MDGs [37] (nutrition, water and sanitation). The MDGs were set up to be achieved by 2015. The post-2015 development agenda prioritizes the idea of sustainability. Therefore, the 17 Sustainable Development Goals (SDGs) were identified for the period 2015–2030, also including the health-related target No. 3 'Ensure healthy lives and promote well-being for all at all ages'. The idea of ensuring health could be seen as a smooth conceptual transition to the term Global Health Security since health security is recognized as one of the most important non-traditional security issues. Not only since the Ebola virus disease outbreak in Western Africa it seems to be general consensus that societies are vulnerable to infectious disease threats that spread across national borders [38]. Thus, Global Health Security is a term used to describe the efforts countries are making to prevent, detect and respond to infectious diseases and other public health threats and to reduce the risk of these threats crossing borders. To enhance the health security according to collective and world's health security, legally binding regulations have been defined as International Health Regulations (IHR) setting out concrete requirements for countries (see section IHR). Besides this understanding of collective health security, there are also aspects subsumed under the term individual health security. The latter is based on access to safe and effective health services, products and technologies [39].

2.2.3 Global Partnership Against the Spread of Weapons and Materials of Mass Destruction

The Global Partnership against the Spread of Weapons and Materials of Mass Destruction (GP) is a security initiative in response to the attacks in New York and Washington on September 11, 2001. In 2002, the G8 countries Canada, France, Germany, Great Britain, Italy, Japan, Russia[2] and the United States decided during the G8 summit in Kananaskis [40], Canada, on having another forum to discuss about weapons and materials of mass destruction and their reduction or safe handling. The members committed themselves to spend around 20 billion US Dollars to secure materials and weapons of mass destruction from unauthorized or terroristic access and further proliferation. Furthermore, the goal was to destroy remaining stocks of chemical weapons, mainly of the former Soviet Union. By now, the GP has 30 member states. Initially the G8, and subsequently the G7 members, the European Union, Australia, Belgium, Chile, Czech Republic, Denmark, Finland, Hungary Ireland, Georgia, Kazakhstan, Mexico, New Zealand, the Netherlands, Norway, the Philippines, Poland, Portugal, Republic of Korea, Sweden, Switzerland, Spain and Ukraine joined the initiative. The chairmanship rotates every year together with G7 chairmanship.

In 2011, the G8 summit decided in Deauville [41], France, to continue the GP initiative, initially envisaged for 10 years only, for an unlimited time. After initially focusing on the destruction of chemical weapons, the group decided to concentrate more on biological security and strengthening the Security Council Resolution 1540

[2]After political difficulties in 2014, Russia was suspended from being part of the G8 group. Although Russia was welcomed back to the group, they did not attend a further meeting of the G8 so far.

on nonproliferation of weapons of mass destruction [25]. GP member countries started individually to initiate programmes fostering biosafety and biosecurity and preventing pathogenic materials to be misused or reached by non-state actors.

2.2.4 The EU CBRN Risk Mitigation Centres of Excellence Initiative

The EU CBRN Risk Mitigation Centres of Excellence Initiative (EU CBRN CoE) was developed by the European Union as a CBRN security and nonproliferation approach based on EU programmes and instruments such as the Instrument for Stability (IFS). The EU CBRN CoE was launched in 2010, representing an innovative methodology and platform for collaboration with partner countries outside the EU.

The main goal of the EU CBRN CoE is the mitigation of CBRN risks by capacity-building in partner countries on national level, linking regional security with international objectives. In this context, strengthening national capacities such as prevention, preparedness and response measures against CBRN risks is of special interest, including natural and accidental or intentional incidents with focus on needs and already existing national structures of the countries which participate voluntarily. The EU CBRN CoE seeks to support a sustainable bottom-up procedure, aiming at increased local ownerships and expertise.

National activities like identification and involvement of national stakeholders, establishment of national CBRN teams or facilitation of project proposals and subsequent projects are coordinated by national focal points (NFP) of the EU CBRN CoE. They also collaborate with CoE regional secretariats which host national CBRN experts. The CoE regional secretariats are led by a national Head of Secretariat and were supported by members of the UN Interregional Crime and Justice Research Institute (UNICRI). Technical assistance, networking and collaboration with partner countries are provided by the Directorate-General for International Cooperation and Development (DG DEVCO). Cooperation agreements with further international and regional organizations and programmes like WHO or BWC with regard to biological issues are put in place if necessary with regard to their special fields of activity and mandates. Currently, the EU CBRN CoE network comprises 55 partner countries across 8 regions: African Atlantic Façade, Central Asia, Eastern and Central Africa, Gulf Cooperation Council countries, Western Asia, Northern Africa and Sahel, Southeast Asia and Southeast and Eastern Europe [42–44].

2.2.5 Global Health Security Agenda

The Global Health Security Agenda (GHSA) was launched in 2014 by the United States, and the term nowadays stands for a partnership of over 50 nations, international organizations and non-governmental stakeholders to foster a multilateral and multisectoral approach to strengthen both the global capacity and nations' capacities to prevent, detect and respond to human and animal infectious disease threats, naturally occurring, accidentally or deliberately spread [45]. The GHSA intends to accelerate the implementation of the World Health Organization (WHO) International Health Regulations (IHR) especially regarding global actions to prevent, detect and respond to infectious disease threats. Essential components of this effort include strengthening laboratory systems, achieving real-time disease surveillance, establishing emergency operations centres, reaching national biosafety and biosecurity systems, preventing the spread of emerging zoonotic disease threats and developing a robust workforce to investigate disease outbreaks [45]. In this context, 11 discrete GHSA Action Packages were identified with the main emphasis on prevention, detection or response.

International Health Regulations[3]

Infectious diseases are an important issue for mankind and have been documented throughout history. Triggered by cross national epidemics and subsequent international public health cooperation, the WHO and its member states adopted the International Sanitary Regulations in 1951 that were imbibed and superseded in 1969 as the International Health Regulations (IHR) (1969), comprising requirements for notification to the WHO with regard to three specific infectious diseases: cholera, plague and yellow fever (and, in addition, smallpox until its eradication).

Ongoing globalization effects, further epidemics and new emerging infectious diseases revealed the limits of the IHR (1969), such as a very narrow scope of included diseases and a lacking formal internationally coordinated containment mechanism, resulting in a resolution at the 48th World Health Assembly in 1995 for revision of the IHR (1969). Following this call, a revised version was adopted in May 2005 and entered into force in June 2007 as the IHR (2005) [46].

Thus, the IHR (2005) are a legally binding instrument aiming to "prevent, protect against, control and provide a public health response to the international spread of disease in ways that are commensurate with and restricted to public health risks, and which avoid unnecessary interference with international traffic and trade" (IHR 2005, Art. 2). The main amendments comprise transition from a disease-specific scope to a multi-hazard and agent-independent strategy linked to obligations and action-oriented WHO recommendations, intending to cover and efficiently contain and mitigate any event that could be constituted as a public health emergency of international concern (PHEIC). Nevertheless, some diseases are still designated in the IHR (2005) algorithm to be notified to the WHO, such as smallpox, pneumonic plague, viral haemorrhagic fevers and cases of human influenza caused by a new virus subtype. Notification is one important function of the IHR (2005) for early identification of potential PHEICS, their verification and subsequent initiation of countermeasures and international response, coordinated by WHO. According to IHR (2005), the notification has to take place within 24 hours, with the ongoing obligation to respond to follow-up requests. Within this context, national IHR focal points with 24/7 availability have to be assigned for communication with WHO and coordinative function. IHR (2005) also obligates each state party to develop core capacities for "their surveillance, reporting, notification, verification, response and collaboration activities and their activities concerning designated airports, ports and ground crossings" (IHR (2005) Annex 1), using own structures and resources. Currently, 196 states are parties to the IHR (2005).

---

[3]Due to their connection to the GHSA, IHR are not included in Chapter I although they are legally binding (binding in terms of international law).

2.3 Conclusion

With regard to the selection of international agreements, regulations and multilateral programmes and initiatives being presented within this chapter, it becomes apparent that there is a multitude of activities fostering national and international biosafety/security, nonproliferation and health situation. All programmes have in common that they are dependent on a critical mass of participating parties and on their commitment and compliance for successful implementation, at best supported by a legal framework that has to be integrated into national law to build an effective and sustainable impact, especially in case trusted informal networks have not been established yet. Since most of the activities only focus on very specific topics, associated/overlapping aspects of other programmes often are not considered. In addition, detailed information about current outcomes and running or planned activities are not accessible in many cases and therefore do not enable an exchange between implementing actors which could avoid redundancies and generate synergy effects, for example, by sharing experiences and lessons identified/lessons learnt. Another challenge is the consideration of ethical and dual-use aspects, in addition to communication and technical issues. Therefore, multidisciplinary forums and cooperation cannot be overestimated to share ideas and to increase efficiency of the programmes, even if legal regulations are applicable. Acknowledging the interdependency of ongoing activities and strengthening biosafety and security, bio-nonproliferation and health might enhance the prognosis for the global success in a sustainable manner.

References

1. Definition of dual use research of concern (DURC). 2013 [cited 2017]. Available from: http://www.who.int/csr/durc/en/.
2. John Hart RT. Science, technology, and the biological weapons convention. 2012 [cited 2017]. Available from: https://www.armscontrol.org/act/2012_10/Science-Technology-and-the-Biological-Weapons-Convention#hart.
3. Geneva Protocol. [cited 2017]. Available from: https://unoda-web.s3-accelerate.amazonaws.com/wp-content/uploads/assets/WMD/Bio/pdf/Status_Protocol.pdf.
4. Biological and toxin weapons convention. [cited 2017]. Available from: http://www.unog.ch/80256EDD006B8954/%28httpAssets%29/C4048678A93B6934C1257188004848D0/$file/BWC-text-English.pdf.
5. Becker-Jakob U. Balanced minimalism – The biological weapons convention after its 7th review conference. PRIF-Report No. 120. 2013.
6. Regional groups in the biological weapons convention. Available from: http://www.unog.ch/80256EE600585943/%28httpPages%29/73C3C49BEA5621A0C12572DB00477B4A?OpenDocument.
7. Milton Leitenberg RAZ. The Soviet Biological Weapons Program – a history. Cambridge: Harvard University Press; 2012.
8. Chandré Gould PF. Project Coast: Apartheid's Chemical and Biological Warfare Programme. Geneva: United Nations Publication; 2002.
9. Zilinskas RA. Iraq's biological weapons. The past as future? JAMA. 1997;278(5):418–24.

10. The confidence building measures (CBMs). [cited 2017]. Available from: https://www.unog.ch/80256EE600585943/(httpPages)/5E2E8E6499843CCBC1257E52003ADED4?OpenDocument.
11. Findlay T. Verification and the BWC: last gasp or signs of life? 2006 [cited 2017]. Available from: https://www.armscontrol.org/act/2006_09/BWCVerification.
12. Bansak KC. Enhancing compliance with an evolving treaty: a task for an improved BWC intersessional process. 2011 [cited 2017]. Available from: https://www.armscontrol.org/print/4902.
13. Gerstein DM. National security and arms control in the age of biotechnology: the biological and toxin weapons convention. Lanham, MD: Rowman & Littlefield; 2013. p. 234.
14. BWC meeting of states parties (10–14 December 2012). [cited 2017]. Available from: https://www.unog.ch/80256EE600585943/(httpPages)/89835CB0A2DAA4A0C1257B6E003415C5?OpenDocument.
15. Eighth Review Conference of the States Parties to the Convention on the Prohibition of the Development, Production and Stockpiling of Bacteriological (Biological) and Toxin Weapons and on Their Destruction. 2016 [cited 2017]. Available from: http://www.unog.ch/80256EDD006B8954/%28httpAssets%29/F277FA6A2B96BA98C125807A005B2F59/$file/2016-1129+Final+report+adv+vers.pdf.
16. Seventh Review Conference of the States Parties to the Convention on the Prohibition of the Development, Production and Stockpiling of Bacteriological (Biological) and Toxin Weapons and on Their Destruction. 2011 [cited 2017]. Available from: http://www.unog.ch/80256EDD006B8954/%28httpAssets%29/C2BAA955E58E15C7C1257D01005260D3/$file/BWC_CONF.VII_07.pdf.
17. Robertson SP, et al. Frontometaphyseal dysplasia: mutations in FLNA and phenotypic diversity. Am J Med Genet A. 2006;140(16):1726–36.
18. Convention on biological diversity. Available from: https://www.cbd.int/.
19. Bagley MA, Rai AK. The Nagoya protocol and synthetic biology research: a look at the potential impacts. Washington, DC: Wilson Centre; 2013.
20. Nagoya Protocol. Available from: https://www.cbd.int/abs/nagoya-protocol/.
21. Access and benefit-sharing clearing-house. Available from: https://absch.cbd.int/.
22. Smallpox. [cited 2017]. Available from: http://www.who.int/csr/disease/smallpox/en/.
23. Damon IK, Damaso CR, McFadden G. Are we there yet? The smallpox research agenda using variola virus. PLoS Pathog. 2014;10(5):e1004108.
24. WHO Advisory Committee on Variola Virus Research: Report of the Eighteenth Meeting, Geneva. Geneva: World Health Organization; 2017.
25. Security Council Resolution 1540. 2004 [cited 2017]. Available from: http://www.un.org/en/ga/search/view_doc.asp?symbol=S/RES/1540%282004%29.
26. Security Council Resolution 1540 (2004) [on non-proliferation of nuclear, chemical and biological weapons]. 2004 [cited 2017]. Available from: http://unbisnet.un.org:8080/ipac20/ipac.jsp?profile=voting&index=.VM&term=sres1540%20(2004)#focus.
27. Australia Group. [cited 2017]. Available from: http://www.australiagroup.net/en/.
28. CBW Events. [cited 2017]. Available from: http://www.cbw-events.org.uk/.
29. Global Health Security Initiative (GHSI). [cited 2017]. Available from: http://www.ghsi.ca/english/index.asp.
30. Global Health Security Initiative. Available from. https://ec.europa.eu/health/preparedness_response/cbrn_threats/ghsi_en.
31. Global Health Security Initiative (GHSI). Available from: https://www.phe.gov/Preparedness/international/ghsi/pages/default.aspx.
32. De Cock KM, et al. The new global health. Emerg Infect Dis. 2013;19(8):1192–7.
33. Lee K, Collin J. Global change and health. Maidenhead. Open University Press; 2015.
34. Rowson M, et al. Conceptualising global health: theoretical issues and their relevance for teaching. Glob Health. 2012;8:36.
35. Kickbusch I. Global health – a definition. Available from: https://www.ilonakickbusch.com/kickbusch/global-health/.

36. Omran AR. The epidemiologic transition. A theory of the epidemiology of population change. Milbank Mem Fund Q. 1971;49(4):509–38.
37. Nations U. Millennium development goals report 2015. New York: United Nations; 2015.
38. McInnes C, Lee K. Health, security and foreign policy. Rev Int Stud. 2006;32:5–32.
39. Heymann DL, et al. Global health security: the wider lessons from the west African Ebola virus disease epidemic. Lancet. 2015;385(9980):1884–901.
40. The G8 global partnership against the spread of weapons and materials of mass destruction, in G8 summit. Kananaskis, Canada; 2002.
41. Report on the G8 global partnership against the spread of weapons and materials of mass destruction, in G8 summit. Deauville, France; 2011.
42. EU CBRN risk mitigation centres of excellence initiative. Available from: http://www.cbrn-coe.eu/.
43. EU CBRN CoE initiative: 5th round table meeting for Central Asia.
44. CBRN Centres of Excellence. Available from: http://www.unicri.it/topics/cbrn/coe/.
45. Global Health Security Agenda. [cited 2017]. Available from: https://www.ghsagenda.org.
46. WHO, editor. International Health Regulations (2005). Geneva: World Health Organization; 2008.

The Role of Event-Based Biosurveillance in Biodefense

3

David M. Hartley, Wai-Ling Mui, and Courtney D. Corley

3.1 Background

According to Merriam-Webster online, *biodefense* is "the means or methods of preventing, detecting, or managing an attack involving biological weapons." In practice, the term typically implies defense against threats to human health beyond biological terrorism and warfare. Threats posed by infectious agents are both numerous and compelling: pathogens can emerge anywhere in the world and spread via many pathways, including food, water, insects, and the air we breathe; vaccines and drug therapies do not exist for many infectious agents; and, especially when pandemics occur, pathogens can cause widespread morbidity and mortality. Moreover, the growing and increasingly mobile world population elevates the potential for new pathogens to emerge and travel long distances rapidly through the normal course of trade, commerce, and travel. Adding to the natural threat spectrum is the specter of intentional bioterrorism perpetrated by state or non-state actors attempting to advance their economic, social, or political objectives.

Regardless of where, when, or why outbreaks and epidemics occur, they can grow rapidly, and mitigation requires timely action on the part of a range of responders. This requirement imposes several epidemic intelligence needs [1],

D. M. Hartley (✉)
University of Cincinnati College of Medicine, Cincinnati, OH, USA

Cincinnati Children's Hospital, Cincinnati, OH, USA
e-mail: david@no-infection.net

W.-L. Mui
Synertex, LLC, McLean, VA, USA
e-mail: ling@synertex.com

C. D. Corley
Pacific Northwest National Laboratory, Richland, WA, USA
e-mail: court@pnnl.gov

S. K. Singh, J. H. Kuhn (eds.), *Defense Against Biological Attacks*,
https://doi.org/10.1007/978-3-030-03053-7_3

including comprehensive and continuous situational awareness [2] of infectious disease activity; early warning and rapid identification of emergent bioevents; and effective dissemination of this knowledge to governments, humanitarian organizations, businesses, and private citizens alike. Both indicator-based and event-based surveillance [3] play roles in epidemic intelligence. Indicator-based surveillance, an approach utilized for decades, aims to describe patterns of disease and link that information to public health action, study the epidemiology and natural history of disease, and provide baseline data [4]. Because the indicator-based approach typically analyzes structured data in the form of disease case reports from health care providers or hospital laboratories [5], surveillance systems built around this approach can provide highly specific and detailed data in terms of time, geography, microbe, and host. However, they can be slow, since they depend upon one or more interactions with the healthcare system.

More recently, the internet, portable computers, and mobile devices (tablets and smart phones) have made it possible for groups and individuals throughout the world to communicate instantly to global audiences, enabling the event-based approach, also known as "internet-based biosurveillance", "digital disease detection", or simply "biosurveillance" [3, 6–10]. Event-based biosurveillance systems analyze news reports and stories, rumors, and other information related to infectious disease published online [11], and most often using information in the public domain. Table 3.1 illustrates several past and existing biosurveillance systems. In contrast to traditional systems, the data gleaned using this approach are typically non-specific and unstructured, since they are obtained from dynamic, non-standardized, and often informal and subjective sources. While the event-based approach does not normally provide highly specific information, the chief advantage is that events portending or related to infectious disease outbreaks and epidemics may be detected earlier than in indicator-based systems, since it does not require ill persons to have sought medical attention [5]. This supports the timeliness requirement of epidemic intelligence.

This chapter provides an overview of the process of event-based biosurveillance as it relates to infectious disease broadly, as well as future directions and research needs.

3.2 Objectives, Aims, and Process of Event-Based Biosurveillance

3.2.1 Objectives and Aims

The United States Department of Homeland Security (DHS) defines biosurveillance as "the science and practice of managing human, animal, plant, food, and environmental health-related data and information for early warning of threats, early detection of events, and rapid characterization of events so that effective actions can be taken to mitigate adverse health, social, and economic effects," [17]. This broad definition highlights the overall objective of biosurveillance as enabling action to mitigate a broad range of biological threats. The US National Biosurveillance

Table 3.1 Examples of biosurveillance systems

System name	Years active	Regions covered	Languages	Access	Sources	User interface
ProMED[a]	1994–present	Global	5	Public	Media reports, official reports, online summaries, local observers, others	Web portal, email
GPHIN[b]	1997–present	Global	8	Public	20,000+ online news reports	Web portal, email
MedISys[c]	2004–present	Global	20	Public and private	90,000 news articles	Web portal, email RSS, RNS (rapid news service)
BioCaster[d]	2006–2012	Global, with Asia-Pacific regional concentration	8	Public	1700 RSS feeds	Web portal, email
Argus[e]	2006–2012	Global	~40	Private	Several thousand local, native-language Internet media articles daily	Web portal, email
HealthMap[f]	2006–present	Global	9	Public	Aggregated news	Web portal, mobile app
BSVE[g]	2012–present	Global	Unknown	Private	Open-source, RSS feeds, de-identified diagnostic results, historic outbreak data, zoonotic data non-health data	Web portal

[a]https://www.promedmail.org/aboutus/; [b]https://www.canada.ca/en/public-health/services/reports-publications/canada-communicable-disease-report-ccdr/monthly-issue/2015-41/ccdr-volume-41-9-september-3-2015-data/ccdr-volume-41-9-september-3-2015-data-1.html; Mykhalovskiy and Weir [12]; [c]Linge et al. [13]; [d]Collier et al. [14]; [e]Nelson et al. [15, 16]; [f]http://www.healthmap.org/site/about; [g]https://www.defense.gov/News/Article/Article/681832/dtra-scientists-develop-cloud-based-biosurveillance-ecosystem/

Strategy [18] explicitly identifies biosurveillance aims and core functions as (1) scanning and discerning the environment, (2) identifying and integrating essential information, (3) alerting and informing decision makers, and (4) forecasting and advising users on impacts. Based on the DHS and White House documents, and extensive expert input from providers and consumers of biosurveillance [19], researchers have identified four distinct goals for biosurveillance: (1) early warning, (2) early detection, (3) situational awareness, and (4) consequence management support.

These goals are general; when developing a biosurveillance system, it is vital to undertake a detailed process of gathering, defining, and documenting system user and stakeholder requirements. Requirements must be specific and focus on the business, mission, and operational needs of system stakeholders and users. Requirements, for example, should describe needs for viewing and interacting with data and information, the geographic region/s of interest, how frequently information is to be collected, the diseases of highest interest, the hours of system operation, and needed system reliability among many other functional issues. Many tools and approaches are available [20–23] that are adaptable to biosurveillance requirements management. Once compiled, requirements determine critical design and evaluation parameters against which system performance can be judged later. In the next section, we elucidate common processes by which goals, aims, and requirements can be met.

3.2.2 The Biosurveillance Process

Based on the experience of several past and existing operational biosurveillance systems, Hartley et al. [9] have described the process of event-based biosurveillance. Here, we revise and extend their process model, as described in Fig. 3.1. While biosurveillance systems are designed to meet different user requirements, in general, the process of biosurveillance begins with data source identification, collection and processing, analysis, and dissemination leading to action.

3.2.2.1 Data and Source Identification

The validity of biosurveillance depends largely on the appropriateness, integrity, and relevance of the data collected and of the sources providing them [24]. Data needs are often identified subjectively through a process considering, for a specific set of

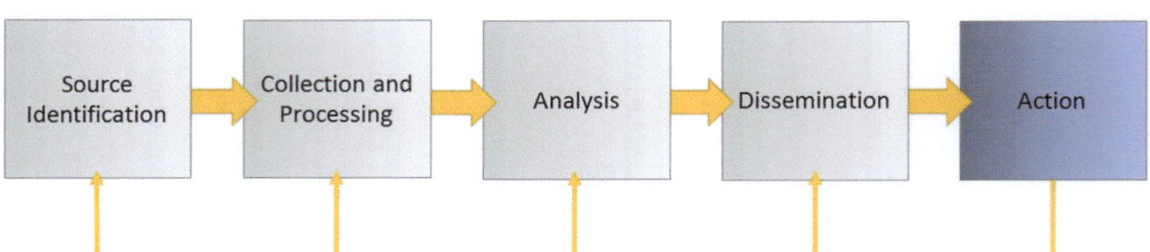

Fig. 3.1 The process of event-based biosurveillance

system requirements, the type/s of the disease/s of interest (e.g., human, plant, animal); the geographic area/areas of interest (e.g., local, regional, national, or international); characteristics of population/populations of interest (e.g., language/s spoken; rate of internet usage); required update frequency (e.g., rapid/frequent versus slow/infrequent); and the needed accuracy or confidence of the information each source provides.

Commonly utilized data sources include text-based news media websites (which may cover local, regional, national, or international events), social media (including blogs and microblogs such as Twitter) [25–28], and interactive sites and platforms soliciting and publishing public input (e.g., participatory epidemiology platforms [29–32]). Audio- and video-based sources, while less frequently utilized due to the need for either manual review or automated transcription to text, may also provide useful information [33]. In addition to information from informal sources, reports from official sources such as ministries and departments of health, the World Health Organization and its regional offices, and similar organizations dedicated to agricultural health, are often collected.

It is becoming feasible to utilize non-subjective techniques for identifying and prioritizing biosurveillance data streams. Generous et al. [34] have proposed a methodology based on multi-attribute utility theory for identifying essential information to include in a biosurveillance system, and Margevicius et al. [19] have described an approach for characterizing and evaluating data streams that relates data to biosurveillance goals. As these approaches are further developed and validated, they are likely to provide a rigorous alternative to the subjective approach often employed in data and source selection.

3.2.2.2 Collection and Processing

Having identified relevant types and sources of data, they are next collected and processed into a form in which analysis is possible. The unstructured nature of data collected from such dynamic and non-standardized sources presents challenges to all processes in this step.

Data are collected from internet sources via one of two general pathways: aggregator news sites (e.g., Yahoo!, Google) or web monitoring of a collection of URLs [9]. Web monitoring typically takes the form of scheduled web scraping (the automated access and storage of specified web pages) or web crawling (e.g., accessing specified web pages, links on those pages, and links on linked pages). Monitoring schedules can vary from several times daily to more infrequent collection. Markup language (e.g., HTML, XML, XHTML) codes and undesired page elements such as advertisements are removed, and the data are converted into a format supporting efficient storage, sharing (if appropriate), and content mining. Other data (e.g., Twitter, Facebook, Wikipedia) may be retrievable via an application programming interface to store and search on these data; technologies such as

Lucene[1] or ElasticSearch[2] are often employed. Regardless of the source, all data must be handled in accordance with applicable copyright laws and provider terms of service. Moreover, especially in the case of social media data, privacy concerns may arise, even though data are publicly available [35, 36].

Once data have been acquired, they are processed. Common data processing tasks include translation, relevancy determination and ranking, ontology, event extraction, and de-duplication [9, 37]:

- *Language translation.* When needed, systems can process and analyze material as native vernaculars or translate to a common language.
- *Relevancy ranking.* Relevancy guidelines are typically identified in the system requirements process, though user interests can and do evolve throughout the lifetime of a system. Approaches to determining and ranking the relevancy of collected material include Boolean keyword searches, machine classifiers (e.g., Naive Bayes or Support Vector Machines [38]), and clustering.
- *Ontology.* By defining key concepts and the relationships between them [39], ontologies enable inferences to be made based on incomplete knowledge. Though commonly used throughout medicine and public health (e.g., ICD, SNOMED, UMLS), few ontologies exist for biosurveillance. Concepts relevant for biosurveillance include time, place, etiologic agent, and signs and symptoms among others. Open-source ontologies developed especially for biosurveillance include BioCaster [40], the Pacific Northwest National Laboratory Medical Linked Dataset [41], and the Anthology of Biosurveillance Diseases [42].
- *Event extraction.* Identifying specific biological events in data is accomplished via different human analyst-based and machine-based approaches. Human analyst-based approaches may identify or postulate bioevents based on direct or indirect indicators of disease, which may be subtle and specific to a specific group under surveillance [15, 16, 43]. The latter approach is known as indications and warning of biological events. Machine-based approaches include simple keyword recognition algorithms and more sophisticated event meta-data extraction, in which types of information related to bioevents are identified [44, 45].
- *Document de-duplication.* Identifying collected documents that are similar or identical (e.g., the same news story carried by multiple news outlets) is often desirable so that users may skip documents that do not add value. Avoiding errors in the de-duplication process can be difficult; collected documents can be nearly the same and differ only by updated specific information such as case numbers or reports of symptoms that may be of prime importance. Overcoming these difficulties can require a combination of machine and manual review.

[1]https://lucene.apache.org

[2]http://www.elastic.co

3.2.2.3 Analysis

After data have been processed, analytic tools identify events of potential interest to users, e.g., those that are abnormal in a sense meaningful to system users or events potentially caused by select agents. Two general classes of algorithms are commonly employed to achieve such objectives: trend analysis and anomaly detection.

- *Trend analysis.* Patterns and trends in data, e.g., keywords corresponding to environmental precursors (e.g., floods) or human behavior (e.g., protests or evacuations), can indicate bioevents. Patterns observed in multiple bioevents define event signatures. Signal processing algorithms can discriminate signatures from seasonal patterns of keywords or noise not associated with bioevents.
- *Anomaly detection.* Anomaly detection assigns a level of significance to an observation or set of observations, which can be in the form of temporal trends, geographic clusters, or a mixture of spatiotemporal observations. Significance is assessed based on one or more features of the data (e.g., disease counts, geographic locations, and human behavior).

3.2.2.4 Dissemination

Biosurveillance systems utilize a variety of different platforms and modalities to disseminate information. Systems often provide a website or portal where users view (typically via maps or tables) and interact with (e.g., search and filter) data and information. Typically, a map illustrates geographic disease clusters over a time interval (e.g., the previous 4–6 weeks). Web sites may be entirely open to the public, have password-protected areas, or be completely closed to those not possessing credentials. In contrast to websites, which are passive and require a user to visit the site to receive information, biosurveillance systems may also disseminate information actively via email, mobile apps, or social media.

3.2.2.5 Action

One of the main goals of biosurveillance is to detect threats to public health at the earliest times possible so that stakeholders can act. For public health authorities, biosurveillance often triggers a verification phase in which observations are confirmed and additional information is gathered [46]. Other actions based on biosurveillance can include the issuance of alerts, cuing of intensified surveillance, and formation of task forces.[3] For the general public, biosurveillance can provide individuals with information allowing them to evaluate travel plans and the appropriateness of prevention measures.

---

[3] A detailed case study illustrating the application of biosurveillance related to an outbreak of dengue fever can be found in [9].

3.2.2.6 Feedback and Optimization

Finally, it is important for users of biosurveillance to communicate successes and failures to system owners, analysts, and administrators so that system performance can be revised and optimized if needed. Barboza et al. [47] and Velasco et al. [48] have discussed the strengths and limitations of event-based biosurveillance systems, and Corley et al. [49] have provided a set of metrics for characterizing system performance. Tracking of such performance metrics, combined with post-event reviews of the performance of biosurveillance, provides mechanisms for optimizing systems to meet user and stakeholder needs.

3.3 Theoretical Underpinnings

Biosurveillance systems aim to collect material containing indicators of outbreaks or epidemics. By "indicator" we mean an observable (e.g., a behavior, laboratory test, or action) that is related to (i.e., *indicates*) a recent or ongoing biological event. "Indicator" used in this context does *not* imply or relate to indicator-based surveillance.

Indicators can be classified as "direct" or "indirect" [43]. Direct indicators of disease are observations and reports related to a disease itself, such as the results of diagnostic tests, detection and verification of etiologic agents, and clinical and epidemiological reports. Additional direct indicators often of interest include unique clinical presentation, excess morbidity and mortality, appearance of a seasonal disease outside the expected time, unusual geographic occurrence of disease, and whether discrete populations or risk groups are affected. Relevant geographic features include the distance and speed of spread, whether spread is observed in unexpected patterns, and the observation of disease in new areas where outbreaks have not occurred historically.

Indirect indicators are social responses to epidemic disease (e.g., reports that individuals or groups and organizations have altered their normal daily routines and activities in response to a real or perceived bioevent). Such responses have been studied by McGrath [50], and more recently by Collmann et al. [51]. The latter work provides detailed case studies of cholera in Venezuela in the 1990s [52] and of war and biological attacks in Ningbo, China, in 1937–1941 [53]. Indirect indicators can be classified as reports of official acknowledgment of a bioevent, official action, demand for medical services, local perception of threat, changes in business practice, and integrity of medical infrastructure [15, 43].

3.4 Computer Methods for Analyzing Biosurveillance Data

One emergent source for indirect indicators of disease, social media, has become a resource for studying different social, emotional, health and economic conditions of communities through their online activities and shared content. One of the challenges noted above (Sect. 3.2.2.2 on collection and processing) is that these data are often messy, incomplete, and sparse. However, applying machine learning and natural language processing (NLP) techniques to web and social media can identify, characterize and monitor health and wellbeing. Using these tools researchers have studied the emotions and behavior in different groups of people through their social media footprints [54, 55] and investigate social issues and phenomena existing in communities through their online behavior [56, 57]. Of particular relevance is the use of signals from social media to detect food poisoning within certain subpopulations and geographic regions [58], and to study subpopulations of smokers and drug addicts [59].

Another valuable computer method is automated linguistic analyses of affective content facilitated by text mining. These methods may reveal emotion, mood, behavior, attitude, cognitive state, physical state, hedonic signal, and sensations. In addition, cultural artifacts extracted from such media may point to useful demographic indicators that have potential utility for the detection of sociocultural and psychosocial signatures [60]. Such signatures can vary considerably in complexity. For example, Pennebaker [61] uses the occurrence of function words to characterize texts authored by terrorist groups as compared to non-terrorist radical groups. Volkova et al. [28] tracked sentiment and emotion in online social media with clinical records of influenza like illness of over 800 medical facilities in the United States and abroad. These techniques are important for the analysis process (Sect. 3.2.2.3 above) to strengthen confidence in assessments based on the integration of direct and indirect indicators.

Traditional computer methods for biosurveillance relying on statistical first principles or classic machine learning techniques, while valuable, can fail in the face of sparse data and social complexities such as language and behavior. Corley et al. [62] systematically reviews these issues within the context of biosurveillance models for forecasting or prediction. To overcome such limitations, deep learning is enabling a paradigm shift in how biosurveillance collection, processing, and analysis steps are conducted. Deep learning (i.e., multi-layered artificial neural networks) has existed for several decades, but the technology convergence of computing power, data and algorithms has moved it from obscurity to the forefront of industry use and scientific discovery. For a detailed history and review of deep learning, see the study of Schmidhuber [63].

Deep neutral networks have also demonstrated promising results for a variety of NLP tasks critical for accurate biosurveillance collection and processing, such as event detection [64], question answering [65], dependency parsing [66], named entity recognition [67, 68], sentiment analysis [69], paraphrase identification [70] and machine translation [71, 72]. Deep learning is also relevant to the analysis phase of biosurveillance (e.g., trend analysis, anomaly detection, forecasting). Recurrent Neural Networks (RNNs) are designed to model dynamic temporal behaviors; they have an internal memory to process arbitrary sequences of inputs. Important RNNs include Hopfield networks, Ekman networks, Jordan networks, and Long Short Term Memory networks [73]. RNNs are most commonly used for text and sequence analysis, and are showing promise for time series forecasting.

3.5 Discussion and Future Directions

The term "biosurveillance" refers to the collection, analysis, and dissemination of different data for early warning, early detection, situational awareness, and in support of consequence management of biological events. This broad view encompasses both traditional and recent approaches to surveillance as they relate to natural and intentional outbreaks and epidemics in human and agricultural populations. While this chapter focuses on event-based biosurveillance utilizing data available from the internet, comprehensive epidemic intelligence requires the fusion of different surveillance approaches, each of which has its strengths and limitations in terms of timeliness, specificity, and actionability.

Biosurveillance systems generally collect reports of disease or human behavior related to disease in the early stages of a bioevent. As such, they do not usually discriminate low-importance events from significant public health threats. Identification of events posing significant threats to public health generally requires human analysts possessing expertise and additional information (including more specific information derived from laboratory data, syndromic and sentinel physician surveillance data, intelligence threat assessments, and other sources). In this sense, biosurveillance can be thought of as generating hypotheses to be tested through additional investigation.

Part of the difficulty in discriminating low- versus high-impact bioevents is a paucity of validated event-based biosurveillance signatures. Case studies exist of the role of biosurveillance in the detection and response to different bioevents (e.g., pandemic influenza [15, 16, 74]), but translating case studies into reliable signatures (e.g., in terms of temporal and/or geographic distributions of indicators) is challenging due to event-to-event variability and the difficulty in identifying indicator patterns that generalize across events.

In the absence of validated event signatures based on past observation, biosurveillance analysts often must rely on theoretical or anticipated signatures [75]. This is especially true for the early detection of bioterrorism [76], when presumed signatures are frequently based on hypothetical scenarios. Typically, such scenarios are constructed for specific pathogens, affected populations, social customs and healthseeking behaviors of populations, and reporting habits of available data sources; the characteristics of biosurveillance data are postulated based on the specific scenario. The accuracy, sensitivity, and specificity of such hypothetical signatures are uncertain. Burkom et al. [75] has an excellent discussion of signatures and tradeoffs between assumptions and requirements within the context of syndromic surveillance; similar considerations apply to event-based biosurveillance.

The sustainability of biosurveillance systems can be complex to manage. Engineering platforms and analytic tools become outdated quickly due to the rapidly evolving information technology, underscoring the importance of flexible system design to accommodate new innovations. Moreover, the trans-disciplinary nature of biosurveillance requires individuals with expertise in many areas (e.g., anthropology and sociology, epidemiology and medicine, computer and data science, software and systems engineering, and project management) for successful system design, development, and operation. Balancing the technical, analytic, and human resource dimensions of biosurveillance against the constantly changing landscape of federal, state, and local programmatic focus and funding is a perennial challenge to system sustainability.

Nonetheless, biosurveillance provides value to a variety of stakeholders. At present, however, measures of the value of information produced by biosurveillance are incompletely developed. Although the broader medical domain has developed methodologies for valuing information within the context of diagnosis and insurance, for example, it is unclear that the value of surveillance information is well understood. Hartley [77] has noted that, in the case of environment- and climate-based surveillance for Rift Valley fever in Africa, the value of early warning and forecasting depends on many scenario-specific parameters, including the cost of control and intervention activities (if they exist), the cost of morbidity and mortality of affected populations, and the costs associated with lost production among others. Public health agencies, who have bounded budgets that must address a range of important activities, must also balance the costs of supporting and/or acting upon biosurveillance against the cost of not investing those monies in support of competing objectives. The literature is virtually silent on such economic dimensions of biosurveillance.

Macauley and Laxminarayan [78] and Laxminarayan and Macauley [79] have enunciated criteria for information providing value more generally. Applying several of these to biosurveillance, we propose that: (1) Biosurveillance has value if it enables decisions regarding action. (2) Biosurveillance has value if it decreases uncertainty surrounding bioevents or if it reveals ambiguity that is not evident otherwise. (3) Although biosurveillance does not provide perfect information, it may be valuable when more-perfect information is not available or is too costly. (4) In any given situation, some dimensions of biosurveillance may be more valuable than others (e.g., early warning, provision of indirect and contextual knowledge, etc.). (5) The growing record of bioevents observed by biosurveillance systems has value for different activities, including signature development and impact assessment among others. Developing appropriate, meaningful, and robust measures of the value of biosurveillance is an area meriting future research.

Another area worthy of development is the operational application of standard statistical methods to biosurveillance data. A wealth of statistical methods can, in theory, be applied to such data. The syndromic surveillance community has developed the application of time series, spatial, and spatiotemporal analytic methods to observational data originating from indicator-based surveillance systems [80, 81]. While these methods can, in principle, be applied to biosurveillance data, in practice sophisticated statistical or signal processing methods tend not to be used in biosurveillance systems.

Perhaps the most exciting area under development concerns the fusion of large, disparate data sets ("big data") collected for purposes unrelated to public health to improve biosurveillance and infectious disease surveillance more generally. Previous efforts have utilized massive data streams from public internet queries to infer influenza incidence locally over large geographic areas in real time [82]. While these experiments have demonstrated limited predictive ability [83, 84], they have also produced insights [85]. At present, researchers are investigating hybrid approaches combining traditional surveillance and big data sets [86]. Such approaches aim to exploit the complementary nature of small and large data sets, and recognize that traditional surveillance data contains important information not necessarily present in "big data" [85]. Methodologies combining disparate data sources (such as Twitter, electronic health records, and crowd-sourced surveillance platforms among others) have shown promising results for influenza surveillance [27, 28, 87] and forecasting. Integrating disparate data is an important step toward more advanced capabilities, such as surveillance-based models forecasting epidemic impact and algorithm-based hypothesis generation [88].

Regardless of the nature of a bioevent, the capacity to act promptly is crucial for minimizing impacts. Internet-based biosurveillance is an important tool in the armamentarium supporting early warning, early detection, situational awareness, and consequence management. It will undoubtedly increase in terms of capability and application as technology and analytic methodology advance and mature.

References

1. Paquet C, Coulombier D, Kaiser R, Ciotti M. Epidemic intelligence: a new framework for strengthening disease surveillance in Europe. Euro Surveill. 2006;11(12):212–4.
2. Endsley MR. Toward a theory of situation awareness in dynamic systems. Hum Factors. 1995;37(1):32–64.
3. O'Shea J. Digital disease detection: A systematic review of event-based internet biosurveillance systems. Int J Med Inform. 2017;101:15–22. https://doi.org/10.1016/j.ijmedinf.2017.01.019.
4. Declich S, Carter AO. Public health surveillance: historical origins, methods and evaluation. Bull World Health Organ. 1994;72(2):285–304.
5. CDC. Global disease detection operations center: event-based surveillance. 2016. https://www.cdc.gov/globalhealth/healthprotection/gddopscenter/how.html. Accessed 26 Aug 2017.
6. Brownstein JS, Freifeld CC, Madoff LC. Digital disease detection – harnessing the Web for public health surveillance. N Engl J Med. 2009;360(21):2153–7. https://doi.org/10.1056/NEJMp0900702.
7. Choi J, Cho Y, Shim E, Woo H. Web-based infectious disease surveillance systems and public health perspectives: a systematic review. BMC Public Health. 2016;16(1):1238. https://doi.org/10.1186/s12889-016-3893-0.
8. Hartley D, Nelson N, Walters R, Arthur R, Yangarber R, Madoff L, Linge J, Mawudeku A, Collier N, Brownstein J, Thinus G, Lightfoot N. Landscape of international event-based biosurveillance. Emerg Health Threats J. 2010;3:e3. https://doi.org/10.3134/ehtj.10.003.
9. Hartley DM, Nelson NP, Arthur RR, Barboza P, Collier N, Lightfoot N, Linge JP, van der Goot E, Mawudeku A, Madoff LC, Vaillant L, Walters R, Yangarber R, Mantero J, Corley CD, Brownstein JS. An overview of internet biosurveillance. Clin Microbiol Infect. 2013;19(11):1006–13. https://doi.org/10.1111/1469-0691.12273.
10. Milinovich GJ, Williams GM, Clements AC, Hu W. Internet-based surveillance systems for monitoring emerging infectious diseases. Lancet Infect Dis. 2014;14(2):160–8. https://doi.org/10.1016/S1473-3099(13)70244-5.
11. WHO. Early detection, assessment and response to acute public health events: implementation of early warning and response with a focus on event-based surveillance. 2014. http://www.who.int/ihr/publications/WHO_HSE_GCR_LYO_2014.4/en/. Accessed 26 Aug 2017.

12. Mykhalovskiy E, Weir L. The global public health intelligence network and early warning outbreak detection: a Canadian contribution to global public health. Can J Public Health. 2006;97(1):42–4.
13. Linge J, Steinberger R, Fuart F, Bucci S, Belyaeva J, Gemo M, Al-Khudhairy D, Yangarber R, van der Goot E. MedISys: medical information system. In: Asimakopoulou E, Bessis N, editors. Advanced ICTs for disaster management and threat detection: collaborative and distributed frameworks, Chapter 9. New York: IGI Global; 2010. p. 131–42.
14. Collier N, Doan S, Kawazoe A, Goodwin RM, Conway M, Tateno Y, Ngo QH, Dien D, Kawtrakul A, Takeuchi K, Shigematsu M, Taniguchi K. BioCaster: detecting public health rumors with a Web-based text mining system. Bioinformatics. 2008;24(24):2940–1. https://doi.org/10.1093/bioinformatics/btn534.
15. Nelson NP, Brownstein JS, Hartley DM. Event-based biosurveillance of respiratory disease in Mexico, 2007–2009: connection to the 2009 influenza A(H1N1) pandemic? Euro Surveill. 2010;15(30)
16. Nelson NP, Yang L, Reilly AR, Hardin JE, Hartley DM. Event-based internet biosurveillance: relation to epidemiological observation. Emerg Themes Epidemiol. 2012;9(1):4. https://doi.org/10.1186/1742-7622-9-4.
17. DHS. National Biosurveillance Integration Center Strategic Plan. 2012. https://www.dhs.gov/sites/default/files/publications/nbic-strategic-plan-public-2012.pdf. Accessed 26 Aug 2017.
18. National Strategy for Biosurveillance. https://obamawhitehouse.archives.gov/sites/default/files/National_Strategy_for_Biosurveillance_July_2012.pdf. Accessed 12 Dec 2018
19. Margevicius KJ, Generous N, Taylor-McCabe KJ, Brown M, Daniel WB, Castro L, Hengartner A, Deshpande A. Advancing a framework to enable characterization and evaluation of data streams useful for biosurveillance. PLoS One. 2014;9(1):e83730. https://doi.org/10.1371/journal.pone.0083730.
20. Gottesdiener E. Good practices for developing user requirements. CrossTalk J Defense Software Eng. 2008;21(3):13–7.
21. HHS. Enterprise life cycle artifacts. 2014. https://www.hhs.gov/ocio/eplc/Enterprise%20Performance%20Lifecycle%20Artifacts/eplc_artifacts.html. Accessed 26 Aug 2017.
22. Mitre. Eliciting, collecting, and developing requirements. 2017. https://www.mitre.org/publications/systems-engineering-guide/se-lifecycle-building-blocks/requirements-engineering/eliciting-collecting-and-developing-requirements. Accessed 13 Aug 2017.
23. Software Engineering Institute. A framework for software product line practice, Ver. 5.0, Requirements engineering. (2017). https://www.mitre.org/publications/systems-engineering-guide/se-lifecycle-building-blocks/r. Accessed 13 Aug 2017
24. Walters RA, Harlan PA, Nelson NP, Hartley DM, Voeller JG. Data sources for biosurveillance. In: Voeller JG, editor. Wiley handbook of science and technology for homeland security, vol 4: Wiley; 2010. p. 2431–47. https://doi.org/10.1002/9780470087923.hhs15.
25. Charles-Smith LE, Reynolds TL, Cameron MA, Conway M, Lau EH, Olsen JM, Pavlin JA, Shigematsu M, Streichert LC, Suda KJ, Corley CD. Using social media for actionable disease surveillance and outbreak management: A systematic literature review. PLoS One. 2015;10(10):e0139701. https://doi.org/10.1371/journal.pone.0139701.
26. Hartley DM. Using social media and internet data for public health surveillance: the importance of talking. Milbank Q. 2014;92(1):34–9. https://doi.org/10.1111/1468-0009.12039.
27. Hartley DM, Giannini CM, Wilson S, Frieder O, Margolis PA, Kotagal UR, White DL, Connelly BL, Wheeler DS, Tadesse DG, Macaluso M. Coughing, sneezing, and aching online: Twitter and the volume of influenza-like illness in a pediatric hospital. PLoS One. 2017;12(7):e0182008. https://doi.org/10.1371/journal.pone.0182008.
28. Volkova S, Charles LE, Harrison J, Corley CD. Uncovering the relationships between military community health and affects expressed in social media. EPJ Data Sci. 2017;6:9. https://doi.org/10.1140/epjds/s13688-017-0102-z.

29. Bach M, Jordan S, Hartung S, Santos-Hövener C, Wright MT. Participatory epidemiology: the contribution of participatory research to epidemiology. Emerg Themes Epidemiol. 2017. doi: https://doi.org/10.1186/s12982-017-0056-4
30. Paolotti D, Carnahan A, Colizza V, Eames K, Edmunds J, Gomes G, Koppeschaar C, Rehn M, Smallenburg R, Turbelin C, Van Noort S, Vespignani A. Web-based participatory surveillance of infectious diseases: the influenzanet participatory surveillance experience. Clin Microbiol Infect. 2014;20(1):17–21. https://doi.org/10.1111/1469-0691.12477.
31. Smolinski MS, Crawley AW, Baltrusaitis K, Chunara R, Olsen JM, Wójcik O, Santillana M, Nguyen A, Brownstein JS. Flu near you: crowdsourced symptom reporting spanning 2 influenza seasons. Am J Public Health. 2015;105(10):2124–30. https://doi.org/10.2105/AJPH.2015.302696.
32. Nguyen A, Brownstein JS. Flu near you: crowdsourced symptom reporting spanning 2 influenza seasons. Am J Public Health. 2015;105(10):2124–30. https://doi.org/10.2105/AJPH.2015.302696.
33. Garimella VRK, Alfayad A, Weber I. Social media image analysis for public health. In: Proceedings of the 2016 CHI Conference on Human Factors in Computing Systems. New York: ACM; 2016.
34. Generous N, Margevicius KJ, Taylor-McCabe KJ, Brown M, Daniel WB, Castro L, Hengartner A, Deshpande A. Selecting essential information for biosurveillance – a multi-criteria decision analysis. PLoS One. 2014;9(1):e86601. https://doi.org/10.1371/journal.pone.0086601.
35. Collmann J, Matei SA, editors. Ethical reasoning in big data: an exploratory analysis, Computational social sciences. Series. Basel: Springer; 2016.
36. Collmann J, Robinson A. Designing ethical practice in biosurveillance: the project Argus doctrine. In: Zeng D, et al., editors. Infectious disease informatics and biosurveillance, integrated series in information systems, vol. 27. New York: Springer; 2011.
37. Collier N. What's unusual in online disease outbreak news? J Biomed Semantics. 2010;1(1):2. https://doi.org/10.1186/2041-1480-1-2.
38. Torii M, Yin L, Nguyen T, Mazumdar CT, Liu H, Hartley DM, Nelson NP. An exploratory study of a text classification framework for Internet-based surveillance of emerging epidemics. Int J Med Inform. 2011;80(1):56–66. https://doi.org/10.1016/j.ijmedinf.2010.10.015.
39. Studer R, Benjamins R, Fensel D. Knowledge engineering: principles and methods. Data Knowl Eng. 1998;25(1–2):161–98.
40. Collier N, Kawazoe A, Jin L, Shigematsu M, Dien D, Barrero RA, Takeuchi K, Kawtrakul A. A multilingual ontology for infectious disease surveillance: rationale, design and challenges. Lang Resour Eval. 2006;40:405–13.
41. Smith WP, Chappell AR, Corley CD. Medical and Transmission Vector Vocabulary Alignment with Schema.org. In Proceedings of the International Conference on Biomedical Ontology (ICBO 2015), July 27-30, 2015, Lisbon, Portugal, edited by FM Couto and J Hastings. Aachen: CEUR Workshop Proceedings. http://ceur-ws.org/Vol-1515/regular11.pdf
42. Daughton AR, Priedhorsky R, Fairchild G, Generous N, Hengartner A, Abeyta E, Velappan N, Lillo A, Stark K, Deshpande A. A globally-applicable disease ontology for biosurveillance. Anthology of Biosurveillance Diseases (ABD). 2016. https://arxiv.org/abs/1609.05774
43. Wilson JM, Polyak MG, Blake JW, Collmann J. A heuristic indication and warning staging model for detection and assessment of biological events. J Am Med Inform Assoc. 2008;15(2):158–71. https://doi.org/10.1197/jamia.M2558.
44. Keller M, Freifeld CC, Brownstein JS. Automated vocabulary discovery for geo-parsing online epidemic intelligence. BMC Bioinf. 2009;10:385. https://doi.org/10.1186/1471-2105-10-385.
45. Yangarber R, Jokipii L, Rauramo A, Huttunen S. Extracting information about outbreaks of infectious epidemics. In: Proceedings of the Human Language Technology Conference/Conference on Empirical Methods in Natural Language Processing: HLT/EMNLP-2005. Vancouver, Canada, 2005; pp. 22–23. Available at: http://aclweb.org/anthology/H/H05/H05-2012.pdf.

46. Grein TW, Kamara KB, Rodier G, Plant AJ, Bovier P, Ryan MJ, Ohyama T, Heymann DL. Rumors of disease in the global village: outbreak verification. Emerg Infect Dis. 2000;6 (2):97–102.
47. Barboza P, Vaillant L, Mawudeku A, Nelson NP, Hartley DM, Madoff LC, Linge JP, Collier N, Brownstein JS, Yangarber R, Astagneau P, Early Alerting Reporting Project of The Global Health Security Initiative. Evaluation of epidemic intelligence systems integrated in the early alerting and reporting project for the detection of A/H5N1 influenza events. PLoS One. 2013;8 (3):e57252. https://doi.org/10.1371/journal.pone.0057252.
48. Velasco E, Agheneza T, Denecke K, Kirchner G, Eckmanns T. Social media and internet-based data in global systems for public health surveillance: a systematic review. Milbank Q. 2014;92 (1):7–33. https://doi.org/10.1111/1468-0009.12038.
49. Corley CD, Lancaster MJ, Brigantic RT, Chung JS, Walters RA, Arthur RR, Bruckner-Lea CJ, Calapristi A, Dowling G, Hartley DM, Kennedy S, Kircher A, Klucking S, Lee EK, McKenzie T, Nelson NP, Olsen J, Pancerella C, Quitugua TN, Reed JT, Thomas CS. Assessing the continuum of event-based biosurveillance through an operational lens. Biosecur Bioterror. 2012;10(1):131–41. https://doi.org/10.1089/bsp.2011.0096.
50. McGrath JW. Biological impact of social disruption resulting from epidemic disease. Am J Phys Anthropol. 1991;84(4):407–19.
51. Collmann J, Blake J, Bridgeland D, Kinne L, Yossinger NS, Dillon R, Martin S, Zou K. Measuring the potential for mass displacement in menacing contexts. J Refug Stud. 2016;29(3):273–94. https://doi.org/10.1093/jrs/few017.
52. Briggs CL, Mantini-Briggs C. Stories in the time of cholera: racial profiling during a medical nightmare. Berkeley: University of California Press; 2004.
53. Crouch A. Japanese biological warfare in China: one family's encounter. In the inventory of the Sheldon H. Harris Papers, Box 7, Folder 1. Hoover Institution Archives, Stanford University, Stanford, California. http://www.oac.cdlib.org/findaid/ark:/13030/kt8f59r20s/entire_text/. Accessed 12 Dec 2018
54. Ammari T, Schoenebeck S. Understanding and supporting fathers and fatherhood on social media sites. In: Proceedings of the 33rd Annual ACM Conference on Human Factors in Computing Systems. New York: ACM; 2015. p. 1905–14.
55. De Choudhury M, Counts S, Horvitz E. Predicting postpartum changes in emotion and behavior via social media. In: Proceedings of the SIGCHI Conference on Human Factors in Computing Systems. New York: ACM; 2013. p. 3267–76.
56. Lin YR. Assessing sentiment segregation in urban communities. In: Proceedings of the 2014 International Conference on Social Computing. SocialCom '14. New York: ACM; 2014. p. 9:1–8. https://doi.org/10.1145/2639968.2640066.
57. Valdes JMD, Eisenstein J, De Choudhury M. Psychological effects of urban crime gleaned from social media. In: Proceedings of ICWSM; 2015. http://www.aaai.org/ocs/index.php/ICWSM/ICWSM15/paper/view/10563.
58. Harris J, Mansour R, Choucair B, Olson J, Nissen C, Bhatt J, Brown S. Health department use of social media to identify foodborne illness-Chicago, Illinois, 2013-2014. MMWR. 2014;63 (32):681.
59. Paul MJ, Dredze M. Drug extraction from the web: summarizing drug experiences with multi-dimensional topic models. In: Proceedings of HLT-NAACL. Atlanta: Association for Computational Linguistics; 2013. p. 168–78.
60. Ortony A. On making believable emotional agents believable. In: Trappl R, Petta P, Payr S, editors. Emotions in humans and artifacts. Cambridge, MA: MIT; 2003.
61. Pennebaker J. Using computer analyses to identify language style and aggressive intent: the secret life of function words. Dynamics of Asymmetric Conflict. 2011;4(2):92–102. https://doi.org/10.1080/17467586.2011.627932.
62. Corley CD, Pullum LL, Hartley DM, Benedum C, Noonan C. Disease prediction models and operational readiness. PLoS One. 2014;9(3):e91989. https://doi.org/10.1371/journal.pone.0091989.

63. Schmidhuber J. Deep learning in neural networks: an overview. Neural Networks. 2015;61:85–117.
64. Nguyen TH, Grishman, R. Event detection and domain adaptation with convolutional neural networks. 2015. Volume 2. Short Papers, 365.
65. Dong L, Wei F, Zhou M, Xu K. Question answering over freebase with multi-column convolutional neural networks. In: Proceedings of the 53rd Annual Meeting of the Association for Computational Linguistics and the 7th International Joint Conference on Natural Language Processing, vol. Volume 1. Beijing: ACL; 2015. p. 260–9.
66. Lei T, Xin Y, Zhang Y, Barzilay R, Jaakkola T. Low-rank tensors for scoring dependency structures. In: Proceedings of the 52nd Annual Meeting of the Association for Computational Linguistics, vol. Volume 1; 2014. p. 1381–91.
67. Godin F, Vandersmissen B, De Neve W, Van de Walle R. Multimedia Lab@ ACL W-NUT NER shared task: named entity recognition for Twitter microposts using distributed word representations. ACL-IJCNLP. 2015;2015:146.
68. Wu Y, Schuster M, Chen Z, Le QV, Norouzi M, Macherey W, et al. Google's neural machine translation system: bridging the gap between human and machine translation; 2016. arXiv preprint arXiv:1609.08144.
69. Tang D, Wei F, Yang N, Zhou M, Liu T, Qin B. Learning sentiment-specific word embedding for Twitter sentiment classification. In: Proceedings of the 52nd Annual Meeting of the Association for Computational Linguistics, vol. 1; 2014. p. 1555–65.
70. Yin W, Schütze H. Convolutional neural network for paraphrase identification. In: Proceedings of the 2015 Conference of the North American Chapter of the Association for Computational Linguistics: Human Language Technologies. Denver: Association for Computational Linguistics; 2015. p. 901–11.
71. Bahdanau D, Cho K, Bengio Y. Neural machine translation by jointly learning to align and translate. In: Proceedings of the International Conference on Learning Representations (ICLR); 2015. arXiv preprint arXiv 1409.0473.
72. Cho K, Dzmitry Bahdanau D, Bougares F, Schwenk H, Bengio Y. Learning phrase representations using RNN encoder-decoder for statistical machine translation. In Proceedings of the Empirical Methods in Natural Language Processing (EMNLP) Conference; 2014. arXiv preprint arXiv:1406.1078.
73. Srivastava N, Mansimov E, Salakhutdinov R. Unsupervised learning of video representations using LSTMs; 2015. arXiv preprint arXiv:1502.04681.
74. Stoto MA. Biosurveillance capability requirements for the global health security agenda: lessons from the 2009 H1N1 pandemic. Biosecur Bioterror. 2014;12(5):225–30. https://doi.org/10.1089/bsp.2014.0030.
75. Burkom HS, Loschen WA, Mnatsakanyan ZR, Lombardo JS. Tradeoffs driving policy and research decisions in biosurveillance. Johns Hopkins APL Tech Dig. 2008;27(4):299–312.
76. Buehler JW, Berkelman RL, Hartley DM, Peters CJ. Syndromic surveillance and bioterrorism-related epidemics. Emerg Infect Dis. 2003;9(10):1197–204.
77. Hartley DM. Space imaging and rift valley fever vectors. In: Laxminaryan R, Macauley MK, editors. The value of information: methodological frontiers and new applications for realizing social benefit. Springer; 2012.
78. Macauley MK, Laxminarayan R (2010) The value of information: methodological frontiers and new applications for realizing social benefit. Resources for the future conference summary. http://www.rff.org/research/publications/value-information-methodological-frontiers-and-new-applications-realizing. Accessed 22 Aug 2017.
79. Laxminarayan R, Macauley MK. The value of information: methodological frontiers and new applications for realizing social benefit. Springer; 2012.
80. Chen H, Zeng D, Yan P. Infectious disease informatics: syndromic surveillance for public health and biodefense, integrated series in information systems 21. New York: Springer; 2010.

81. Wong W-K, Moore AW. Classical time-series methods for biosurveillance. In: Wagner MM, Moore AW, Aryel RM, editors. Handbook of biosurveillance. MA: Elsevier Academic Press; 2006.
82. Ginsberg J, Mohebbi MH, Patel RS, Brammer L, Smolinski MS, Brilliant L. Detecting influenza epidemics using search engine query data. Nature. 2009;457:1012–4. https://doi.org/10.1038/nature07634.
83. Pollett S, Boscardin WJ, Azziz-Baumgartner E, et al. Evaluating Google Flu trends in Latin America: important lessons for the next phase of digital disease detection. Clin Infect Dis. 2017;64:42–3. https://doi.org/10.1093/cid/ciw657.
84. Santillana M. Perspectives on the future of internet search engines and biosurveillance systems. Clin Infect Dis. 2017;64:34–41. https://doi.org/10.1093/cid/ciw660.
85. Lazer D, Kennedy R, King G, Vespignani A. The parable of Google Flu: traps in big data analysis. Science. 2014;343(6176):1203–5. https://doi.org/10.1126/science.1248506.
86. Bansal S, Chowell G, Simonsen L, Vespignani A, Viboud C. Big data for infectious disease surveillance and modeling. J Infect Dis. 2016;214(suppl_4):S375–9. https://doi.org/10.1093/infdis/jiw400.
87. Santillana M, Nguyen AT, Dredze M, Paul MJ, Nsoesie EO, Brownstein JS. Combining search, social media, and traditional data sources to improve influenza surveillance. PLoS Comput Biol. 2015;11:e1004513. https://doi.org/10.1371/journal.pcbi.1004513.
88. Parker J, Yates A, Goharian N, Frieder O. Health-related hypothesis generation using social media data. Soc Net Anal Min. 2015;5(1) https://doi.org/10.1007/s13278-014-0239-8.

Trust and Transparency in Biodefense

4

Filippa Lentzos

4.1 Introduction

In early 2000, a series of secret projects were reportedly underway in the United States to improve biodefense capabilities. The Pentagon was acquiring commercially available equipment to build a small-scale factory capable of producing *Bacillus anthracis* simulants—*Bacillus thuringiensis*, the biopesticide made at the main Iraqi bioweapons facility before it was destroyed by United Nations inspectors in 1997. Another US project involved genetically modifying *Bacillus anthracis* to make a vaccine-resistant "superbug". Meanwhile the CIA, in one of its projects, was building Soviet-style bio-bomblets and testing them for dissemination characteristics and performance in different atmospheric conditions.[1]

Pentagon and CIA lawyers stated the projects were legitimate defensive activities: Building and operating a bioweapons facility helped uncover the telltale clues of distinctive patterns of equipment buying; genetically modifying *Bacillus anthracis* was essential to test whether the current vaccines administered to soldiers were effective; and building and testing bomblets was a defensive response to specific intelligence about a possible adversary. Others disagreed, arguing the projects were not permitted under the Biological Weapons Convention (BWC), signed and ratified by the United States in 1975.[2]

[1]Miller J, S Engelberg and W Broad (2001) *Germs: The Ultimate Weapon*. New York: Simon & Schuster.

[2]Ibid.

F. Lentzos (✉)
Department of War Studies, Department of Global Health and Social Medicine, King's College London, London, UK
e-mail: filippa.lentzos@kcl.ac.uk

S. K. Singh, J. H. Kuhn (eds.), *Defense Against Biological Attacks*,
https://doi.org/10.1007/978-3-030-03053-7_4

The BWC permits almost any kind of research in the name of defense. Some of this work is justifiable. Other research edges closer to the blurred line between defensive and offensive work. The trouble with distinguishing permitted biodefense projects from non-permitted projects is that one cannot just assess the facilities, equipment and activities involved, but must also examine and interpret the purpose, or intent, of those activities. Article I of the BWC—through which states "agree to never under any circumstances acquire or retain biological weapons"—is therefore vague in demarcating the borders of prohibited and legitimate activities. Article I merely refers to biological agents "of types and in quantities that have no justification for prophylactic, protective or other peaceful purposes." An essential part in judging states' compliance with the Convention is therefore an analysis of justifications provided for the activities in question.

Undertaking secret borderline projects, such as building mock bioweapons factories, creating "superbugs" or testing germ bomblets, undermines the BWC because such activities, regardless of their legitimacy, will inevitably stir suspicion when they come to light. While there is defensive research that a nation might legitimately keep secret—such as experiments exploring the vulnerabilities of existing vaccines—the existence of such research and its general outlines should be disclosed whenever possible to allay fears and suspicions.

States with biodefense programs have a special responsibility to demonstrate that their defensive activities are not used as cover for offensive programs, and, importantly, that their programs are not perceived as such. It is particularly important to proactively counter the perception that a biodefense program may be used to disguise an offensive program, or elements of an offensive program, because such a perception may provide other states with justification for initiating, or continuing, their own offensive biowarfare programs.

Unusually for a disarmament and non-proliferation treaty, however, the BWC was agreed without including verification mechanisms to ensure compliance. The decade-long effort to introduce a legally binding verification protocol for the BWC failed in 2001, and it is highly unlikely that political agreement can be reached on such a protocol in the near- to mid-term. Yet, this does not mean that states cannot demonstrate responsible biodefense, or that it is impossible for states to be assured other countries are abiding by their treaty obligations. There are a number of actions and activities that cumulatively may give a reasonable indication of a state's intent and compliance status over time. These include implementing national biosafety and biosecurity legislation, developing effective oversight processes for dual use research of concern (DURC), working with open publication policies on research at biodefense facilities, efforts to promote a culture of responsibility in the scientific community, on-going and constructive participation in BWC meetings, and regular engagement with academia and civil society on issues related to the Convention.[3]

[3]See e.g. 2013 BWC, Meeting of Experts, 'We need to talk about compliance: a response to BWC/MSP/2012/WP.11', Working paper submitted by the United Kingdom, BWC/MSP 2013/MX/WP.1.

Paramount for states with biodefense programs, however, is the nurturing of trust through transparency. Trust between states, and between nations and their citizens, is enhanced through openness and communication.

This chapter considers the information-sharing mechanisms currently operating in the BWC that are aimed at demonstrating that biological activities are solely undertaken for peaceful purposes.[4] The chapter first describes and appraises the confidence-building measures (CBMs) that constitute the principal information exchange that formally applies to all member states. The second part of the chapter presents and analyzes a series of new and informal information-sharing mechanisms that have gained currency with a number of states in the last few years: 'compliance assessment', 'peer review', 'implementation review', and 'facility visits'. In conclusion, the chapter argues that these more informal exchange mechanisms have the potential to achieve, in the longer term and in conjunction with other efforts, what the verification protocol set out to do, but they would so in a different way, better suited to engendering trust today.

4.2 Confidence-Building Measures

The BWC CBMs constitute the core, formal mechanism whereby member states exchange information on a regular basis. Submitted annually, the set of six agreed forms is comprehensive in scope and includes data on biodefense programs. The forms cover the objectives and funding of biodefense programs, the principal research and development activities, the facilities involved, the organizational structure and the reporting relationships of the facilities, and details of any sub-contracted parties from industry, academia or other non-defense institutions. The forms also cover information on maximum containment facilities outside defense, the national oversight framework implementing the treaty and regulating biological research, published articles or reports detailing results from research directly related to the Convention, unusual outbreaks of infectious diseases, vaccine production facilities, and details of any past offensive and/or defensive activities.[5]

---

[4]Much of the material in this chapter is based on previous publications, including Koblentz G and F Lentzos (2016) '21st Century biodefence: Risks, trade-offs and responsible science' ILPI BWC Review Conference Series Paper No. 3 available at http://nwp.ilpi.org/wp-content/uploads/2016/11/03-21st-century-biodefence-gold.pdf [accessed 17 July 2017]; Lentzos F (2016) 'Increasing transparency in biodefence: A 2016 visit to a German military medical biodefence facility,' EU Non-proliferation Consortium Paper No. 52; Lentzos, Filippa (2015) '3D Bio: Declare, Document & Demonstrate,' EU Non-proliferation Consortium Paper No. 45; and Lentzos, Filippa (2011) 'Strengthening the BWC confidence-building measures: Toward a cycle of engagement' *Bulletin of the Atomic Scientists* Vol. 67(3):26–33.

[5]For more details on the information requested as part of the CBM exchange, see the blank form templates and the 'Guide to participating in the confidence-building measures of the Biological Weapons Convention' (Revised edition 2015) prepared by the United Nations Office for Disarmament Affairs, available from www.unog.ch/bwc [accessed 17 July 2017].

The CBMs emerged in the early 1980s following the crisis of confidence among states that resulted from unresolved allegations of non-compliance, rapid developments in science and technology, and other pressures. They were conceived, developed, and agreed upon at a time when it seemed plausible that a verification mechanism was going to be put in place that resembled the declarations and on-site inspections of the Chemical Weapons Convention (CWC) then under negotiation. The measures were therefore not conceived of as a verification tool, but merely as a layer within a larger "regime of compliance."[6]

Anchored in Article V of the BWC on consultation and cooperation,[7] the CBMs were agreed at the Second Review Conference in 1986, elaborated at a meeting of scientific and technical experts in 1987, and modified and considerably expanded at the Third Review Conference in 1991. The emphasis on seeking verification between the end of the Cold War and the Fifth Review Conference in 2001 resulted in decreased interest in further developing the measures. The Sixth Review Conference in 2006 resulted in agreement on various improvements to the mechanism for submission and distribution of CBMs, but political differences meant the CBMs remained substantively unchanged until minor modifications to the reporting forms were adopted at the Seventh Review Conference in 2011.

Although the CBMs are not derived directly from the text of the Convention itself, the Second Review Conference resulted in a decision by consensus that BWC member states were "to implement [the CBMs] on the basis of mutual co-operation". This decision meant that participation in the CBMs became a requirement for all BWC member states. The majority of states with biodefense programs recognize their responsibility to ensure high standards of transparency, and they submit information about their programs under the CBM mechanism to reassure other states that their activities are solely for peaceful purposes.

Initially, in the first round of submissions in 1992, 13 states declared biodefense programs.[8] The number of states declaring biodefence programs has grown over the years, with the last two decades witnessing a dramatic increase in biodefence activities and in the number of facilities and researchers working with dangerous pathogens around the world. In the most recent set of submissions of 2016, 29 countries declared biodefense programs.[9] Concerns about bioterrorism and emerging infectious disease have also triggered a construction boom around the world in maximum biocontainment [or biosafety level 4 (BSL-4)] laboratories, in which the most dangerous pathogens are handled. There are 13 BSL-4 laboratories in operation or under construction in the United States and at least 24 such

[6]Sims N.A. (2001) *The Evolution of Biological Disarmament.* Oxford: Oxford University Press.

[7]BWC Article V states: The States Parties to this Convention undertake to consult one another and to cooperate in solving any problems which may arise in relation to the objective of, or in the application of the provisions of, the Convention. Consultation and cooperation pursuant to this article may also be undertaken through appropriate international procedures within the framework of the United Nations and in accordance with its Charter.

[8]Op. cit. Koblentz G and F Lentzos (2016).

[9]Ibid.

laboratories in operation in the rest of the world.[10] In addition, there are more than 1400 BSL-3 laboratories registered to work with dangerous pathogens in the United States.[11] Such laboratories are increasingly common around the world as countries upgrade their public health infrastructure, but exact numbers are unknown.

The equipment, material, and science behind much of the work on dangerous pathogens can be considered as "dual-use". This means that while intended for peaceful purposes, they may also be used for nefarious purposes. Distinguishing between permitted and prohibited activities is difficult at the level of basic biological research, for which the same techniques used to gain insight and understanding about fundamental life processes for the benefit of human health and welfare may also be used for the development of biological weapons agents. Examples of the sorts of experiments that are of particular concern include those that increase capacity to: manipulate the pathogenicity, virulence, host-specificity, transmissibility, resistance to drugs of a replicative entity; synthesize pathogens and toxins without cultivation of microorganisms or using other natural sources; identify new mechanisms to disrupt the healthy functioning of humans, animals, and plants; and develop novel means of delivering biological agents and toxins. The US State Department has noted in its annual compliance report to Congress that China, Iran, and Russia are engaged in dual-use activities, but that available information does not indicate whether these activities are conducted for purposes inconsistent with the BWC.[12] The same logic, of course, applies to the United States and any other country with a developed bioscience base.

The CBMs provide states engaged in such dual-use activities, especially within their biodefense programs, with an opportunity to provide rationales and justifications for their activities, which form a key component of compliance judgements.

On a national level, collating national biodefense information and combining projects from different funders, like the Pentagon and the CIA, offers a state the opportunity to review the full range of biodefense efforts it undertakes countrywide and thereby provides an additional layer of oversight. Had the White House known about the entire package of secret projects performed in early 2000, it could, for instance, have initiated a legal review to ensure US compliance with the BWC. The process of collecting and submitting information for CBM submissions provides a mechanism for individual governments to draw domestic stakeholders together, to focus internal inter-agency or inter-departmental coordination, and to increase their awareness and oversight of relevant national biological activity. Through complete, accurate, and annual submissions, national patterns of normal activity can be

[10]National Research Council (2011) *Biosecurity Challenges of the Global Expansion of High-Containment Biological Laboratories*. Washington, DC: National Academies Press.

[11]Kaiser J. (2011) 'Taking stock of the biodefence boom', *Science* Vol. 333, No. 6047: 1214.

[12]US Department of State (2015) *Adherence to and Compliance with Arms Control, Nonproliferation, and Disarmament Agreements and Commitments*. Bureau of Verification, Compliance and Implementation. In 2016 and 2017, these reports limited themselves to Russia only on the grounds that "there is insufficient information to support the inclusion of other countries".

established, and this can also make a significant contribution to a compliance judgement. CBM submissions demonstrate to peers in government, and to peers in other governments, that states have their house in order.

Unfortunately, in practice, many states, some of which have biodefense programs, fall far short of this ideal, only submitting partially completed returns, submitting returns in some years but not in others, or, in the worst case, failing to submit returns altogether. Over the years, there have been various proposals to address these challenges, and to improve and strengthen the CBM mechanism. More specifically, these proposals have sought to:

1. Review the questions asked on the CBM forms, ensuring that they are clear, relevant, and that they contribute to enhancing transparency and building trust between states;
2. Improve the format of the CBM forms, making their completion more intuitive and user-friendly, while also addressing the need to make CBMs available in a wider number of languages, ensuring universal accessibility;
3. Modernize the reporting process, making greater use of electronic CBM forms and online resources, moving towards a comprehensive web-based information management system that is accessible to all member states;
4. Improve national data collection processes, encouraging improved collator rotation through handover notes, and offering guidelines, completed CBM forms, data collection, and collation workshops and one-to-one assistance;
5. Strengthen the role of the Implementation Support Unit (ISU), moving towards an administrative office that will have the appropriate authority and the resources to facilitate the implementation of the CBM mechanism;
6. Promote cooperation between member states, encouraging bilateral and multilateral dialogue, allowing states in a position to do so to assist other states struggling to fulfill their CBM obligations; and to
7. Invite civil society groups and international organizations to play a role in the CBM information exchange, drawing on their expertise and energy to help address problems with the CBM mechanism and seek possible solutions.[13]

At the most recent Review Conference in 2016, there were renewed calls to generate more useful information by further refining the type and range of information requested in the CBMs. The United States, for instance, emphasized that clarification was needed to confirm that the information requested on biodefence programs includes both military and civilian programs. It noted: "At present, roughly one-third of the States Parties declaring national biodefense research programs report civilian biodefense research. For the remaining two-thirds of States Parties, it is not clear whether they have construed the request for information to apply only to

[13]Lentzos F and RA Hamilton (2010) *BWC Confidence Building Measures: Preparing for a Comprehensive Review of the CBM mechanism at the Seventh Review Conference*, Report prepared for BWC MX 2010.

military programs, or whether they do not have biodefense research programs conducted by civilians aimed at protecting the civilian population."[14]

The United States has, for example, by one estimate, spent on average $1.79 billion annually during FY2010–FY2016 on federal civilian biosecurity to prevent, prepare for, and respond to attacks on civilians with biological agents and accidental releases of biological material.[15] Capturing this information as part of the CBM information exchange is crucial to provide an accurate reflection of the national biodefense capacity and of the research, projects, and programs carried out. It is also particularly important because, although the primary aim of the CBMs is to enhance transparency about military biodefense programs, their main use today is more likely to be in ensuring that the rapidly expanding civil biodefense programs built up in a number of countries over the last decade are not 'creatively interpreting' the treaty, and the line between offensive and defensive activities, to their own ends.

To maximize their transparency, an increasing number of states are now also making their CBM submissions publicly available and open to civil society analysis. Making these submissions public can greatly enhance their function. The knowledge, experience, and expertise of civil society can contribute to the communication process and to enhancing transparency between states in several ways, including through: assisting states to collect and collate information for and on the CBMs; monitoring states' activities; collecting data from open sources; processing the data submitted to generate accessible information; and, ultimately, by bringing this information into the public sphere. Restricting access to CBM returns risks building suspicion rather than confidence among important stakeholders and misses an opportunity to engage these same stakeholders in processes that might actually enhance the quality and completeness of the submitted information. At the 2011 Review Conference, 22 states made their CBMs publicly available. By the 2016 Review Conference, 30 states made their CBM submissions publicly available, 18 of which had biodefense programs.[16]

4.3 Compliance Assessment

Recently, a small number of states have voluntarily gone further in their efforts to be transparent and to allay any potential suspicions about the status of their biodefense programs.

[14]2016 BWC Review Conference, 'Strengthening confidence-building and consultative mechanisms under the Biological Weapons Convention,' Working Paper submitted by the United States, BWC/CONF.VIII/PC/WP.6, 21 April 2016.

[15]Watson C, M Watson and T Kirk Sell (2017) 'Federal funding for health security in FY2018,' *Health Security* Vol. 15(4).

[16]www.unog.ch/bwc [accessed 17 July 2017].

Canada, in the lead-up to the Seventh Review Conference in 2011, proposed a BWC 'compliance assessment' initiative.[17] The initiative approached the concept of compliance verification from the broad perspective of examining national legislation and implementation programs, rather than from the more traditional and focused perspective of inspecting facilities. Canada sought to demonstrate that options to evaluate compliance and implementation of the treaty exist outside of a legally binding verification protocol.

Canada viewed national implementation programs as including both the implementation measures a state party has taken and its efforts to administer and enforce those measures on an ongoing basis. Submissions under the compliance assessment initiative are therefore comprised of two sections: (1) a detailed description of national legislation and regulations supporting the national implementation of the BWC, including section-by-section analysis of how the statutory measures work, the scope of the statutory measures, and the penalties associated with contraventions; and (2) a detailed description of how the program operates on a national level, including the yearly budget for running the program, process flow diagrams and organizational charts of the implementing program, process and standard operating procedures, as well as clear indications of the inspection program, frequency of inspections, the number of announced versus unannounced inspections, and how major and minor non-compliances are handled.

States deemed to be effectively regulating facilities that handle pathogens would enhance confidence that the state is in compliance with its non-proliferation obligations under Article I (as well as Articles III and IV) of the Convention.

Canada put itself forward to act as a test case and made an initial compliance assessment submission in the form of a Working Paper to the 2012 BWC meeting.[18] The submission provided more in-depth reporting on national implementation as well as operational details to supplement and add to the information Canada provides through its annual CBM submissions and its quinquennial review conference compliance reports. Other interested states were invited to join in the initiative, and Switzerland, the Czech Republic, and France all made Working Paper submissions to recent BWC meetings.[19]

Given the diversity of methods to implement the Convention, the participating states did not follow a common template. The Canadian submission was built to resemble a compliance report and involved an article-by-article review of Canada's implementation measures. Canada has a large and centralized oversight system in

[17]2010 BWC Meeting of States Parties, 'National implementation of the BTWC: compliance assessment: a concept paper', Working Paper submitted by Canada, BWC/MSP/2010/WP.3/Rev.1.

[18]2012 BWC Meeting of Experts, 'National implementation of the BTWC: compliance assessment', Working Paper submitted by Canada and Switzerland, BWC/MSP/2012/MX/WP.17.

[19]2012 BWC Meeting of States Parties, 'National implementation of the BTWC: compliance assessment: update', Working Paper submitted by Canada, the Czech Republic and Switzerland, BWC/MSP/2012/WP.6; and 2013 BWC Meeting of Experts, 'National implementation assessment report of the Biological Weapons Convention', Working Paper submitted by France, BWC/MSP/2013/MX/WP.16.

which the Public Health Agency of Canada is responsible for almost all aspects of BWC implementation in the country. This means that the federal oversight programs that have been developed can easily be described and scrutinized in detail. The Swiss and Czech submissions more closely resembled a very detailed CBM Form E, with concise descriptions of the legal authorities and programs arranged by ministry. In Switzerland, where implementation of the BWC is more decentralized and many measures are implemented at the cantonal level, a smaller, though broader, submission was more manageable and effective. The submission by the Czech Republic was designed to reflect a smaller centralized system.

Preparation of the initial submissions were seen to require significant effort, but it was highlighted that the included level of detail would only need to be submitted once. Subsequent submissions would only require updating of the 'tombstone' (initial) information with annual compliance and enforcement metrics and any new policy initiatives. In 2012, Canada and Switzerland submitted sample annual declarations to demonstrate the ease of subsequent submissions.

4.4 Peer Review

A more interactive arrangement was put forward by France around the same time as Canada proposed its compliance assessment initiative.[20] Here, participating states were to physically come together to make mutual assessments of national implementation standards based on common understandings reached during the intersessional process. Like the 'compliance assessment' submissions, the 'peer review' mechanism was also aimed at building confidence and providing transparency between states. Additionally, this mechanism also aimed at improving national implementation and, importantly, providing an opportunity to share experiences and best practices among experts.

The mechanism was presented as voluntary, with all assessments being initiated at the request of the state concerned. Whether the assessment report was to be published was up to the state, as was the nature of the follow-up process. The main advantages identified of peer review were enhancing national implementation, improving confidence among states through increased transparency, sharing good practices, and providing the opportunity to develop international cooperation.

France organized the first peer review exercise in 2013.[21] National experts from nine states—Canada, China, Germany, India, Mexico, Morocco, Switzerland, the United Kingdom, and the US—as well as a representative from the BWC Implementation Support Unit as an observer, were given presentations on French biosafety

[20] 2011 BWC Review Conference, 'A peer review mechanism for the Biological Weapons Convention: enhancing confidence in national implementation and international cooperation', Working Paper submitted by France, BWC/CONF.VII/WP.28.

[21] 2014 BWC Meeting of States Parties, 'Peer review pilot exercise held from 4 to 6 December 2013 in Paris', Working Paper submitted by France, BWC/MSP/2014/WP.3.

and biosecurity measures, export controls, and awareness-raising policy. The information provided in the presentations was then illustrated through on-site visits to two civilian laboratories (Maisons-Alfort Laboratory for Animal Health and the Pasteur Institute) that included exchanges with personnel involved in research activities. France encouraged other states to organize similar exercises to develop their own approaches to peer review and broaden understanding of the concept.

In 2015, Belgium, Luxembourg, and the Netherlands conducted a Benelux peer review exercise among themselves based on a mutually developed and agreed format.[22] In their interpretation of 'peer review,' the CBMs, and in particular Forms A and E, were given a central role as the basis on which to perform the national implementation evaluation. The Benelux peer review consisted of a first phase during which written consultations took place between the three states, and a second phase during which each state organized an event and hosted visiting peers from the other two states for a review through presentations and on-site visits.

4.5 Implementation Review

A third informal arrangement, labelled 'implementation review', was put forward by Canada, Chile, Ghana, Mexico, and the United States in 2016.[23] This arrangement aimed to voluntarily enhance transparency and to exchange information, experiences and best practices on national implementation of the BWC by mixing elements from both 'compliance assessment' and 'peer review'. The five participating states jointly developed a national reporting form that was to operate as the basis for information exchange on national laws, regulations and other measures implementing BWC obligations; legal prohibitions; export controls; biosafety and biosecurity; and oversight of life sciences DURC. Each state then completed the national reporting form by consulting relevant ministries, departments, and agencies. The completed form was circulated to the other participating states for review, and areas were identified for which further information was desired.

The next step in the implementation review process involved 2-day visits to the capitals of each participating state for in-depth conversations on laws, regulations, and implementation in practice. In each country, the one or two BWC experts invited to the capital would meet with a larger number of local practitioners directly responsible for some aspect of implementation (e.g., specialists in biosafety, biosecurity and biocontainment; lawyers; export and import control officers). To

[22] 2016 BWC Preparatory Committee, 'Strengthening the BWC: reflecting on the peer review concept', Working Paper submitted by Belgium, Luxembourg and Netherlands, BWC/CONF. VIII/PC/WP.26; and 2016 BWC Preparatory Committee, 'Peer review: an innovative way to strengthen the BWC', Working Paper submitted by Belgium, France, Luxembourg and Netherlands, BWC/CONF.VIII/PC/WP.13.

[23] 2016 BWC Review Conference, 'BWC Implementation Review Initiative', Working Paper submitted by Canada, Chile, Ghana, Mexico and the United States of America, BWC/CONF. VIII/WP.22.

minimize travel costs, all four visits in the Americas were done sequentially from North to South over a 2-week period; the visit to Accra was conducted later. Facility visits were also part of the program. In Canada, the invited experts were given a tour of the Ottawa Hospital Research Institute. In the US, the visit included a tour of the BSL-3 and BSL-4 training laboratories at the National Institutes of Health (NIH) in Bethesda, Maryland. Four facilities were visited in Mexico: The National Service for Food Safety and Quality Safety, the Federal Commission for Protection Against Sanitary Risk, the National Institute for Nuclear Research, and the BSL-3 lab of the Institute of Epidemiological Reference and Diagnosis. In Chile, the invited experts were given a demonstration of the national border control mechanism at the International Airport Arturo Merino Benítez, in Santiago. In Ghana, the participants toured the BSL-2 and BSL-3 facilities of the Noguchi Memorial Institute for Medical Research located on the University of Ghana campus in Accra.

Each partner state completed a summary report of its capital visit and circulated it among the invited experts to allow them to review it for accuracy. Participants were encouraged to share their observations, but the aim was not to grade or criticize others' progress in implementing the BWC.

The participating states noted that one of the strengths of the implementation review process was the use of a common reporting form. This commonality helped the assessment of the pros and cons of the different approaches, such as the varying approaches to which Department or institution is assigned leadership on BWC implementation within the participating states' governments, and facilitated a direct comparison of the implementation systems. Participants also observed that a discussion would have been helpful at the beginning of each visit on the government structure and organizational relationships in that country as they apply to BWC implementation. This sort of introduction would provide a framework for the subsequent presentations and discussions. After the visits to the Western Hemisphere capitals, Ghana took advantage of this lesson and provided such information early in its briefings in Accra. The participating states also noted that they operate with different definitions of biosafety and biosecurity. For example, for some, biosafety referred to the means of protecting laboratory workers, the public, and the environment from inadvertent exposure to dangerous biological agents used in laboratories. Other countries took a broader approach, including the import of foodstuffs and the use of genetically modified organisms. With regards to biosecurity, some countries limited their definition to protecting infectious biological agents from loss, theft or misuse, whereas others used a more expansive definition that included monitoring food for infectious agents that could impact the agricultural sectors.

4.6 Facility Visits

In 2011, the US emphasized its commitment to building an environment of openness and collaboration in its biodefence activities, and outlined a set of unilateral, voluntary activities for a 'bio-transparency and openness' initiative.[24] In 2012, the US reported on these activities.[25] A cross section of Geneva-based ambassadors, including the BWC chair, had visited the National Interagency Biodefense Campus at Fort Detrick, Maryland. The ambassadors toured the US Army Medical Research Institute of Infectious Diseases and the National Biodefense Analysis and Countermeasures Center of the US Department of Homeland Security. They received briefings from senior representatives from the Departments of Defense, Homeland Security, Health and Human Services, and Agriculture, as well as the Federal Bureau of Investigation. The visit provided insights into US programs and facilitated greater understanding of the nature and scope of US biodefense activities. The visit also demonstrated the 'whole-of-government' approach that the US follows for biodefense and provided an opportunity to share ongoing international cooperation activities and assistance.

Similar initiatives to host visits to maximum containment laboratories and to organize international conferences on BWC topics have also been performed by other states to encourage transparency and foster the exchange of views. Switzerland, for instance, has on a number of occasions invited Geneva-based BWC delegations to visit its BSL-4 laboratory in Spiez. This laboratory also hosts annual open days for locals and other interested parties.

From 2015 to 2016, Spain, along with Chile and other Latin American countries, launched a series of 'voluntary visits' with the aim of exploring strengths and weaknesses related to biosafety, and in particular biosecurity, at the national level in each of the participating states. In each country, a number of bio-facilities were visited and seminars organized. To follow up on the process, workshops were run in Madrid with the goal of fostering regional cooperation and to support Latin American countries in improving their biosafety and biosecurity legislations.

To focus compliance assessment more specifically on biodefense, military activities and on-site visits, Germany organized what was described as a 'peer review compliance visit exercise' in 2016.[26] This exercise was a joint project organized by the German Ministry of Foreign Affairs (MFA) and the Ministry of Defense (MOD), and part of the country's long-standing efforts to pragmatically and incrementally strengthen the BWC to contribute to the longer-term national and

[24] US Secretary of State Hillary Rodham Clinton 'Remarks at the 7th Biological and Toxin Weapons Convention Review Conference,' 7 December 2011. Available at: www.state.gov/secretary/20092013clinton/rm/2011/12/178409.htm [accessed 17 July 2017].

[25] 2012 BWC Meeting of States Parties, 'The United States Government's Bio-transparency and Openness Initiative,' Working Paper submitted by the United States, BWC/MSP/2012/WP.3.

[26] 2016 BWC Review Conference, 'Confidence in compliance: peer review visit exercise at the Bundeswehr Institute of Microbiology in Munich, Germany', Working Paper submitted by Germany, BWC/CONF. VIII/WP.11.

European Union objective of strengthening treaty compliance and verification capacities. The German MOD opened the Bundeswehr Institute of Microbiology in Munich to 20 international experts, with the stated objective to demonstrate that the facility complied with the provisions and obligations of the BWC.

The experts were asked to evaluate the information gained during the visit and to develop a compliance assessment report.[27] Their approach was to assess the visit based on the facilities, equipment and materials, and procedures and personnel. The information provided in Germany's CBM submission was taken into account, as was the information provided and observed during the visit. The summary generally followed the structure of the assessment meeting discussion.

By looking at the facility, the visiting experts found that the size of the institution was consistent with the information provided in the CBM. The experts also noted that although located within an old building, the facility had been adapted to meet the purpose of its activities and that the physical security included several layers of access control to prevent unauthorized access to sensitive areas. In terms of equipment and infrastructure, the assessment report stated that everything the visitors had seen was relevant for the prophylactic, protective, and other peaceful research and diagnostic purposes stated by the Institute. The report continued: "With regard to pathogen and toxin material, particularly that held in the BSL3 strain collection, the process for access control was explained in detail to us. Although not state-of-the-art access control, the database methodology was suitable for the purpose and an example of good practice. There was also evidence of a strongly developed culture of trust between the staff. The pathogens and toxins were of types and in quantities consistent with the stated purposes in the CBM and the briefings during the visit."

The experts noted that in addition to the procedures explained during the initial briefing, they were given further and more detailed elaborations on request. Documentation was provided to demonstrate that specific processes were in place to assess projects considered DURC both at startup and during conduct of the work. Documented procedures for the transfer and export of pathogens and toxins from the Institute were also provided. The experts noted that safety incident reporting could have been made available if needed; that access to laboratory notebooks was permitted on specific request; and that the Institute followed international standards (ISOs).

The observations continued: "Information was provided on training programs for staff both on initial employment and annual refresher training covering technical, safety and procedural issues. Records of staff training for specific protocols were shown. All staff underwent security vetting, and senior staff developed relationships of trust with those in their teams. Although not requested, the opportunity for interviewing additional staff members would have been available under the facility provisions. Throughout the visit, all local personnel and escorts were very helpful

[27] Lentzos F (2016) 'Increasing transparency in biodefence: A 2016 visit to a German military medical biodefence facility,' EU Non-Proliferation Consortium Paper No. 52.

and provided open responses to all requests for access, documentation and information."

The visiting experts concluded that they had been provided with good insight into the research and diagnostic activities of the Institute, and that the hosts' cooperation had "helped to promote transparency and confidence with regards to the BWC and reassurance that all activities are within the permitted prophylactic, protective and other peaceful purposes in Article I." The procedures for biosafety and biosecurity, DURC issues, transfer and export of pathogens and toxins, and other measures had demonstrated commitment to implementing the obligations under Articles III and IV of the BWC, and that the information provided in the CBM submission had added to this commitment. In addition, some information had been provided on international assistance and cooperation activities of relevance to Articles VII and X.

In closing, the summary of observations maintained that as well as providing an opportunity for the Institute and Germany to demonstrate consistency with their CBM submission and transparency of their implementation of the BWC, the exercise provided a useful platform for exchange of good practices among all participants.

4.7 Conclusion

Conflicting views on the verifiability of the BWC have endured through the decades since the treaty was first negotiated and remain one of the fundamental challenges in the field of biological disarmament and non-proliferation. Some states maintain that the nature of biological weapons is such that it is inherently impossible to verify that there are no biological weapons: not only can significant quantities of biological agents be produced in small and readily concealable facilities, but most of the equipment required (e.g. fermenters, centrifuges, freeze-dryers) is ubiquitous in public, private, and commercial laboratories. Others argue that, although the same level of accuracy and reliability as the verification of, for example, nuclear arms control treaties is unattainable, it is possible to build a satisfactory level of confidence that biology is only used for peaceful purposes.

There are many states that support strengthening the Convention on the basis of a legally binding document. Yet there are few who believe that a return to the draft verification protocol of the 1990s is the way forward. The draft protocol, however, was a package of measures, only part of which involved verification. The fact that the protocol as a package is 'dead' does not necessarily mean that all the constituent parts, including verification, are also 'dead'.[28] Indeed, as indicated by the recent informal information-sharing initiatives described in this chapter, there is growing appetite among BWC member states for meaningful dialogue on how best to enhance transparency and build confidence that states are in compliance with their obligations under the Convention.

[28]Lennane R. (2011) 'Verification for the BTWC: If not the protocol, then what?' *Disarmament Forum* No. 1, pp. 39–50.

It is possible to extract and develop useful components from the draft protocol that could conceivably operate (at least initially) without a legally binding framework and without the institutional support of an organization like the Organisation for the Prohibition of Chemical Weapons (OPCW). A bottom-up approach could result in the development and implementation of individual components on a small scale, their refinement and improvement in operation, gradually expanding participation and scope, and then—once everyone knows what is involved and is confident the measures work in practice—in bringing them together in a legally binding instrument.

The draft protocol had four complementary components: declarations, visits, consultation, and clarification provision, and investigations. Both 'declarations' and 'investigations' are similar to mechanisms that already exists: the CBMs and the UN Secretary-General's Mechanism for Investigation of Alleged Use of Chemical and Biological Weapons (UNSGM). Both the CBMs and the UNSGM require strengthening, but the mechanisms themselves are in place. The 'consultation and clarification provisions' of the protocol can be reproduced independently, and to some extent exist in the form of the procedures agreed at the Second and Third Review Conferences for consultations under Article V of the Convention.

Arguably, in the longer term and alongside the CBMs, UNSGM, and consultation and clarification provisions, the informal information-sharing initiatives described in this chapter could potentially be developed to supplant the final 'visits' component of the protocol. Although compliance assessment, peer review, implementation review, and facility visits are distinct from one another, they all share an overarching objective: to move the verification debate on in a workable, pragmatic way. The initiatives also share many similar characteristics: (1) the participation of foreign, highly qualified experts, allowing for substantive exchanges among practitioners and the construction of a network of qualified experts; (2) on-site visits of facilities with opportunities to share experience with local staff, essential for participating states to compare approaches in a way that goes beyond examining the text of laws and regulations and to see what the implications are in the field; (3) the involvement of various national agencies, which can help foster interagency efforts within the host nation, inform relevant agencies about BWC-related issues, and demonstrate how agencies work together to implement the Convention; and (4) feedback from participants with a view to enhancing the way the BWC can be implemented by the host country and to informing other states about how the host country implements the Convention.[29] In general, the information-sharing initiatives were deemed to add real value to states' compliance judgements by all participating states.[30] Their emphasis on interaction and flexibility, on expert level exchanges of

[29]2016 BWC Review Conference, 'Building confidence through voluntary transparency exercises,' Working Paper submitted by Belgium, Canada, Chile, Czech Republic, France, Ghana, Germany, Luxembourg, Mexico, the Netherlands, Spain, Switzerland and the United States, BWC/CONF. VIII/WP.35.

[30]Ibid.

best practices rather than just on-site monitoring, and on a broad conception of relevant laboratories and facilities, arguably make them even better suited than the protocol 'visits' to enhancing transparency and building trust in twenty-first century biodefense.

The Science of Forensic Epidemiology in Detecting a Biological Attack (Bioterrorism)

5

Steven A. Koehler

5.1 Introduction

Historically the role of health departments and their staff of epidemiologists has been to monitor the overall health of the community they serve. This monitoring is conducted by a number of surveillance systems that track the number and types of disease, illness, injury, and death within that community. The department responds to food outbreaks; conducts investigations related to exposure risks such as the release of chemicals into the water, atmosphere, or soil; and informs the population of other public health risks. In addition, they conduct studies demonstrating the risks associated with certain types of exposures and behaviors and resulting diseases and health risks. They also serve the community by conducting educational programs addressing health risks and methods to reduce disease, illness, injury, and death.

The current and growing threats from either an overt or covert biological attack have resulted in health departments around the country and its staff to take on a new and unfamiliar role. Health departments typically investigate diseases, illnesses, injuries, and deaths that occurred naturally or accidentally, but now they are involved in the investigations of criminal or terroristic events. The investigation of health-related criminal acts, such as the deliberate poisonings of water systems or a biological terrorist attack, marks a new era for public health officers and highlights the increasing role and function of the new specialist familiar with the science of epidemiology, forensics, and law enforcement: the forensic epidemiologist. The emergence of this new public health threat has resulted in the field of forensic epidemiology and forensic epidemiologists to expand their role and function from civil to criminal investigations. The specialty of forensic epidemiology utilizes the foundations of epidemiology, public health methods such as surveillance and outbreak investigation techniques, forensic techniques, and law enforcement

S. A. Koehler (✉)
Forensic Medical Investigations, Pittsburgh, PA, USA

S. K. Singh, J. H. Kuhn (eds.), *Defense Against Biological Attacks*,
https://doi.org/10.1007/978-3-030-03053-7_5

investigation protocols in the setting of a potential criminal/bioterrorism investigation [1]. The forensic epidemiologist bridges the gap between public health and the law enforcement communities. During an investigation into a possible bio-attack, the public health officials, forensic epidemiologists, and law enforcement agencies work side-by-side, each possessing a unique set of skills and background.

Typically, a food outbreak is considered a naturally occurring event. Therefore, during the health department's investigation, epidemiologists look for unintentional breaches in food preparation or food handling protocols as the cause for the outbreak. However, recognizing early on that a criminal intent (bio-attack) is involved in a health-related incident would result in preserving evidence and increased odds of solving the crime.

There are basically two types of terroristic attacks, overt and covert. In a covert attack, no group or individual takes responsibility for the attack, and the incident may not be initially recognized as an attack. The large *Salmonella* Typhimurium outbreak that occurred in 1984 in Oregon is an example of a covert attack. It was almost a year after the outbreak that authorities discovered that members of a religious commune had deliberately contaminated a number of restaurants causing the outbreak. When a well-planned covert attack occurs, a forensic epidemiologist is provided with surveillance data; case information such as medical history, signs, symptoms, and laboratory results; and an understanding of the issues and motives of area groups are organizations to be able to differentiate between a natural outbreak from a terroristic attack. Therefore, it is critical to have a sophisticated surveillance system that is linked to area hospitals, emergency rooms, laboratories, pharmacies, real-time sales of over-the-counter (OTC) medications, and information from private healthcare providers. Individuals exposed to a possible biological attack may present in the emergency room (ER) or at their primary care physicians (PCP) or simply try to treat themselves with OTC medications.

Even after all those affected by the exposure have been identified within the community, it may still be challenging to immediately confirm that a bioterrorist incident had occurred. Therefore, when a large number of cases present that are off-season, unexpected, exceed normal background levels, or there are unexplained diseases or deaths, the local health department should immediately notify the state health department. In such cases, a forensic epidemiologist should be consulted to perform a preliminary investigation. If this investigation discovers strong suspicion that a crime such as biological attack might have occurred, it would trigger a joint investigation. This investigation is composed of a forensic epidemiologist, law enforcement with assistance from the health department and local and state laboratories, and access to surveillance system data [1].

This chapter has six objects: First, to examine the changing roles of the health department in an era where the possibility of a bioterrorism attack is very real; second, to examine the genesis of the field of forensic epidemiology, the roles and functions of forensic epidemiologists, and structure of a forensic epidemiological approach in a joint investigation; third, to examine the steps of a standard food outbreak investigation conducted by public department epidemiologists; fourth, to provide historic background information on the Rajneeshee Cult in Oregon, the cult's motives for engaging in a biological attack; the methods used in the selection, production, and dispersal of the biological agent; and the outcome of the biological attacks. The biological attacks involved *Salmonella* Typhimurium and represented the first recognized, thoroughly investigated, documented, and reported instance of domestic bioterrorism in the United States (US) [1]; fifth, to examine the steps used by the public health department to investigate the covert biological attack by the Rajneeshee Cult in 1984; and, finally, to illustrate some of the shortfalls and missed warning signs during the initial investigation of the biological attack by the Rajneeshee Cult and highlight the lessons learned from that attack and how a forensic epidemiological approach differs from that of an epidemiological as it relates to an investigation into a covert bio-attack that may present as a naturally occurring outbreak.

5.2 The Changing Roles of the Health Department and Epidemiology

5.2.1 The Changing Roles of the Health Department

The traditional role of a health department epidemiologist in the United States is to monitor the health of a community, through the surveillance of diseases, illnesses, injuries, and deaths, identify risks, and provide information and alerts regarding situations and hazards within a community. For example, when a food outbreak is suspected, epidemiologists from the local health department conduct a standard food outbreak investigation. Once a specific food item has been identified as causing the outbreak, the investigation then focuses on (1) the history of the food from farm to restaurant, (2) the food handlers, and (3) the storage and preparation of the food. Epidemiologists focus on the personal hygiene of food harvesters, possible food contamination or cross-contamination during transport, improper storage temperatures, equipment failure, or breaches in the preparation of the food item. The local health departments throughout the United States have the legal authority to cite restaurants, impose fines for health code violations, and, if necessary, close the affected restaurant. The role and level of authority of the health department in other countries may vary. All these actions are conducted without the involvement of law enforcement.

Prior to the 1990s, there was limited interaction between public health investigator and law enforcement. However, in today's society, with easy access to the Internet, the expansion of homegrown and international terrorist groups, the deliberate contamination of food, water, air, or the land as an act of bioterrorism is more real today than in the past. In isolation, neither law enforcement nor the health department alone possesses all the expertise required to recognize and investigate certain forms of covert biological terrorism attacks. One form of a covert attack involves the deliberate biological attack that mimics a common/typical food outbreak. This type of attack presents a number of challenges both for law enforcement and for the health department epidemiologists investigating the outbreak. The challenge faced by law enforcement personnel is that they are not trained to recognize these types of attacks. The presentation of unusual, off-season increases in a particular symptomology presenting in the local ER or a sudden increase in the sales of OTC medications for gastrointestinal symptoms or illness would not raise suspicion among law enforcement officers. A challenge also exists for the epidemiologist(s) investigating the outbreak. Although the epidemiologist would accurately assess that an outbreak has in fact occurred, they are not trained to investigate acts of terrorism. As a consequence, their investigation may not identify the "true" cause of the outbreak, fail to collect key pieces of forensic evidence, or preserve evidence to the legal requirements regarding issues such as chain of custody for admissibility in a court of law.

5.2.2 Traditional Epidemiology

The Epidemiologist

To carry out the many functions of the health department, the agency is staffed by a large number of epidemiologists. The field of epidemiology is the foundational study of the distribution and determinants of health-related states or events in specific populations and the identification of risk factors associated with specific diseases and activities, to identify and control epidemic and monitor the rates, distribution, and seasonal patterns of disease, illness, injury, and death. All these actions are designed to identify, reduce, and control health-associated issues. Epidemiologists are also involved in a number of areas of investigations. Below is a list of some of these areas [2]:

- *Study Risk Associated with Exposures*

  Epidemiological studies have provided evidence for the association between exposure and the resulting outcome. The majority of the investigations focus on cause-and-effect questions regarding natural disease or accidental exposure to noxious substances or pathogens. Examples include the relationship between the use of fluoride and better dental health, cigarette smoking and lung cancer, incorrect tampon use and the increase risk of toxic shock syndrome, lead in the water and developmental and behavioral problems, and human papillomavirus exposure and cervical cancer and the risk associated with being overweight and a number of health-related diseases.
- *Identify and Control Epidemics*

  Epidemiologists are charged with the collection and analysis of health data and to alert the health department if there is epidemiological evidence of an atypical spike in a specific disease, an outbreak, or any other situation that represents a public health emergency. The local and state health departments have the legal authority to protect public health and have the legislative powers if necessary to control an epidemic, isolate individuals, and/or quarantine regions of a state [3].
- *Monitor (Surveillance) Reports*

  There are a number of public health surveillance systems operating in the United States that collect data. These data are used to monitor the general health of a community, city, or state and disease trends and to detect outbreaks [4]. These surveillance systems include vital statistics, morbidity data, sentinel surveillance, zoonotic disease surveillance, syndromic surveillance, laboratory data, hospital ER data, and the sales of OTC medications. Vital statistic includes birth and death certificates and represents the critical cornerstone of public health surveillance and has been used as an indicator of a population's overall health. Morbidity surveillance requires the reporting of certain types of disease. Sentinel surveillance involves collecting data from a number of reporting sites (called sentinel sites). One of the most common sentinel surveillance systems used in the United States monitors influenza. Selected healthcare providers report the number of influenza-like illness to their state health department on a weekly basis. Surveillance of zoonotic diseases (an infectious disease that is transmitted from animals to human such as rabies) involves a system for detecting infected animals. In 2001, Florida conducted surveillance for West Nile virus infections using a variety of strategies, most involving animals [5]. Syndromic surveillance is a relatively new surveillance method that uses clinical information about diseases (signs and symptoms) before a diagnosis is made. The syndromic surveillance system uses electronic data from hospital emergency rooms. New York City operates a syndromic surveillance system using emergency department chief complaint data from approximately 44 hospitals [6]. The data are monitored electronically for signals that might indicate the beginning of a disease outbreak. In 2002, the system detected a higher-than-usual number of diarrheal and vomiting symptoms. Based on this information, the health department notified hospital emergency departments of a possible outbreak and ordered the collection of stool specimens, several of which were positive for norovirus infection. Another source of surveillance data are public health laboratories, which routinely conduct tests for viruses, bacteria, and other pathogens. In the United States, public health laboratories participate in the National Salmonella Surveillance System that electronically reports *Salmonella* isolates. In 2006, more than 40,000 isolates of salmonellae throughout the United States were reported through this system [7]. Laboratory serotyping provides information about salmonellosis cases that are likely to be linked to a common source. For this reason, salmonella serotypes are useful for detecting local, state, or national outbreaks. Other laboratory systems that play an important role in surveillance include PulseNet, developed by the CDC and the Association of Public Health Laboratories which is used to monitor foodborne illness outbreaks. PulseNet enables public health laboratories across the United States to compare pulsed-field gel electrophoresis (PFGE) patterns of bacteria isolated from ill persons and determine whether they are similar. This comparison allows scientists to determine whether an outbreak is occurring, even at geographically distant locations, and can decrease the time required to identify outbreaks of foodborne illness and their causes [8].

Epidemiologists use years of information from various surveillance systems to create "background profiles," "base levels," and seasonal patterns for different diseases, illness, and deaths. This is a critical piece of information to differentiate between a true outbreak, normal background levels, and an atypical outbreak. For example, a food outbreak involving *Salmonella*, based on past data, has a known and expected pattern. Salmonellosis has an incubation period of 12–72 h and clinical signs that include diarrhea, fever, abdominal cramps, and vomiting. The number of individuals that presents with these signs increases during the summer months corresponding with events such as picnics, weddings, and other outdoor events (concerts and fairs). Typically, one food item, such as potato salad, becomes contaminated due to a lack of maintaining a safe temperature of the food, causing the organism to multiply resulting in illness among those that consumed that food item. Therefore, based on epidemiological data collected over many years, the epidemiologist can conclude whether the current outbreak is consistent with the seasonal pattern for a particular food consumption seen in the past or whether the occurrence is atypical. Once the laboratory confirms the specific etiology of an outbreak and matches the outbreak to those recorded in the past, the outbreak can be considered a natural event.

5.2.3 The Epidemiological Investigation of a Biological Attack

The basic epidemiologic approach and evaluation of a potential bioterrorist or biological attack are not different from a standard epidemiological investigation. The first step is to collect field, laboratory, and clinical data to confirm that an outbreak has occurred. A case definition is constructed to determine the number of cases and the attack rate. Objective criteria are then used for the development of a case definition; these criteria are used to determine accurate case numbers, add new cases, and exclude others. The estimated rate of illness is then compared with recorded past rates to determine if the rate constitutes a deviation from the norm. Once the case definition and attack rate have been determined, the outbreak can be characterized in the conventional context of time, place, and person. These data will provide crucial information for the determination of the potential source of the outbreak [9]. Epidemiologists use a number of methods to better understand, visualize, and differentiate a natural from a man-made attack. Methods used involve examining the epidemic curve and the examination of epidemiological patterns of clues for the outbreak.

- *The Epidemic Curve*

  Case data is used to construct an epidemic curve, and this pattern is used in differentiating between a natural outbreak and an intentional attack. In most naturally occurring outbreaks, the number of cases gradually increases as a progressively larger number of people come in contact with other patients, fomites, and vectors that can spread disease. Eventually, most of the vulnerable population has been exposed, and the number of cases, or epidemic curve, gradually decreases. Conversely, a bioterrorism attack is most likely to be caused by a point source, with everyone coming in contact with the agent at approximately the same time. The epidemic curve in this case would be compressed, with a peak in a matter of days or even hours. If the biological agent is contagious, it is possible to see a second curve peak after the first, as original cases expose originally unexposed persons to the agent. The steep epidemic curve expected in a bioterrorism attack is similar to what would be seen with other point source exposures, such as foodborne outbreaks. Therefore, the compressed epidemic curve is still not pathognomonic for a bioterrorism attack.

If a specific group has been exposed, the epidemic curve may indicate the time of exposure. From this information, a possible incubation period can be calculated, which can assist in determining the potential cause of illness, thereby suggesting a possible intentional attack (e.g., if the incubation period is shorter than usual as a result of an unusually high inoculum or more effective exposure route). Calculating the incubation period may also help determine if the pathogen is spread from person to person, which is extremely important knowledge to achieve effective disease control measures.

– *Epidemiologic Clues: Natural Outbreak Versus Bioterrorism*

A bioterrorist attack may be difficult to distinguish from a naturally occurring infectious disease outbreak or a food outbreak. Investigators must first examine the etiology and epidemiological features of an outbreak to identify its source, mode of transmission, and the population at risk. Certain clues may indicate whether an outbreak is the result of purposeful release of microorganisms. As steep epidemic curves can be seen in natural point source exposures, additional characteristics of the outbreak should be investigated to help with the determination whether the outbreak is natural or the result of a biological attack [10, 11]. None of the following clues alone constitute proof of intentional use of a biological agent, but together they may indicate whether further investigation is warranted:

(1) The presence of a large epidemic, with greater caseloads than expected, especially in a discrete population
(2) More severe disease than expected for a given pathogen, as well as unusual routes of exposure [12]
(3) A disease that is unusual for a given geographic area, is diagnosed outside the normal transmission season, or occurs in the absence of the normal vector for transmission of the etiologic agent
(4) Multiple simultaneous epidemics of different diseases
(5) A disease outbreak with zoonotic as well as human consequences, as many of the potential threat agents are pathogenic for animals
(6) Unusual strains or variants of organisms or antimicrobial resistance patterns disparate from those common in a given region
(7) Higher attack rates in those exposed in certain regions (e.g., inside a building if the agent was released indoors) or lower rates in those inside a sealed building if an aerosol was released outdoors
(8) Intelligence that an adversary has access to a particular agent or agents
(9) Claims by a terrorist of the release of a biologic agent

(10) Direct evidence of the release of an agent, with findings of equipment, munitions, or tampering of products associated with the attack

Naturally occurring diseases are endemic to certain areas and involve traditional cycles of transmission; seasonal patterns and sentinel cases are not uncommon. In contrast, a disease outbreak due to bioterrorism could occur in a non-endemic disease area, at any time of year, without warning, and depending on the etiologic agent and mode of transmission affect large numbers, and thousands of cases might occur abruptly. Consequently, public health officials must be appropriately sensitized to the possibility of bioterrorism when investigating outbreaks. Therefore, it is important to have suspicious outbreaks investigated by a forensic epidemiologist. If the forensic epidemiologist investigation reveals or suspects bioterrorism, they contact the FBI, which is responsible for coordinating interagency investigations of such episodes. FBI agents and forensic epidemiologists are well trained in forensic methods for criminal investigations and are supported by local and state health departments and laboratories [15].

5.3 Forensic Epidemiology

5.3.1 History of Forensic Epidemiology and the Functions of a Forensic Epidemiologist

In the past, epidemiologist has served in the fields of medicine, pharmacy, and emergency medicine and in public health. A new public health threat has emerged: the threat of biological attacks or terroristic groups targeting food, water, and agriculture. Terrorist attacks not only represent overt but also covert attacks, some not even recognized as attacks at first. These attacks are designed to mimic natural outbreaks so as not to create too much attention and are conducted to determine the feasibility of future larger-scale attacks, evaluate the agent and method of dissemination, and gauge the response of the health department, law enforcement, and the public. In response to this new public health threat, a new field has emerged that merges public health methods, the science of epidemiology, the fields and techniques of forensic science, and protocols of law enforcements in the investigation of acts of bioterrorism. This field is called forensic epidemiology.

– *Early History of Forensic Epidemiology*

  During the 1970s and through the early 1990s, there was no major differentiation between an epidemiologist and a forensic epidemiologist. Both were trained in epidemiology and epidemiological methodology and techniques. Forensic epidemiologists are employed in the fields of medicine, pharmacy, and emergency medicine and in public health setting. One function of forensic epidemiologists was that they more frequently served as expert witness in civil trials. Their role was to determine the cause of specific types of illness, injuries, and deaths, evaluate and interrupt the strengths and weakness of the data and the epidemiological methodology, and review the medical and scientific literature. A forensic epidemiologist would testify to issues relating to the strengths/weaknesses of the methodology of research and medical studies. They evaluated factors such as the number and types of subjects, length of the study, and interruption of the results and highlighted limitations. They were frequently used to resolve such suppositions as, "Does exposure to a certain chemical relate to an increased risk of specific birth defects?"

– *Recent History of Forensic Epidemiology*

  Starting in the late 1990s, the field of forensic epidemiology underwent a major transformation. The term "forensic epidemiology" was first associated with bioterrorism in October 1999 when the term was used by Dr. Ken Alibek, the former first deputy chief of Biopreparat, part of the USSR bioweapon program. Forensic epidemiology was defined then as the activity that would help distinguish natural from made-make epidemics [1]. After the anthrax attacks in September and October of 2001, Dr. Gerberding, a senior official of the Centers for Disease Control and Prevention, Infectious Disease National Centers, defined a forensic epidemiologist as an epidemiologist that would be trained to respond to bioterrorism attacks [1].

  The increasing risk for covert biological attacks underscored the need for law enforcement and public health personnel to work together in conducting joint investigations, especially those centered on a possible bioterrorism event. This new generation of forensic epidemiologist combined the sciences of medical, public health, forensic, and law enforcement information to create situational awareness of the events surrounding illness, injury, and death that could indicate a natural event, an accidental event, a criminal event, or a national security matter such as a terrorist attack. The field of forensic epidemiology differs from epidemiology in that it includes law enforcement information as it relates to terrorists groups, motives, capabilities, and features of previous attacks in its analysis of an event. Forensic epidemiologists are trained in various techniques of forensic science, such as the identification, documentation, and evidence collection, and an understanding of legal techniques such as interviewing, chain of custody, and evidence admissibility protocols. Results of the forensic epidemiological analysis can lead to actions by law enforcement, future investigations, or criminal charges.

– *Definitions*

  In 2002 several definitions of a forensic epidemiology emerged [1]. They included (1) the use of epidemiological methods as part of an ongoing investigation of a health problem for which there are suspicions or evidence regarding possible intentional acts or criminal behavior as factors contributing to the health problem or (2) the use of epidemiologic and other public health methods in conjunction with or as an adjunct to an ongoing criminal investigation. Forensic epidemiology is most commonly understood to refer to that area of practice when public health and criminal investigations overlap, especially in relation to possible instances of bioterrorism [1]. Therefore, a forensic epidemiologist uses public health methods in a criminal investigation and works closely with law enforcement investigators to determine the cause of public health emergencies, such as a *Bacillus anthracis*-laced letter attack or other forms of biological attacks [14].

5.3.2 The Forensic Epidemiological Approach

One of the main differences between a public health epidemiologist and a forensic epidemiologist is that the former is trained to investigate diseases, illnesses, and food outbreaks as caused by a nondeliberate event, whereas the latter approaches such events as very suspicious and a possible bioterroristic attack until proven otherwise, especially those that do not conform to the typical food outbreak pattern. If the features of the outbreak are atypical, the investigation is transferred from the epidemiologist to either a health department forensic epidemiologist or to a forensic epidemiologist consult to conduct the investigation. If there is strong suggestive evidence of a possible biological attack, the forensic epidemiologist join forces with law enforcement with support for local/state health departments and laboratories.

5.3.3 The Joint Investigation Between Public Health, Forensic Epidemiologist, and Law Enforcements

Beginning in the early 1970s, there were a number of joint investigations involving law enforcement and public health authorities. These investigations centered on health problems that were suspected to have been intentionally caused or crimes that had potentially significant public health consequences. Examples include "angel of mercy" syndrome when healthcare providers intentionally assisted patients' death wishes by administering lethal doses of pharmaceuticals. In 2002, the CDC developed the *Forensic Epidemiology: Joint Training for Law Enforcement and Public Health Officials on Investigative Reponses to Bioterrorism.* A joint collaboration between the CDC and the FBI resulted in the *Criminal and Epidemiological Investigation Handbook* [15]. The Department of Criminal Justice Training (DOCJT) had designed courses specifically to aid law enforcement officers to operate closely and more effectively with forensic epidemiologist, during the investigation of a public health problem that may have been intestinally caused or crimes with public health consequence. The goal of this collaboration is to familiarize law enforcement officers with some of the basic principles of public health and the forensic epidemiological approach to investigation of a possible bio-attack.

A comprehensive investigation of a bioterrorism attack includes a forensic epidemiologist, law enforcements, and the assistance of the health departments and local and state laboratories. The role of law enforcement is to enforce the laws of the county, issue warrants and seizure evidence, and make arrests. In the case of a biological attack, the evidence is typically outside the normal type encountered by law enforcement; therefore, law enforcement officers work closely with forensic epidemiologists in the identification, collection, and interruption of the evidence. The forensic epidemiologist utilized the vast amount of surveillance, past outbreak data, and laboratories and served as bridge between the health department and law enforcement. In this joint operation, it is critical that the law enforcement officers have a strong understanding of the skills, training, and role of the forensic epidemiologist within the investigation. Likewise the forensic epidemiologist must understand the limits and powers of the police within the investigation [16].

5.3.4 Issues Related to Evidence Collection

Law enforcement officials gather evidence, while public health officials gather specimens. These are the same under some circumstances, but the purpose for gathering them and the way they are handled differ. One purpose of evidence collection during an investigation is to present that evidence in a court of law. In order to present that evidence in court, two criteria must be met. First, the evidence must be collected during a legitimate criminal/public health investigation. Second, the collected evidence must conform to the admissibility of evidence standards, i.e., for evidence to be admissible in a court of law, it must have a chain of custody form. This form documents the chronological history of a piece of evidence; provides the name (initials) of the individual who collected the evidence and the date of the collection, a case number, the victim's/suspect's name, and a brief description of the item; and contains a list of each person or entity that received the item [17]. A chain of custody is also a routine part for evidence collected by medical examiners and forensic pathologists [18]. Chain of custody is not typically established in a public health investigation [17]. Individuals documented as custodians of a piece of evidence should be able to testify in court that the item was secure, unaltered, and uncontaminated during the time it was in their custody and should be able to explain the procedures they used to store, examine, test, and otherwise process the item [19].

5.4 A Standard Food Outbreak Investigation

It is important to understand the basic steps involved in a standard food outbreak investigation and the thought processes of the epidemiologist conducting the investigation. The investigation involves the health department, epidemiologist, laboratories, and the media.

The health department employs a number of surveillance systems collecting data from a number of sources. The process begins when one of these systems alerts an epidemiologist that an outbreak may have occurred. Signs of an outbreak can include a spike in the number of patients presenting at the local ER with similar symptoms, an increase in the workload of local pathology laboratories, or an increase in the sales of OTC medications for a specific illness such as GI issues. Other warning signs of an outbreak include a large number of patients presenting with similar symptoms to their PCP or an increase in first responder calls with similar complains.

Once the epidemiologist has determined that a food outbreak has occurred, the next step involves identification possible cases and collecting case data. Based on preliminary information from the first few cases, a wide net is cast to locate other possible victims of the outbreak and to contact them. The epidemiologist investigating the outbreak is trying to identify a common point of exposure such as eating at the same restaurant. To identify this common point of exposure, the epidemiologist conducts interviews with all accessible victims to locate a common event that all the victims' attended.

One of the key functions of the food outbreak investigation is to identify the specific food item that was consumed by all those affected. This is accomplished by obtaining a detailed food history that lists all the food and drinks consumed by the victims of the outbreak over a crucial time period. Once the common specific food item has been identified, the investigation shifts focus to ascertain whether the contamination was caused by a breach relating to the harvesting, transport, or processing of the food item or related to human factors. This phase of the investigation takes two paths: one path examines the history of the food item in question, and the second examines all the individuals that handled the food from farm to table. The investigator examines the history of the food item from the farm where it was grown, the route and method of transportation of the food, and finally the environment at the restaurant where the food was served. In addition, the investigation also involves environmental testing of the water and soil of the farm where the food was grown. If the source of the food outbreak was meat, livestock is examined and tested. The goal is to identify where the breach in safe food handling, transportation, preparation, or storage has occurred in order to correct it.

Next, the investigation examines those handling the food, starting with the food handlers at the farm and ending with the food preparers and servers at the restaurant. The investigators collect biological samples, medical history, and level of education regarding hygiene practices from all those that interacted with the food item.

The next step is to collect the following information: the time/date individuals were exposed, the specific food item most likely responsible for the outbreak, the time/date symptoms/signs first appeared, the type of symptoms/signs, severity of the symptoms/signs, actions taken by individuals to address the symptoms/signs (OTC, PC, ER), and the duration of symptoms/signs. The epidemiologist will use this information to create what is called a, "Case definition." This case definition will be used to determine if future cases should be considered related to a specific outbreak. These data are also used to "define the scope of the outbreak." The epidemiologist will describe the features of the outbreak such as the numbers affected, the site where the outbreak occurred, the symptoms/signs, the treatment, using descriptive epidemiology methodology and terminology.

Once the specific contaminated food item responsible for the outbreak is identified, the next step is to identify the etiological agent (specific organism). During this phase of the investigation, biological samples (blood, stool) are collected and sent to local, state, or national laboratories for analysis to determine the agent and identify the specific organism. The next step involves the identification of the mechanism of transmission. Mechanism includes poor hygiene of food handlers at the farms, cross-contamination during transport, or inadequate storage or displaying of food. Cross-contamination can occur when contaminated food is placed on top of other foods, and when the top container is washed, it contaminates the lower containers or when a contaminated shipping container is not disinfected prior to the transport of other food items.

Once the source and method of food contamination are confirmed, the next step is to determine the time period that the public was at risk. This period is referred to as the critical period of exposure. After all the key data related to the outbreak are available, the next step is to inform the public that a food-related outbreak had occurred. Public notification by the health department can include press releases, news conferences, and the use of social media. These notifications provide the site of the outbreak, the period of exposure, the food involved, the symptoms/signs, and medical advice. Below is an example of a public health warning notification:

> If you have recently eaten at Karen's Dinner located on Forest Hills between the dates of November 15–18 and ate Potato Salad you might have been exposed to Staphylococcus bacterium. This organism typically causes mild gastrointestinal infections distress for a few days. If you are experiencing more serious symptoms see your medical provider or your local ER. If you have any questions regarding the outbreak please contact the local health department.

The final role of the epidemiologist is preventative. The health department makes suggestions and initiates implantation control measures to prevent similar outbreaks from occurring in the future.

Once all the facts of the outbreak are available, the epidemiologist will determine if the recent outbreak conforms to past outbreaks, is typical, and is expected. This function is performed by comparing a number of features of the recent outbreak to the characteristics of past outbreaks. For example, epidemiological data reveals that during the summer months, there is an increase in the number of food-related gastrointestinal infection (GI) symptoms among 20–50-year-old individuals. These individuals present with GI-related clinical signs at their general practitioners' (GP) office or ER and increased sales of OTC medication related to GI clinical signs. There is also an increase in the number of *Staphylococcus aureus* (*S. aureus*) isolation at diagnostic laboratories. Typically, these outbreaks are associated with poorly maintained temperature of potato salad at outdoor events. Typically, only a small number of individuals become ill with only mild GI symptoms that are self-treated with OTC medications. Occasionally, an outbreak affects a large number of individuals. The local health department epidemiologist, using years of surveillance data from a wide range of sources, is able to establish the normal background levels for specific diseases and illness by characteristics such as age of victims, seasonal patterns, and the specific agents involved in the outbreaks.

The procedure outlined above represents the standard methodology used to investigate a food outbreak not associated with a bioterroristic attack. However, "What if the food outbreak is in fact a bioterrorists attack?"

5.5 The Bioterrorism Attack in 1984: The Rajneeshee Cult

Until 1984, there had been 222 bioterror-related incidents in the United States with only 24 confirmed attacks. Of these, only the Rajneeshee Cult attack had resulted in mass casualties until the attack of 9/11. Mass casualties can be defined as an event where the number of casualties (with injuries ranging from moderate to severe) exceeds the resources normally available from local resources. Below we will examine "how" and "why" the 1984 Rajneeshee Cult in Oregon succeeded with the first documented bioterrorist incident in the United States.

5.5.1 Background

In 1974, an individual by the name of Bhagwan Shree Rajneesh established the Acharya Rajneesh Ashram, a spiritual organization in India based on Buddha, Krishna, Jesus, Zen masters, Hasidism, and Sufism [20]. In 1981, Bhagwan immigrated to the United States due to growing hostility in India and charges of tax evasion by the government. He established a cult in Oregon in an area known to locals as the Big Muddy Ranch. The property straddled Jefferson and Wasco Counties, with the majority in Wasco. In 1985 the area was largely rural with a population of approximately 20,000, with the county seat located in a small town called The Dalles, which had a population of about 10,000. Bhagwan's plan was to build a "Buddhafield," or agricultural commune, where the "Rajneeshees" could celebrate the enlightened master's philosophy of "beauty, love, and guiltless sex" [21, p. 15].

In 1982, Bhagwan self-imposed a 4-year vow of silence. Consequently, he was not directly involved in the day-to-day operations of the organization. This duty was taken over by Bhagwan's personal secretary, Sheela P. Sliver, otherwise known as Ma Anand Sheela [22]. It was widely understood among the Rajneeshees that Sheela represented Bhagwan. She spoke in his name and had sweeping powers throughout the community. Ma Sheela ruled over a large number of "moms" and the more senior "big moms" for which she relied on to direct the daily operations of the commune. Only one senior member was male besides Bhagwan himself: David Berry Knapp, otherwise known as Krishna Diva or "K.D.". He served as mayor of the municipality of "Rajneeshpuram." Another key member of the inner circle was big mom Dianne Y. Onang, also known as Ma Anand Puja. She was a licensed family nurse practitioner, registered nurse, and vice president of Bhagwan's church and the Rajneesh Foundation International [23, p. 121]. Puja wielded absolute control over all medical facilities within the organization, including the Pythagoras Clinic and Pythagoras Pharmacy of the Rajneesh Medical Corporation. In 1980, she became director of the Shree Rajneesh Ashram Health Center. Puja's interests included death, poisons, viruses such as HIV-1, and a fascination with carrying out various plots. Her fascination with HIV-1 included purchasing a quick freeze dryer (September 1984) to culture the virus, one human trial, and one unconfirmed report that she may have deliberately infected at least one individual with HIV-1 to see if it was possible to transmit the agent [23, pp.126–127].

5.5.2 The Rajneeshees and the Community

To fully comprehend the Rajneeshees' motivation to engage in a bio-attack, it is important to understand the relationship between the Rajneeshees and the surrounding community. The cult was externally rich, especially when compared to the surrounding residences. Their considerable income came from followers turning over all their assets to the cult plus substantial income generated from the sales of books and tapes on Bhagwan's teachings. The Rajneeshees' community had a vast infrastructure complete with a dam; lake; a large number of farms; a network of roads; sophisticated water, sewage, and transportation systems (85 buses); and even an airstrip for the cult's five jet aircrafts and a helicopter [25, p. 16]. The community also included a 160-room hotel, a 2-block-long shopping mall, casino, and a disco [25, p. 16]. The Rajneeshees used the legal system to advance their agenda by suing in response to the smallest provocation and flooding the courts with hundreds of motions to steamroll and intimidate opponents. Deputy county clerk Karen LeBreton estimated that 60% of her work involved responding to Rajneeshees' legal suits and petitions [26, p. 22]. On the other hand, the group disregarded laws that interfered with its objectives. Sham marriages to conceal members brought the cult into conflict with the US Immigration and Naturalization Service and the US Attorney's Office in Portland, Oregon. In addition, the group registered as a religion to leverage tax benefits despite years of condemning organized religion [27, p. 118]. The locals tolerated the Rajneeshee because they brought doctors, lawyers, and engineers to the area [28, p. 17]. Also, their massive spending amounted to a local investment of more than $35 million in construction. This was a tremendous financial boost to The Dalles since its sole industry, an aluminum smelting plant, had shut down years earlier.

5.5.3 The Main Issue of Contention

There was one major problem with the cult's impressive expansion: it conflicted with rural zoning ordinances. The cult continued with the construction anyway, without the approval of the Wasco County commissioners or securing the necessary permits. The Rajneeshees found themselves in a series of protracted political and legal disputes, many of their own making. They had a rational plan to solve all of their problems by fighting politics with politics. To evade zoning restrictions, the Rajneeshees took over the nearby town of Antelope in 1982 along with its school district [27, p. 118]. This was accomplished by moving enough people into the small community of 75 and outnumbering the existing inhabitants by exploiting Oregon's liberal voter registration laws. Shortly after securing electoral control of the town, they renamed Antelope and its school district "Rajneesh" and turned Antelope's sole business into a vegetarian health food café called Zorba the Buddha. The second stage of their plan was to incorporate the commune at Big Muddy Ranch into a township called "Rajneeshpuram." Rajneeshpuram ultimately became the center of a legal dispute between the cult and Oregon's Attorney General David Frohnmayer, who concluded there was no separation between church and state within this newly created township. Therefore, the creation of Rajneeshpuram was considered unconstitutional by Frohnmayer. For the Rajneeshees, diplomacy had failed, and political maneuvering had reached an impasse. With time running out within the courts, desperation was setting in, the cult escalated to the use of violence in order to preserve their community's sense of autonomy, control, and power. These motivations breed intentions that were aimed at taking political control of the county seat and local government.

5.5.4 Plans to Take Control of the Wasco Commission

An examination of the group's motivations and intentions is necessary to understand the type of attack and the selection of the agent used to carry out the attack. Ma Sheela and Bhagwan decided to take control of the Wasco County Commission by winning the upcoming 1984 November election. By some means, the 4000 members of the commune (many of whom were not US citizens and could not vote) needed to take over a county of 20,000 inhabitants including some 15,000 registered voters. Three plans were considered. The first was to rent apartments throughout Wasco County and register Rajneeshees under multiple names to vote several times during the elections. However, the risk of discovery was considered too great, and this plan was rejected. The second proposal involved running a candidate for County Commissioner that would support Rajneeshees' interests. This plan failed when the group was unable to secure the required number of signatures to get its candidate on the ballot. Ma Sheela, influenced by Ma Puja, came up with the third plan to make Wasco citizens sick before the election so they would be unable to vote. Sheela talked with Bhagwan about a plot to decrease voter turnout by making people sick.

An examination of the Rajneeshees' organizational structure revealed a classic terror network model consisting of a motivationally led bureaucracy, containing a small group of self-interested individuals with great power that are supported by a number of subunits [29, pp. 202–229]. A small group of only 14 persons executed the plans conceived by three key players in an organization numbering 4000 members. Bhagwan's vow of silence and hands-off approach gave Ma Sheela power to push her agenda. Ma Puja's dark side, technical expertise, and special connection with Sheela enabled her to coax the small group toward bioterrorism. Finally, K.D.'s public relationship with the members of the cult and as the mayor of Rajneeshpuram allowed him to control and constrain the majority. Although he actively participated in the attacks, it was K.D. who convinced Sheela and Puja not to use deadly pathogens in the planned attack for fear the cult would be immediately suspected.

5.5.5 The Biological Agent

Once the plan to use a biological agent was approved, the next step was to select and obtain a bioagent. The cult leaders considered a number of agents to sicken the residents and lower voter turnout on Election Day. Ma Puja considered agents that cause typhoid fever, hepatitis, and even AIDS [30, pp. 124–127]. AIDS was discarded by Puja because of the difficulty to culture and weaponized HIV-1. Puja's team also considered putting dead beavers into The Dalles' water supply. Beavers were considered because they are natural carriers of *Giardia lamblia*, a protozoan that causes the GI disease giardiasis or "beaver fever" in humans, but the rodents were too big to fit through screens on the town's water tanks. Infected rats and mice were also considered. The cult ordered and received a variety of pathogens, such as *Enterobacter cloacae* (responsible for various infections including bacteremia, lower respiratory tract infections, skin and soft tissue infections, and urinary tract infections), *Neisseria gonorrhoeae* (gonorrhea), and *Shigella dysenteriae* (shigellosis).

The group finally agreed to unseat the local government of The Dalles by using salmonellae to sicken voters and swing the upcoming election in favor of the Rajneeshees [31]. The selected agent was *Salmonella enterica* serotype Typhimurium—a bacterium strain known by American Type Culture Collection (ATCC) under catalog number 14028. *Salmonella* Typhimurium (ATCC 14028) is a commonly used bacterium in medical settings. The Rajneesh Medical Corporation (RMC) had a legal and legitimate need for *Salmonella* Typhimurium because this bacterium is used as a control organism used to meet the requirements for quality assurance of licensed clinical laboratories. Puja and the RMC were required by law to test the proficiency of their technicians by having them identify samples contaminated with known agents. The RMC also uses the organisms to ensure the quality of growth media used in their diagnostic testing. Because of these requirements and needs, Puja could purchase *Salmonella* Typhimurium without raising suspicion from a medical supply company such as VWR Scientific. The RMC legally purchased a set of "bactrol disks" from VWR Scientific, a medical supply company based in Seattle, Washington. RMC laboratory technician Parambodhi provided technical expertise and assisted Puja in culturing the salmonellae. Puja and Parambodhi used the disks in the RMC's state-licensed medical laboratory to culture and mass-produce the final preparation. Carla Chamberlain, the nurse who ran the county public health office, visited with Puja at the RMC to discuss the county's health reporting requirements and to receive assurance that the laboratory would comply with them. She was astonished to find that Rajneeshee facilities were better equipped than those of the county [26, p. 22]. Both Carla Chamberlain and Judge Hulse were suspicious of the cult and their possible role in the outbreaks. Mrs. Chamberlain was raised by the sophistication of the cult's laboratory equipment and the judge as a result of hospitalization after meeting with cult members.

Salmonellae Typhimurium were selected for a number of reasons. Salmonellosis is the second most common cause of bacterial foodborne illness and is underreported by a factor of about 38-fold [32, 33]. The average onset period for salmonellosis is about 12–36 h, and the disease manifests as acute gastroenteritis. Other symptoms/signs include fever, anorexia, and diarrhea that can persist for several days, and more severe manifestations may at times occur, especially among the very young or elderly. Contaminated food is the principal route of agent transmission [32, 34]. It has been estimated that 1.4 million cases of salmonellosis occur annually in the United States, resulting in 15,000 hospitalizations and 400 deaths [35]. Therefore, the index of suspicion for an intentional salmonellosis outbreak was—and remains today—low [34].

Large-scale production of the agent took place in the "Chinese Laundry" of Rajneeshpuram and was later moved to the Alan Watts complex that consisted of approximately two dozen buildings in the northeast section of the commune. The actual facility for producing the *Salmonella* Typhimurium consisted of two A-frame structures connected by a common bathroom. The facility contained a large freeze dryer and "a green incubator the size of a small apartment-type refrigerator" containing petri dishes used to grow the bacteria. The next phase was to evaluate the effectiveness of the agent and experiment with the methods of disposal.

5.5.6 The Attacks, Targets, and Outcome

The attacks targeted a number of individuals and several restaurants. The attacks were known as the "salad bar contaminations" or "restaurant attacks." The attacks occurred over the course of several weeks in 1984.

First Attack The Rajneeshees first attempted to spread *Salmonella* Typhimurium by contaminating commune members' hands to greet outsiders, as well as the doorknobs and urinal handles at the Wasco's county courthouse. These efforts did not result in causing any illness.

Second Attack The next documented use of the biological agent by the Rajneeshees occurred on August 29, 1984. During a routine fact-finding visit to Rajneeshpuram by Wasco County commissioners (Judge William Hulse and Raymond Matthews) [36, p. 128], cult member Ava slipped *Salmonella* Typhimurium into the judge's breakfast at Zorba the Buddha café in Antelope (a.k.a. Rajneesh) the morning the two men were on their way to visit the ranch [26, p. 22]. Cult members also gave the two men water laced with *Salmonella* Typhimurium later that day at the commune. Judge William Hulse and Raymond Matthews were two of the commissioners most unfavorable to the group. Both men became sick, and Judge Hulse required hospitalization. According to hospital records, Judge Hulse was admitted and almost died.

Third Attack The restaurant attacks occurred in two distinct waves. The attacks involved targeting a number of area restaurants. The Dalles is located along Interstate 84, a major east-west transportation route just 90 miles east of Portland, Oregon. Due to its prime location, it contained more restaurants (35 establishments in all) than a town of its size could normally support [37, p. 19]. Approximately 12 individuals were involved in the plot, and up to 8 individuals distributed *Salmonella* Typhimurium cultures to area restaurants [34].

The first wave of attacks occurred on September 9–18, 1984, and involved the Arlo's Restaurant, Portage Inn, the Recreation Café, Johnny's Café, and the Chuck Wagon restaurant [34]. The restaurants were contaminated with salmonella during the cult's trial runs. K.D. and her accomplice Ma Ava directly participated in the attacks, along with Ma Sheela and Ma Puja. Puja gave K.D. "a plastic bag containing a test tube sealed with a cork stopper and filled with a 'mostly clear' light brown liquid" [37, p. 19]. Puja ordered him to spread the content of the vial at restaurants in town. During a trip into town to attend a meeting with another cult member, K.D. went to Portage Inn. Because they arrived after lunchtime, the salad bar was closed, but he poured the contents of the vial into the salad dressing. Then, Puja tainted the salad bar at the Recreation Café. Ava contaminated three additional restaurants at Sheela and Puja's request. She and cult member Swami Satyam Bodhidharma drove to The Dalles with five or six vials of "salsa." They poured the content of their vials into coffee creamers at Johnny's Café and The Chuck Wagon restaurant. They also contaminated blue cheese salad bar dressing at the Arlo's restaurant.

The second wave of cases occurred between September 19 and October 10, 1984, and involved the following restaurants: Arctic Circle Restaurant, Arlo's Restaurant, Burgerville USA, Kopper Kitchen, Pietro's Pizza Restaurant, Portage Inn, Recreation Café, Shakey's Pizza Parlor, Skipper's Seafood Restaurant, and Taco Time.

Fourth Attack (Not Carried Out) The cult had plans to attack a critical infrastructure, the town's water supply. Sheela ordered K.D. to acquire maps of The Dalles' water system. Puja contacted the Rajneeshpuram Resource Manager to obtain raw sewage. Two cult members named Julian and Anugiten conducted two attempts to contaminate the town's water supply with the raw sewage. Julian and Anugiten climbed up a hill to a water tank that overlooked a nearby school. They tried to pry open the screen on the water tank but were unsuccessful.

The Outcome In late September the *Salmonella* attacks ended. In total, the attacks hospitalized 45 people, caused 751 people to fall ill, and would serve as a historical benchmark of bioterrorism in the United States. Two explanations were offered why for the decision to terminate the attacks. According to K.D., "as soon as the Share-A-Home project began, the contamination project seemed to be deemphasized and everyone concentrated on the Share-A-Home project" [38]. The project was difficult to manage, and there was not enough time to continue the bio-attacks. Ma Ava provided this explanation: "The *Salmonella* poisoning was simply an experiment through which Sheela and her associates could assess whether or not they could incapacitate the voting population in The Dalles at the time of voting" [39]. When the Rajneeshees decided that they could not prevail in the local elections, they abandoned both the Share-A-Home and the contamination projects [40].

5.6 The Public Health Departments' Investigations

The initial response to the outbreak involved the Wasco-Sherman Public Health Department who conducted a standard food outbreak investigation. When the outbreak increased to a second wave, the health department was assisted by the CDC Epidemic Intelligence Service (EIS).

5.6.1 The Outbreak Investigations

First Outbreak The first wave of salmonellosis cases occurred on September 11–18, 1984, with the first case reported to the Wasco-Sherman Public Health Department in September 17. Individuals presented with gastroenteritis signs after eating at the Arlo's Restaurant and at the Portage Inn [34]. Twenty more cases from two different eating establishments followed within days. The local and state health officials initiated a standard food outbreak investigation outlined above. The investigation revealed that those patronizing a number of restaurants in the city of The Dalles and eating salad bar food items became ill. Over time, the number of cases increased. Biological samples were sent to pathologists at the Mid-Columbia Medical Center, who identified salmonellae in the patients' stool samples less than 48 h after the outbreak began. *Salmonella* Typhimurium was isolated from a total of 388 patients. Authorities confirmed cases either microbiologically by stool culture of *Salmonella* Typhimurium or using clinical criteria of diarrheal illness and at least three of the following symptoms/signs: fever, chills, headache, nausea, vomiting, abdominal pain, or bloody stools. Four days later, state public health scientists in Portland analyzed additional samples and confirmed the presence of *Salmonella* Typhimurium, a very unusual strain of salmonella for that region. *Salmonella* Typhimurium isolates were eventually obtained from clinical specimens from all affected individuals [35]. The source for this outbreak was puzzling. Epidemiological analysis revealed multiple food items rather than a single suspect item as the cause of the outbreak [34].

Second Outbreak County public health officials thought the outbreak was over when the epidemiological wave began to crest. On September 25, 1984, the county was in the midst of the second wave, and the state sought help from the EIS [41]. During the outbreak, the Mid-Columbia Medical Center's laboratory was immediately overwhelmed with specimens (20 petri dishes to test in a normal week mushroomed to 200 every other day). Mid-Columbia Medical Center's 125-bed hospital was filled for the first time ever, with some patients being kept in the corridors. Doctors struggled to treat panicked and even hostile patients. Violent patients and their families demanded their test results; some even throwing stool and urine samples at hospital employees [41].

Public health authorities conducted an extensive investigation in response to the salmonellosis outbreak. The 38 restaurants located in The Dalles were grouped according to the number of culture-confirmed customer cases with a single restaurant exposure in the week before symptom/sign onset [34]. Additional ill customers were located through laboratory reporting of clinical specimens or clinician reporting to public health authorities (passive disease surveillance). Press releases were issued to encourage disease reporting by patients and clinicians (Stern 2006). Public health officials interviewed ill persons to obtain their symptoms/signs, risk factors, and a comprehensive food history, as well as the names of all persons who had eaten with them at the restaurant.

In order to differentiate a natural from a man-made outbreak, it is critical to know what is referred to as the "background levels." Background data provide the number of cases by organisms, strain, yearly pattern, and pattern of past outbreaks. Based on information from Carla Chamberlain, the director of the counties health department office, she reviewed data between 1980 and 1983 and discovered that 16 isolates of salmonella were detected with only 8 of which were *Salmonella* Typhimurium. On closer examination of the eight cases, none matched the strains isolated during the 1984 outbreak [37, p. 19].

5.6.2 The EIS Investigation

EIS was created in 1951, following the start of the Korean War, as a warning system against biological warfare and man-made epidemics. EIS is composed of medical doctors, researchers, and scientists. Today, the EIS mission includes surveillance and responding to all types of epidemics including chronic disease and injuries. A substantial epidemiological investigation was conducted by EIS and the Wasco-Sherman Public Health Department. The team questioned hundreds of patients, family members, and friends. Investigators interviewed all 325 food handlers who worked at the 10 affected restaurants, approximately 100 of which had been infected, and many of these falling ill before their patrons [42, pp. 20–21]. A detailed investigation was conducted on out-of-state visitors who purchased meals with credit cards, 120 people who ordered home delivery, and those that were served by the restaurants at banquets. Fortunately, none of these people were infected [42, pp. 20–21]. As part of the EIS investigation, the two local water systems and the water at the restaurants were tested, salad bar temperature systems were evaluated, and food-handling practices were scrutinized [42, pp. 20–21]. Investigators visited farms that supplied cucumbers, tomatoes, and melons to the restaurants. They also checked a dairy in neighboring Washington State to test cows, cow feces, raw milk, and the farm's pond water. None of the taken samples contained *Salmonella* Typhimurium. No common source of contamination was even identified. The lettuce and other vegetables came from different suppliers, and the salad dressings were purchased from different wholesalers. The EIS team did find *Salmonella* Typhimurium in the milk of coffee creamers in one café and in the blue cheese dressing of another, but not in the dry mix used to make the dressing. This finding suggested that the dressing had been contaminated during or after its preparation [42, pp. 20–21].

5.6.3 Actions, Results, and Conclusions of the Public Health Investigation

As the number of gastroenteritis cases increased, public health officials imposed a closure of all salad bars in The Dalles on September 25, 1984. The outbreak resulted in 751 salmonellosis cases affecting individuals ranging from newborns to 87-year-olds and was associated with dining in 10 area restaurants. At least 45 persons were hospitalized, but no fatalities occurred [34]. By late 1984, the local public health officials concluded that the outbreaks were accidentally caused by contaminated food handlers. Although dozens of food handlers became ill, their time of symptom/sign onset did not precede those of their customers. It only took 4 days for the local laboratory to identify *Salmonella* Typhimurium as the cause of the outbreak but over a year for CDC to file its final report that confirmed that a single strain had caused all the illnesses [43].

The investigation concluded the following: a failure to locate a common supplier for all of the implicated restaurants; many food items served at the salad bars were associated with illness and differed among the restaurants; illness was associated with eating blue cheese dressing, potato salad, and lettuce; and *S.* Typhimurium was isolated from the blue cheese dressing collected at one restaurant, but not from the dry mix used to prepare the dressing. State public health investigator and CDC reports published on the fall of 1984 and early 1985 both concluded that the Rajneeshees were not to blame for the outbreaks [44, p. 23].

5.7 The Subsequent Criminal Investigations

The perception of the events of 1984 took a dramatic change on September 16, 1985. The cult began to fracture from within, exposing the true nature of the outbreak, revealing for the first time that it was a biological attack by the cult members. The first evidence was derived from public statements by Bhagwan after the completion of his 4-year vow of silence, revealing the activity of the cult members involved in the bio-attack. A press conference was held at the ranch 2 days after Ma Sheela and Puja resigned their posts and flew to Europe. Bhagwan accused his second-in-command and others of creating a "fascist regime" that plotted to kill fellow cult members, stealing money, mismanaging commune affairs, and leaving the cult some $55 million in debt [44, p. 23]. More importantly, he publicly accused Sheela of poisoning his personal doctor and dentist and the district attorney of neighboring Jefferson County and of trying to contaminate The Dalles' water system. Shortly after Bhagwan's statements, the Oregon Attorney General established a joint task force between the state police and the FBI to conduct a criminal investigation. On October 2, 1985, investigators found the bacterial disks and laboratory equipment at the commune [45]. In October 1985, Oregon public health laboratory officials and the FBI discovered an open vial containing the original culture *Salmonella* Typhimurium in the Rajneeshees' clinic laboratory ([46]; Stern 2006). The bacterial strain was indistinguishable from the outbreak strain isolated from food items and clinical specimens. ATCC documents discovered at the commune (ATCC invoices) showed that the agent was purchased before the outbreak (Stern 2006; [34]). The ATCC invoices seized during the FBI and state police search were not shared with the public health officials who would have immediately recognized their significance in light of the criminal investigation. When the public health officials learned of the invoices years later, they considered both the pathogen and timing of their purchases at the RMC laboratory "ominous" [47, p. 27].

5.7.1 Outcome of the Criminal Investigation

Bhagwan fled the commune on October 27, 1985, and was arrested when his jet landed in Charlotte, North Carolina. West German police detained Sheela and Puja at a luxurious German resort hotel, and both were later extradited back to the United States. Sheela and Puja were tried in criminal proceedings divided between state and federal courts. The Oregon Attorney General's Office took the poisoning cases of Judge William Hulse and Commissioner Raymond Matthews, while the US Attorney's Office took the restaurant cases [24, p. 126]. On July 22, 1985, Sheela and Puja both pled guilty in state court to first-degree assault and conspiracy assault for the poisoning of Judge Hulse and pled guilty to second-degree assault charges for the poisoning of Commissioner Matthews. In total, Sheela received three concurrent 20-year sentences, was fined $400,000, and was ordered to pay Wasco County restitution in the amount of $69,353.31 [48, p. 136]. Puja was sentenced to two concurrent 20-year sentences and a concurrent 7.5-year sentence. Sheela was sent to the Federal Correctional Institution in Dublin, California, prison for female criminals. In December 1988, she was released on good behavior after serving 29 months of her 20-year sentence and moved to Switzerland. Sheela married Swiss resident Urs Birnstiel, a fellow Rajneesh follower. In 1999 she was convicted by a Swiss court for "criminal acts preparatory to the commission of murder" in relation to a plot to kill US federal prosecutor Charles Turner in 1985. The Swiss government refused to extradite her to the United States but agreed to try her in Switzerland. She was found guilty of the equivalent Swiss charge and was sentenced to time served. Puja was also released after only serving 2.5 years in a minimum-security federal prison for good behavior. Both women fled to Europe before the US Justice Department was able to notify the state of the early release. Bhagwan received a 10-year suspended sentence for his knowledge of the attacks, was fined, and paid $400,000 in fines. He later left the United States in 1985 [49, p. 32]. Bhagwan Shree Rajneesh died on January 19, 1990, at the age of 58 from heart disease [50]. However, some of his followers suspected that he was poisoned by confidantes who had an eye on his riches.

5.7.2 The Investigation: Missed Warning Signs, Failures, Biases, and Lack of Understanding

A review of the health department and the EIS investigations found that throughout the process, the investigation and investigators missed a number of indicators and warning signs and showed a level of bias and a lack of understanding of the mindset of terror organizations. The overall result was a failure to realize signs that the outbreak was not a naturally occurring event but in fact a biological attack. These signs involved biological, epidemiological, and political. The investigators lacked an understanding of the framework of a terrorist organization and their motivations and ultimate goals. Furthermore, the training of epidemiologist has a built-in level of bias that contributed to the failure in the detection of the attack. Below are some of the indicators that were missed, overlooked, or ignored that suggested that this outbreak was inconsistent with features of a typical natural outbreak. The steps and type of investigations that forensic epidemiologists would have conducted are presented.

Biological Between 1978 and 1983, the local health department identified 16 isolates of *Salmonella* and only 8 of which were *Salmonella* Typhimurium. In addition, no local cases of salmonellosis were reported in 1984 before the August outbreak (Stern 2006; [34]). The rarity of the specific strain of *Salmonella* Typhimurium isolated during the outbreak coupled to the infrequency of salmonellosis should have increased the level of suspicion that this outbreak was atypical. Based on these initial features of the outbreak, a forensic epidemiologist would have suspected that the outbreak was not a natural event. Once the laboratories identified the exact bacterial strain, a forensic epidemiologist would have contacted medical supply companies that offer that strain. Then the forensic epidemiologist would have worked with law enforcement to obtain a list of the locations that were supplied with that bacterium within a 100 radius of the outbreak.

Epidemiological There were several epidemiological features that would indicate an atypical food outbreak. First, the epidemic exposure curves indicated that the salad bars were infected multiple times during a several-week period, suggesting a sustained source of the agent was necessary. The vast majority of natural-occurring outbreaks present with a single exposure curve. The anomaly of the double crest of the epidemiological wave should have suggested foul play. Second, several different food items were contaminated. In almost all cases of a natural food outbreak, the source of the outbreak can be traced back to one specific food item. In this case, several different items were involved. A forensic epidemiologist would have identified the presence of multiple curves and the contamination of several different food items as clear markers of a deliberate contamination and a bio-attack.

Other signs that a forensic epidemiologist would have used to determine that an outbreak was nonnatural are the appearance of an extremely rare disease (such as smallpox), an ordinary or common disease emerging out of season, an ordinary disease with an unusual mode of transmission, or the public health investigation failing to identify the source of the outbreak.

Alternative Hypotheses One of the failures of the health department, staff epidemiologists, and the EIS investigations was their unwillingness to consider alternative theories for the outbreak. There was resistance by the team members to consider the possibility of a bio-attack. Their resistance was based mostly on the lack of documentation of previous biological attacks involving a food outbreak. All those involved in the investigation focused *only* on traditional and common causes of an outbreak. This included breaches involving food harvesting, handling, transport, storage, and preparation. By contract, forensic epidemiologists are trained to consider all possible alternative theories, even those that are unlikely. In the case the forensic epidemiologist would search beyond the typical causes associated with a food outbreak. This includes considering and investigating local organizations that would have motives or would somehow benefit from the outbreak [3].

The Cult All the agencies involved in the investigation had limited to no understanding of the structural organization and motives of the Rajneeshee Cult. In addition, they were not aware of the conflict that existed between the Rajneeshees and the local community. Religiously motivated cults such as Aum Shinrikyo and the Rajneeshees are cut off from the outside world and are often guided by charismatic, all-powerful leaders, making them less subject to societal norms. These cults also have a tendency to escalate the level of terrorist violence over time and to use innovative weapons and tactics. Motives used by terrorist groups include seizing political control over a community, animal rights, ecoterrorism, or targeting a specific company or organization.

Political Issues There were a number of signs that would have suggested that the Rajneeshee Cult could have been linked to the attack. The evidence included the following: It was a well-established fact that there were growing tensions between the cult and the local community. There were concerns expressed by the county's public health official Carla Chamberlain and Commissioner Judge William Hulse, relating to the activity of the cult. These facts should have been sufficient to heighten the level of suspicion. Despite concerns of potential election fraud, the outbreaks in September and October were not linked to the upcoming November elections. Investigators failed to understand that the attacks in September and October represented trail attacks used to evaluate the effectives of the agent/attack and the reaction of the public to the outbreak. The cult assesses the reaction to the bio-attack by examining the response of the local health department and law enforcement. Neither agency believed it was anything other than a natural outbreak. One of the major failures of the investigation was that investigators believed that there was no claim for responsibility or no demands by a terroristic organization that would rule out a terroristic attack. Furthermore, if a terroristic organization did engage in an act of terrorism, a public statement would have been issued to create widespread fear. The investigators/epidemiologists failed to comprehend the underlying motives and goals of the Rajneeshees to lower voter turnout without drawing unwanted attention focused on the outbreak or the foundational features of a covert attack. A forensic epidemiologist would have gathered intel from the local law enforcement agency relating to recent actions of the cult, talked with the commissioners about any conflicts between the Rajneeshee Cult and the community, and gained an understanding of the possible motives and goals of the cult and the importance of the upcoming election. In addition, the forensic epidemiologist would have connected the recent purchase of the ATCC-14028 bacteria, the unprecedented number of *S.* Typhimurium cases, and the upcoming election and how the Rajneeshees could benefit.

Training/Biases The original investigators Laurence Foster and Thomas Török failed to detect a biological attack or even an intentional contamination. Part of the failure was rooted in the traditional training of public health investigators (epidemiologist). Somewhat unique to the 1984 investigation was a level of bias that was introduced by the key investigator Laurence Foster, the most senior state epidemiologist and a widely respected figure in the regional medical community. He was a staunch civil libertarian who "ardently believed that the Rajneeshees were being unfairly harassed because of their strange religious beliefs" [26, p. 22]. Additionally, Foster was also the mentor of Thomas Török—one of the principal EIS team member dispatched to The Dalles to lead the federal response. Foster's preliminary report in November 1984 found no evidence supporting an intentional contamination and concluded it was more likely that food handlers were responsible. Although Foster did acknowledge failing to find no common source for the infections, he deduced their cause "could have occurred where food handlers failed to wash their hands adequately after bowel movements and then touched raw foods". Török's EIS team concurred with these findings in their preliminary report issued January 1985, saying there was "no epidemiological evidence" supporting intentional contamination.

The types of bias demonstrated in this investigation were not just individual bias in the form of personal support for the cult but also institutionalized bias. Epidemiologists are trained first at the graduate level and later by senior investigators at the health department to investigate food outbreaks from the prospective of food, individual, or environmental causes for the outbreak. In this case the personal ideology and training of the investigators affected the investigation and therefore failed to identify the true cause of the outbreak. The investigators' training resulted in never considering a line of investigation forced on the possible of terrorism. Forensic epidemiologists, on the other hand, are trained to consider all possible even remote explanations to explain the outbreak and then look for evidence that supports or refutes the hypotheses of a bioterrorism attack.

Law Enforcement Law enforcement investigators found no questionable activities reported by restaurant patrons, among the restaurant workers, or establish a recognizable pattern of unusual behavior. No disgruntled employees were identified, and the criminal investigation confirmed that employees did not intentionally contaminate any food. The narrowly focused investigation, only on restaurant employees, resulted in missed opportunities to investigate the role of other individuals or organizations. The law enforcement knew that there had only been two reports of a foodborne illness caused by intentional contamination using a biological agent, and neither incident appeared to be politically motivated. Therefore, one of the explanations for why a covert bioterrorism attack was not considered or discovered was that such an event never occurred before. In addition, law enforcement failed to appreciate the significance of the failure of the health department's investigation to locate a common source for the outbreak, the importance of multiple contaminated food items and multiple outbreaks.

5.8 Lessons Learned from the Rajneeshee Cult Investigation

A number of lessons can be learned by evaluating the facts of the bio-attack carried out by the Rajneeshees. These lessons are important to identify areas of vulnerabilities; sectors of the society that can be the target of attacks; the methods that biological agents are obtained, produced, and spread; the mindset of terrorist organizations; the need for better lines of communications between investigative agencies; and the importance of conducting joint investigations involving the health department and law enforcement with the forensic epidemiologist bridging the gap between these very different agencies.

First, the Rajneeshees' attacks demonstrated the inextricable link between organizational objectives, agent selection, and ultimate outcomes. The cult took a deliberate, procedural, and rational approach regarding agent selection. Key players within the organization debated and carefully calculated the costs and benefits of various lethal pathogens, eventually deciding on a mild bacterium as the final choice.

Second, the Rajneeshees' attacks highlighted the ease that certain sectors of a society (hospitals, medical facilities, laboratories, testing clinics) were able to obtain and produce biological agents.

Third, the Rajneeshees' attacks revealed the difficulty in discerning signatures of a biological attack and tracking the acquisition and production of viruses, bacteria, or toxins. There are a wide range of sites to legally acquire biological agents. For example, the World Federation for Culture Collections is composed of 472 specimen repositories in 61 different countries, and there are 46 germ banks located in countries as diverse as Germany, India, and Iran. In addition, the legitimate dual-use equipment and sensitive medical materials (viruses and bacteria) required for hospital laboratories, veterinary clinics, and research facilities are easily obtained, and not to mention the international trade in microbial cultures or "germ commerce" all combine to aggravate an already difficult situation [51]. In this case the equipment used to culture the agent is considered common everyday equipment that would not alarm anyone.

Fourth, the objectives of a biological attack are not always to produce mass casualties, panic, or actions by the government. The Rajneeshee case illustrated that to prepare only for mass casualty scenarios is ill-advised. In fact, both the Rajneeshee and the 2001 US Amerithrax attacks did not aim at achieving mass casualties. The Rajneeshee Cult's successful attack against a localized population is very relevant today. It showed the two US sectors are at great risk: agriculture and public health. Experts warn that bioterrorism is a tool to wage sustained economic warfare. For example, an attack on US agriculture can raise fear of food contamination that would impact the general population and the national economy and cause widespread panic. Concerns about the public health sector focus on the ability of emergency medical services, hospitals, and the community to absorb the result of an attack. The relatively simple Rajneeshee attacks challenged both state and local authorities. During and after the attack, the local hospital and laboratory were stretched to the max. Had the agent used by the Rajneeshees been lethal or a contagious virus, the local public health sector could have been severely crippled. The public health sector is vast and diverse, "consisting of state and local health departments, hospitals, health clinics, mental health facilities, nursing homes, blood-supply facilities, laboratories, mortuaries, and pharmaceutical stockpiles" [52]. The ability of these systems to first survive a biological attack, second provide continuity of operations, and third rebound to acceptable and sustainable levels of operations is paramount [53].

Finally, the Rajneeshees' case underscores the need for effective communication and coordination between intelligence, public health, and law enforcement officials. The relationship of the public health and law enforcement during the epidemiological investigation in 1984 and subsequent criminal investigation in 1985 was described as "rocky" and "a clash of cultures" [54, p. 33]. Information was not shared and opportunities were missed. The lines between a natural outbreak and an intentional contamination were blurred, and it proved difficult to establish that a crime had been committed [54, p. 33]. The Rajneeshee Cult biological attacks showed the vulnerability of civilian populations to biological terrorism [55]. Likewise, it forever changed the US perspectives regarding bioterrorism. It highlighted the need for a joint investigation approach involving a forensic epidemiologist alongside law enforcement and using assets of the health department, laboratories, and criminal justice during the early phase of the investigation if a biological attack is even remotely suspected. The events had an impact and made some changes to the public health and law enforcement perspectives regarding bioterrorism, such as joint training involving the health department with law enforcement and forensic epidemiologist and law enforcement task forces.

5.8.1 The Risk of Future Attacks

The Rajneeshee Cult attacks raise three important questions: (1) "What is the risk of a similar bio-attack today?" (2) Can such an attack be prevented? and (3) What is the best approach to investigate a possible covert attack using an unrestricted and readily available biological agents?

A comprehensive review of the risks linked to biological terrorist attack conducted in 1997 made four observations [56]. First, biological terrorism is more likely than ever before and far more threatening than either explosives or chemicals. This observation may have been true in 1997; however, current international data has shown the increased use of chemical attacks such as those in Syria (gas attacks) and bombing by ISIS. Second, official actions directed at the threat to the civilian population have been only marginally funded and minimally supported [57]. Third, preventing or countering bioterrorism will be extremely difficult. Recipes for making biological weapons are readily available on the Internet, and small groups with modest finances and basic training in biology could develop biological weapons [58] at minimal cost. Fourth, detection or interdiction of those intending to use biological weapons is externally challenging. Thus, the first evidence of a biological attack will almost certainly be a spike in the number of hospital emergency room/doctor's office cases all presenting with similar symptoms, an increase in the number of purchases of similar OTC medications, and an increased number of identical pathological agents being reported from testing laboratories [59].

5.8.1.1 Agents of Bioterrorism

The risk from a biological agent is greater when compared to those from a chemical or nuclear attack, in part because biological agents are relativity easy to obtain, easy to produce in large amounts, safe to transport, and can be disturbed by a number of methods. One of the major advantages of using a biological agent is that it can mimic a natural-occurring disease, thereby appearing as a normally occurring outbreak. The agents of anthrax, plague, brucellosis, smallpox, viral encephalitides, and viral hemorrhagic fevers are of particular concern: however, they are not easily produced, are hazardous to handle, and are dangerous to distribute [13].

The short answer to the second question is "No" for three main reasons. First, it is doubtful that regulations and tighter controls of commercially available pathogens would have prevented this or similar future attacks. Second, the cultures such as the one used in The Dalles attacks are easily obtained from medical supply companies and clinical isolates or isolated from raw foods [45, p. 183]. The production and storage of large quantities of the bacteria are inexpensive and require basic equipment and limited technical skills. Finally, the growth, transportation, distribution, and food handling practices in an open society are inadequate to prevent a deliberate contamination by determined actors. This security-freedom trade-off invites a [32] certain level of unmanageable risk.

As for the third question, the public is best served when healthcare professionals and laboratories communicate and cooperate with local and state health departments in the reporting of unusual disease "clusters" [45, p. 184]. The findings from the Rajneeshee case highlight the need for strong surveillance systems and the reporting of unusual disease clusters or atypical symptom presentations. Routine reporting is essential in disease surveillance at both the local and national level, and efforts to improve surveillance will assist in detecting future outbreaks in general.

To recognize a bioterrorism event early, medical, public health, and law enforcement must share information more efficiently than in the past. Since 1984 there have been a number of improvements in data sharing, but a number of challenges still exist. Structural challenges include the fact that over 90% of healthcare in the United States is delivered by private businesses that employ staff who are largely uneducated about the presentations of chemical, radiological, and biological terrorism. Another challenge is the highly irregular system of municipal, county, state, and federal public health authorities, each with different capabilities and interests in terrorism investigations. Each level possesses information on terrorist group characteristics, but there exist limited procedural guidelines to facilitate combined analysis of protected healthcare data and classified law enforcement information. There are also many operational challenges in sharing and analyzing such information. These can be summed up as issues of "threshold" and of "responsibility and authority." Threshold problems deal with the initial recognition and reporting of an event. Individual practitioner limitations, lack of awareness, and lack of confidence to report a questionable event create these barriers. In the absence of an overt threat (e.g., white powder and a threatening letter), illness patterns that may constitute a "terrorist attack" are unclear to most medical, public health, and law enforcement personnel.

In the long term, we need to expand surveillance systems, forensic epidemiologists embedded in health departments, and joint task forces between the health department and the criminal justice system to detect, diagnose, characterize epidemiologically, and respond appropriately to biological attacks especially covert attacks targeting food and water. There also needs to be greater capacity at the international, state, and local levels relating to greater capacity for surveillance, a more integrated and better network of laboratories, and better diagnostic instruments.

5.9 Conclusions

The methods used by terrorist groups are always evolving, looking for soft targets, different methods to wage attacks, and ways of using easily obtainable biological agents as source materials for weapons construction. Overt attacks such as that of the 2001 anthrax attacks, also known as Amerithrax from its FBI case name, occurred within the United States over the course of several weeks beginning on September 18, 2001, 1 week after the September 11 attacks. However, small covert attacks on "soft targets" such as the food and water infrastructure are a growing threat. These groups, often small in number, often conduct small-scale "trail attacks" designed to evaluate the effectiveness of the biological agent, the method of distribution, and the level of suspicion it raises within the public health and law enforcement communities. These attacks are not designed to result in mass number of fatalities but disrupt society in a way that may not be easily detected at first. This bio-attack illustrates how easily a large number of legally obtainable biological agents are available. Many of these agents can be used in a bio-attack due to easy of production and a wide range of disruption. Most experts in the field of terrorism have concluded that future attacks will target infrastructure such as food supplies, restaurants, and other easily accusable targets.

The face of terrorism changes, so must the role of the public health community, including that of forensic epidemiology. The forensic epidemiologist's unique skills such as understanding surveillance systems, prospective data, epidemiology, forensic, and law enforcement techniques and policies such as evidence identification, collection, and presentation are in a key position to lead the investigation that presents atypical in order to differentiate between a natural outbreak and a covert bio-attack. Once there is sufficient prima facie evidence to support the occurrence of a terrorist attack, the forensic epidemiologist can serve as an interface between the health department and the law enforcement community.

Although the majority of apparent food outbreaks is in fact the result of poor hygiene practices of food handlers, cross-contamination, or a failure to maintain food at safe temperatures, a number are in reality bioterroristic attacks. These covert attacks are being carried out without detection for a number of reasons. These include the traditional training of public health offices; inadequate understanding of terrorism ideally and motives; inadequate training in areas of evidence collection, documentation, and chain of custody protocols; and a level of bias among the investigations.

The historical record suggests that future incidents of bioterrorism will probably involve relatively small-scale attacks, such as food contamination. Nevertheless, the dissemination of dual-use technologies relevant to the production of biological and toxin agents and the potential availability of scientists and engineers formerly employed in sophisticated biological warfare programs such as those of the Soviet Union suggest that the technical barriers to mass casualty terrorism are eroding.

To be classified as a terrorist event, an incident must involve an organization or person that conspires to use violence instrumentally to advance a political, ideological, or religious goal [55]. The categories of terrorist organizations involved in the acquisition and use of chemical, biological, radiological, or nuclear materials have changed over time. In recent years, there has been a rise in cases involving three types of terrorist organizations: (1) single-issue groups such as those dealing with abortion and animal rights; (2) nationalist and separatist groups such as Chechen rebel organizations, the Kurdistan Workers' Party, the Tamil Tigers of Sri Lanka, and ISIS; and (3) apocalyptic religious cults such as Aum Shinrikyo [55].

Bioterrorist attacks could be covert or announced and could be caused virtually by any pathogenic microorganism. The bio-attack by the Rajneeshees' religious cult in The Dalles, Oregon, is an example of well-planned attacks using the infectious organism *Salmonella* Typhimurium. The plan was to infect residents with *Salmonella* on Election Day to influence the results of county elections in favor of the cult. The cult conducted a number of trial attacks contaminating ten area restaurants with *S.* Typhimurium on several occasions before the election. These attacks were used to evaluate the agent, number affected, and most importantly the reaction of the health department and law enforcement. A community-wide outbreak of these attacks resulted in 751 salmonellosis cases in a county that typically reports fewer than 5 cases per year. The initial investigation by the health department, epidemiologists, and the CDC all concluded that the outbreak was not linked to the cult. However, they failed to identify the cause of the outbreak. The "true" cause and source of the outbreak became known a year later during a FBI investigation into the cults other criminal violations. During the investigation a vial of *Salmonella* Typhimurium identical to the outbreak strain was found in a clinical laboratory on the cult's compound, and members of the cult subsequently admitted to contaminating area restaurants and several individuals. This incident, among other recent events, underscores the importance of surveillance and improving preparedness at all levels [13].

The Rajneeshees' attack highlighted several features of a covert biological attack: first, the readily ease of obtaining biological agents, producing large amounts of the agent with easily obtained knowledge, and the lack of attention the purchases of such agents are disturbing; second, the level that terroristic groups engage in the evaluation of potential agents, the selection of the agent, the method of distribution, the level of lethality, and the targets selected. The Rajneeshees carefully selected an agent that would not cause death but simply incapacitate the voters. They also choose an agent that caused disease mimicking common food outbreaks, thereby not raising suspicion during the outbreak; third, although there were a number of features of the outbreak that were atypical and markers of a bio-attack, those investigating the outbreak lacked the training, knowledge, and background to reach that conclusion; and fourth, this case highlights the need for a forensic epidemiologist to be a lead investigator serving as a bridge between the health department and law enforcement. While the bio-attack failed to affect the outcome of the election, it did highlight the vulnerability of the general population to biological attacks.

Appendix

Table 5.1 A sample of confirmed use[1], probable use[2], and threatened use[3] (with confirmed possession) of agents involved in criminal or terrorist events from 1970 to 2003

Date	Place	Agent	Investigation findings
2003		Nicotine	Supermarket notified state and federal agencies of planned recall of ground beef because of customer complaints of acute illness after eating the product; supermarket's laboratory then identified nicotine as ground beef contaminant. At least 92 cases of illness meeting case definition. Investigation discovered high concentrations of nicotine in ground beef. Grand jury indictmented a supermarket employee and was arrested and charged with poisoning meat with insecticide
2001	Washington, DC	*B. anthracis*	Following terrorist attacks in fall 2001, envelopes containing *B. anthracis* spores were mailed to news media, companies, and government officials. Total of 22 cases of anthrax (11 inhalational and 11 cutaneous) and 5 fatal cases
1998–1999		Methomyl (highly toxic pesticide)	County health department recorded two clusters of persons with acute gastrointestinal illness with onset soon after eating at a restaurant. A total of 107 persons became ill. Methomyl was identified in table salt; the risk of illness was proportional to the amount of salt added to the food. Epidemiologist concluded an international contamination resulting in an arrest of one cook associate

(continued)

Table 5.1 (continued)

Date	Place	Agent	Investigation findings
1997[1]	New Zealand	Rabbit hemorrhagic disease virus (RHDV)	Farmers used RHDV as an animal (pest) control tool
1996[1]	Dallas, TX	*Shigella dysenteriae* Type 2	Over a 3-day period, laboratory workers of a large medical center developed severe gastroenteritis caused by a rarely identified organism. A laboratory stock culture was used by a disgruntled employee to contaminate pastries of laboratory staff; 12 of 45 workers contracted severe diarrheal illness. A laboratory worker was indicted and charged with first-degree felony tampering with consumer product
1996[3]	England	*Yersinia enterocolitica*	An extortion attempt was directed at British dairies by threatening to contaminate milk
1990[2]	Scotland	*Giardia*	Feces containing agent were allegedly placed in water tank
1990–1995[1]	Japan	*Bacillus anthracis*, botulinum toxin, sarin	Aum Shinrikyo, a religious cult, allegedly released agents on several occasions using cars/trucks, a sprayer on a roof, and briefcases/jars in subway; only one attempt (with sarin) was successful in causing death and injuries
1989[2]	Namibia	*Vibrio cholerae*, yellow fever virus	The Civilian Cooperation Bureau allegedly contaminated the water supply in a refugee camp
1985[2]	Mexico	Screwworm	Mexican workers allegedly spread the parasite to protect jobs in an eradication program
1984[1]	The Dalles, OR	*Salmonella* Typhimurium	The Rajneeshees, a religious cult, contaminated salad bars at restaurants to influence voter turnout in local election; 751 people became sick
1978[1]	London	Ricin	A Bulgarian dissident was fatally stabbed in the leg with an umbrella laced with agent
1977–1980[1]	Norway	Curacit	A nursing home worker used the agent to kill 22 patients
1975		Pancuronium bromide	Over 6-week period, there was a striking increase in incidence of cardiopulmonary arrests in a VA hospital (51 episodes of cardiac arrest in 35 patients). The arrests were more likely during one specific shift. Two nurses were indicted for attempted murder and conspiracy to commit murder
1972[3]	Chicago, IL	*Salmonella typhi*	Teenagers plotted to contaminate the municipal water system

Adapted from Carus (2001), [2]

References

1. Goodman RA, Munson JW, Dammers K, Lazzarini Z, Barkley JP. Forensic epidemiology: law at the intersection of public health and criminal investigations. J Law Med Ethics. 2003;31: 684–700.
2. Mountcastle S. Focus on forensic epidemiology. NC Center for Public Health Preparedness School of Public Health, University of North Carolina, vol 2(5). https://nciph.sph.unc.edu/focus/vol2/issue5/2-5ForensicEpi_issue.pdf. Accessed 25 Nov 2003.
3. Schofield AR. Forensic epidemiology: legal considerations for public health practitioners. Center for Public Health Law Partnerships. www.publichealthlaw.info. Accessed 27 Nov 2003.
4. Anderson M. Focus on field epidemiology. Public Health Surveillance Systems, vol 5(6); 2008. https://nciph.sph.unc.edu/focus/vol5/issue6/5-6SurveillanceSystems_issue.pdf
5. Blackmore CGM, Stark LM, Jeter WC, Oliveri RL, Brooks RG, Conti LA, Wiersma ST. Surveillance results from the first west Nile virus transmission season in Florida, 2001. Am J Trop Med Hyg. 2003;69(2):141–50.
6. Steiner-Sichel L, Greenko J, Heffernan R, Layton M, Weiss D. Field investigations of emergency department syndromic surveillance signals – New York City. Morb Mortal Wkly Rep. 2004;53(Suppl):190–5.
7. Centers for Disease Control and Prevention. Salmonella surveillance: annual summary, 2006. 2008. http://www.cdc.gov/ncidod/dbmd/phlisdata/salmonella.htm. Accessed 14 Oct 2008.
8. Centers for Disease Control and Prevention. What is PulseNet? 2006. PulseNet Website. http://www.cdc.gov/pulsenet/whatis.htm. Accessed 3 Dec 2008.
9. Pavlin JA. Epidemiology of bioterrorism. Emerg Infect Dis. 1999;5:528–30.
10. Noah DL, Sobel AL, Ostroff SM, et al. Biological warfare training: infectious disease outbreak differentiation criteria. Mil Med. 1998;163:198–201.
11. Weiner SL. Strategies of biowarfare defense. Mil Med. 1987;152:25–8.
12. Meselson M, Guillemin J, Hugh-Jones M, et al. The Sverdlovsk anthrax outbreak of 1979. Science. 1994;266:1202–8.
13. Federal Bureau of Investigation, Centers for Disease Control and Prevention, Department of Justice, U.S. Army Soldier Biological Chemical Command. Criminal and Epidemiological Investigation Handbook. 2006. Edition. http://www2.cdc.gov/phlp/docs/CrimEpiHandbook2006.pdf
14. Scott MJ. What is a forensics epidemiologist? http://www.chow.com/fact_6174467_forenensic-epidmiologist_html. Accessed 30 Oct 2012.
15. McDade JE, Franz D. Bioterrorism as a public health threat. Emerg Infect Dis. 1999;4:493–4.
16. Richards EP. Plagues, police, and posse comitatus: legal issues in forensic epidemiology and public health emergency response. http://biotech.law.lsu.edu. Accessed 2 Aug 2017.
17. Technical Working Group on Crime Scene Investigation, National Institute of Justice. Crime scene investigation: a guide for law enforcement. 2000, January. http://www.ncjrs.org/txtfiles1/nij/178280.txt. Accessed 31 Aug 2005.
18. Fierro MF. Chief Medical Examiner. Virginia Department of Health. 2005, November 17.
19. Scenario 1 – Suspicious letter in DeKalb County. Working group exercise. Presented at: forensic epidemiology training course; November 2–5, 2002; NC: Chapel Hill.
20. Answers.com. Rajneesh. http://www.answers.com/topic/rajneesh. Accessed 20 Sept 2006.
21. Miller J, Engelberg S, Broad W. Germs: biological weapons and America's Secret War. New York: Simon & Schuster; 2001. p. 15.
22. The New York Times. Former aides to guru in Oregon plead guilty to numerous crimes. http://topics.nytimes.com/top/reference/timestopics/subjects/i/immigration_and_refugees/index.html?query=ONANG,%20DIANNE%20Y&field=per&match=exact. Accessed 20 Sept 2006.
23. Carus SW. The Rajneeshees-Chapter 8 (1984). In: Tucker JB, editor. Toxic terror: assessing terrorist use of chemical and biological weapons. Cambridge: MIT Press; 2000. p. 121.
24. Carus SW. The Rajneeshees-Chapter 8 (1984). In: Tucker JB, editor. Toxic terror: assessing terrorist use of chemical and biological weapons. Cambridge: MIT Press; 2000. p. 126–7.

25. Miller J, Engelberg S, Broad W. Germs: biological weapons and America's Secret War. New York: Simon & Schuster; 2001. p. 16.
26. Miller J, Engelberg S, Broad W. Germs: biological weapons and America's Secret War. New York: Simon & Schuster; 2001. p. 22.
27. Carus SW. The Rajneeshees-Chapter 8 (1984). In: Tucker JB, editor. Toxic terror: assessing terrorist use of chemical and biological weapons. Cambridge: MIT Press; 2000. p. 118.
28. Miller J, Engelberg S, Broad W. Germs: biological weapons and America's Secret War. New York: Simon & Schuster; 2001. p. 17.
29. Stern J. Terrorist motivations and unconventional weapons. In: Lavoy PR, Sagan SD, Wirtz JJ, editors. Planning the unthinkable: how new powers will use nuclear, biological, and chemical weapons. Ithaca: Cornell University Press; 2002. p. 202–29.
30. Carus SW. The Rajneeshees-Chapter 8 (1984). In: Tucker JB, editor. Toxic terror: assessing terrorist use of chemical and biological weapons. Cambridge: MIT Press; 2000. p. 124–7.
31. Zilinskas RA. Bio criminality and bioterrorism: assessing the threats (Monterey Institute of International Studies, November 4, 2005), slides 25–29. Presentation given as part of Workshop 548, "Simulating a Deliberately Caused Foodborne Disease Outbreak Using the Table-top Exercise (TTX)," held at Monterey Institute of International Studies on 4 and 12 November, 2005.
32. Buehler JW, et al. Syndromic surveillance and bioterrorism-related epidemics. Emerg Infect Dis. 2003;9(10). http://www.cdc.gov/ncidod/EID/vol9no10/03-0231.htm. Accessed 15 Oct 2006.
33. Widdowson MA, et al. Automated, laboratory-based system using the Internet for disease outbreak detection, the Netherlands. Emerg Infect Dis. 2003;9(9). http://www.cdc.gov/ncidod/EID/vol9no9/02-0450.htm. Accessed 15 Oct 2006.
34. Dembek ZF. Epidemiology of biowarfare and bioterrorism. In: Medical aspects of biological warfare. Washington, DC: Bordon Institute, Walter Reed Army Medical Center; 2007.
35. Stern J. Confronting biological terrorism: global epidemiological monitoring. Harv Int Rev. 2001;23(1). http://hir.harvard.edu/articles/1008/. Accessed 15 Oct 2006.
36. Carus SW. The Rajneeshees-Chapter 8 (1984). In: Tucker JB, editor. Toxic terror: assessing terrorist use of chemical and biological weapons. Cambridge: MIT Press; 2000. p. 128.
37. Miller J, Engelberg S, Broad W. Germs: biological weapons and America's Secret War. New York: Simon & Schuster; 2001. p. 19.
38. Knapp DB. Interview. November 13, 1985. AG files. p. 7.
39. Avalos AK. Interview. October 22, 1985. AG files. p. 48.
40. Deva SK. Report of Interview Swami Krishna Deva [K.D.], AKA David Berry Knapp, November 25, 1985. AG files. p. 76.
41. Centers for Disease Control and Prevention. Epidemic intelligence service—about EIS. http://www.cdc.gov/eis/about/about.htm. Accessed 20 Sept 2006.
42. Miller J, Engelberg S, Broad W. Germs: biological weapons and America's Secret War. New York: Simon & Schuster; 2001. p. 20–1.
43. Anderson M. The Rajneeshee Cult-Passagen. http://hem.passagen.se/jan.olofsson/biowarfare/history/rajneeshee.html. Accessed 26 Nov 2005.
44. Miller J, Engelberg S, Broad W. Germs: biological weapons and America's Secret War. New York: Simon & Schuster; 2001. p. 23.
45. Török TJ, et al. A large community outbreak of Salmonellosis caused by intentional contamination of restaurant salad bars. In: Lederberg J, editor. Biological weapons: limiting the threat. Cambridge: MIT Press; 1999.
46. Levy BS, Sidel VW. Terrorism and public health: a balanced approach to strengthening systems and protecting people. New York: Oxford University Press; 2003.
47. Miller J, Engelberg S, Broad W. Germs: biological weapons and America's Secret War. New York: Simon & Schuster; 2001. p. 27.
48. Carus SW. The Rajneeshees-Chapter 8 (1984). In: Tucker JB, editor. Toxic terror: assessing terrorist use of chemical and biological weapons. Cambridge: MIT Press; 2000. p. 136.

49. Miller J, Engelberg S, Broad W. Germs: biological weapons and America's Secret War. New York: Simon & Schuster; 2001. p. 32.
50. Pace E. Baghwan Shree Rajneesh, Indian guru, dies at 58. https://www.nytimes.com/1990/01/20/obituaries/baghwan-shree-rajneesh-indian-guru-dies-at-58.html. Accessed 2 Apr 2018.
51. Barletta M, Sands A, Jonathan B, Tucker JB. Keeping track of Anthrax: the case for a biosecurity convention. Bull At Sci. 2002;58(3):57–62.
52. The White House. National strategy for the physical protection of critical infrastructures and key assets. Washington, DC: The White House; 2003.
53. Parker HS. Agricultural bioterrorism: a federal strategy to meet the threat. McNair Paper 65. 2002, March. http://www.au.af.mil/au/awc/awcgate/ndu/mcnair65.pdf. Accessed 26 Nov 2005.
54. Miller J, Engelberg S, Broad W. Germs: biological weapons and America's Secret War. New York: Simon & Schuster; 2001. p. 33.
55. Tucker JB. Historical trends related to bioterrorism: an empirical analysis. Emerg Infect Dis. 1999;5(4). http://www.cdc.gov/ncidod/EID/vol5no4/pdf/tucker.pdf. Accessed 26 Nov 2005.
56. Bioweapons and bioterrorism. JAMA 1997;278:351–370, 389–436.
57. Tucker JB. National health and medical services response to incidents of chemical and biological terrorism. JAMA. 1997;285:362–8.
58. Danzig R, Berkowsky PB. Why should we be concerned about biological warfare? JAMA. 1997;285:431–2.
59. Henderson DA. Bioterrorism as a public health treat. Emerg Infect Dis. 1998;4:488–92.

Remediation of a Biological Weapons Attack

6

Crystal R. Watson and Francisco J. Cruz

6.1 Introduction to Biological Remediation

Remediation is a critical component of the response to and recovery from a biological contamination event, whether naturally occurring, accidental, or intentional. Regardless of the etiology, remediation steps are needed to enable recovery and reoccupation of contaminated areas when a biological agent with the potential to cause human infections persists in the environment.

One component of remediation—decontamination—is undertaken every day in hospitals and laboratories around the country and the world. Protocols for infection control to decontaminate hospital rooms and medical tools and prevent hospital-acquired infections are well-established, if not always well-adhered to in the US. In addition, accidental biological contamination events, including minor laboratory accidents and infections caused by importation and use of biologically contaminated

---

Crystal Watson is a Senior Scholar and Assistant Professor at the Johns Hopkins Center for Health Security.

Francisco J. Cruz is a Biologist with the US Environmental Protection Agency. He holds a BA in Biological Sciences from the University of Delaware and a MS in Biodefense from George Mason University. Mr. Cruz was also a 2016 Fellow in the Emerging Leaders in Biosecurity Initiative through the Center for Health Security.

This work is not a product of the United States Government or the United States Environmental Protection Agency, and the author (Mr. Cruz) did not conduct this work in any governmental capacity. The views expressed are those of the authors only and do not necessarily represent those of the United States or the US EPA.

C. R. Watson (✉)
Johns Hopkins Center for Health Security, Baltimore, MD, USA
e-mail: cwatson@jhu.edu

F. J. Cruz
Environmental Protection Agency, Washington, DC, USA

S. K. Singh, J. H. Kuhn (eds.), *Defense Against Biological Attacks*,
https://doi.org/10.1007/978-3-030-03053-7_6

materials such as animal hides occur sporadically. In large part, these events are limited in size and scope and are remediated quickly and effectively by responsible laboratorians or local environmental and public health authorities.

An intentional attack using biological agents as weapons is different. Although remediation steps would be similar to other events, an attack will be much higher profile and more complex than an accident or other more routine contamination event. An attack with a biological weapon would likely involve greater amounts of biological material, and potentially "weaponized" agents that are more stable and persistent in the environment. The location of the release will be important for remediation, and may involve a variety of contaminated areas and surfaces, and potentially a much larger area of contamination. There may also be concerns about re-aerosolization of biological agents from surfaces, leading to additional infections, and there will be a need for safe disposal of large amounts of contaminated waste.

An attack with a biological weapon would be an area denial attack in addition to causing illness and death. Whole buildings or city blocks could require remediation, evacuation, and suspension of all economic activity. Therefore, remediation is exceptionally important in a bioweapons attack, both for reducing morbidity and mortality and for ensuring that an affected area is habitable and economically viable into the future.

6.1.1 What Is Remediation?

Remediation following an attack with a biological agent will involve a number of important components from identification and characterization of the biological agent through reoccupation of decontaminated areas.

Identification and Characterization of the Biological Agent

The first step in remediation is to identify the agent that was used. Identification could be made through interdiction of a perpetrator, by an astute healthcare provider making a clinical diagnosis, or environmentally through air sampling, such as via the BioWatch air monitoring system, which operates in over 30 jurisdictions around the US to provide early detection of an aerosolized attack [1]. Once the biological agent has been confirmed, it will also need to be characterized to understand its environmental stability and re-aerosolization potential, as well as properties such as infectious dose and the dose response relationship in humans, to aid in risk assessment. Defensive biological threat characterization research, done prior to an emergency to better understand the agents that might be used as weapons, can help reduce the time required to do this assessment in an emergency situation.

Defining the Boundaries of Contamination

Once an attack has been recognized, the geographic extent of contamination, sometimes called the "hot zone" or "exclusion zone," is established through environmental sampling, testing, and analysis. This process may involve hundreds or thousands of samples from a wide variety of indoor and outdoor surfaces collected by environmental health or public health personnel. These samples would be processed at laboratories with detection capabilities such as the EPA's Environmental Response Laboratory Network (ERLN), and analyzed to identify areas in need of decontamination [2].

Assessing the Risks to Human Health

Scientists will next need to assess the risks that contamination poses to human health, in order to set clearance goals for remediation. This involves the classic risk assessment steps of problem definition, hazard identification, dose-response assessment, exposure assessment, and risk characterization [3]. This well-defined process is intended to lead to identification of acceptable risk/levels of contamination or viable biological material in the environment. Clearance levels or "clean up goals" can then be established to guide decontamination.

Selecting a Decontamination Approach

Decontamination tools and technologies are numerous, and not all approaches to decontamination work in every setting. There is no single tool or approach that will accomplish decontamination following a biological attack. Selection of a decontamination strategy will need to be guided by a number of factors including the location of the attack (e.g., indoors, outdoors, subway), the types of surfaces contaminated (e.g., concrete, glass, carpet, wood, foliage, etc.), levels of contamination, public risk perception of the contaminant and the decontamination method, time available for remediation, and clearance goals established through risk assessment.

Prioritizing Decontamination

Once the boundaries of contamination have been defined and clearance goals have been set, it will likely be necessary to prioritize decontamination efforts, especially in a wide area attack where the problem may be overwhelming to tackle all at once. Priorities might include remediating critical infrastructure first, decontaminating homes to clear living space for residents, ensuring that businesses are decontaminated to enable reestablishment of economic activity, or other options. This process should be empirically based, but will ultimately come down to a political decision about the values and goals of the affected community.

Identifying Responsible Parties

Remediation of large areas will need to be conducted quickly, and will require many partners to accomplish. The federal government will likely not have enough resources to conduct the remediation directly, so federal, state, and local governments, will need to work jointly in partnership with the private sector to operationalize decontamination. In the 2001 "Amerithrax" attacks [4, 5], private companies, employed by both the government and private sector, carried out decontamination. Having a verified list of qualified companies that can perform biological decontamination prior to an emergency will be critical to ensuring that decontamination is conducted swiftly and effectively [6].

Source Reduction
Source reduction is conducted to reduce the levels of contamination prior to the main decontamination effort in order to make decontamination more efficient and effective. This is done through removal and subsequent treatment of contaminated items and pretreatment of surfaces to prepare them for final decontamination [7].

Decontaminating Affected Areas
Decontamination consists of either removal or inactivation of the biological agent from contaminated environments and surfaces. It is an iterative and sometimes lengthy process involving confirmatory sampling, testing, analysis in the laboratory, and further decontamination until determined levels of safe residual contamination and clearance goals are met. Multiple approaches to decontamination may be needed and may include fumigation or surface treatment using liquids or other agents [8].

Safe Disposal of Contaminated Materials
Massive amounts of waste will be generated in the process of decontamination from a biological attack. This will include liquid and solid waste, contaminated equipment and belongings from offices and homes, contaminated materials like carpet and wood, personal protective equipment and other personal waste, and potentially animal waste and carcasses (depending upon the scenario). These materials will need to be decontaminated themselves either through deactivation or incineration, and be ultimately placed safely in a landfill or other site for final disposition [9].

Reoccupation of Contaminated Sites
Upon completion of decontamination efforts, the final phase in remediation is reoccupation of formerly contaminated areas. This will likely involve a large community—and perhaps national—effort to engage businesses and residents and encourage them to return to their homes and workplaces. If residual contamination is likely to persist for long periods, reoccupation may include consideration of a vaccination campaign (if a vaccine exists) to protect community members from future exposures, particularly within vulnerable/immunocompromised populations. Reoccupation and recovery may be a long process, and some contaminated areas may never be useable again.

6.1.2 Biological Agents and Environmental Persistence

Biological agents vary in environmental persistence, with more environmentally stable agents being of greater concern for remediation in the event of a biological weapons attack. Length of persistence in the environment can range from brief minutes to years in certain circumstances, with bacteria being generally more stable, viruses generally less so, and other agents including fungi, protozoa, toxins, and prions with variable stability. Agent persistence also differs based on the medium, with some agents or pathogens being more stable in the open air, others in food or water, and still others in soil or on other surfaces (Table 6.1).

Table 6.1 Selection of NIH priority pathogens and their environmental stability [10]

Biological agent (disease)	Category	Environmental stability
Bacillus anthracis (anthrax)	Spore-forming bacteria	Very stable in most environments, potentially subsisting for years in some cases. Inactivation of spores is very difficult, and success varies according to environmental conditions (e.g., temperature and humidity). The risks of secondary aerosolization are not well-understood, but may be significant in both indoor and outdoor environments [11–13]
Brucella spp. (brucellosis)	Non-spore-forming bacteria	Somewhat stable in the environment, particularly in moist conditions. Contamination attenuates naturally over time, with an estimated maximum of just over 2 months of viability on multiple surfaces [14]
Burkholderia mallei (glanders)	Non-spore-forming bacteria	May survive in moist, warm conditions for multiple weeks, but is thought to not survive for long in dry conditions or in the presence of sunlight or other UV sources [15]
Burkholderia pseudomallei (melioidosis)	Non-spore-forming bacteria	Can survive for months to years in moist, warm environments, and particularly in water and soil. It is less able to survive in dry environments or outside of soil, and is similar to other bacteria in its susceptibility to sunlight or other UV sources [15, 16]
Yersinia pestis (plague)	Non-spore-forming bacteria	Unstable in the open air, but potentially viable for months to years in soil, water, and live tissues [15]
Francisella tularensis (tularemia)	Non-spore-forming bacteria	Can survive for days to weeks on a variety of fomites, and for hours to days as an aerosol. The bacteria may also be able to survive for multiple weeks in water, including tap water. However, less is known about its persistence in soil or its susceptibility to UV radiation [15]
Coxiella burnettii (Q fever)	Spore-forming bacteria	These bacteria are fairly stable in aerosol form for multiple weeks. They are also stable on fomites, potentially for years, and can survive in moist soil for weeks to months [15]
Variola virus (smallpox)	Virus	The virus is unstable in the environment, remaining viable on most surfaces at room temperature for only hours. However, Variola virus can remain viable for years or even decades in smallpox scabs or in cold storage [17, 18]
Hemorrhagic fever viruses	Viruses (e.g., Ebola virus, Marburg virus, etc.)	These viruses are generally unstable in aerosol form—viability lasting only a matter of hours. On fomites, at room temperature, the viruses are not expected to persist for more than a few hours. However, in dried blood and other bodily fluids, in corpses, or in colder temperatures, these viruses may remain viable for days or even weeks in some cases [15]

(continued)

Table 6.1 (continued)

Biological agent (disease)	Category	Environmental stability
Botulinum toxin (botulism)	Spore-forming and toxin-producing bacteria	Relatively unstable, and will degrade naturally in outdoor environments within a few days. The toxin is heat sensitive and can be deactivated with bleach. However, in food matrices or untreated water sources, it may persist for weeks or longer [19, 20]
Ricin	Toxin	May persist in the environment for up to 2 weeks, depending on the surface, but the toxin is heat sensitive and can be deactivated by sodium hypochlorite solution [21]

Stability in the environment is a natural characteristic of biological agents, but agents and pathogens can also be intentionally engineered to have greater environmental stability through weaponization. Engineered stability has been a goal of past offensive biological weapons programs, which aimed to make agents both easier to disseminate in aerosol form, with smaller particle sizes, less clumping, and less susceptibility to the elements, including ultraviolet light and precipitation. In addition, more effective aerosols tend to be more likely to re-aerosolize, and thus are more dangerous as residual contaminants after an attack.

6.2 Biological Decontamination Examples

6.2.1 The 2001 "Amerithrax" Letter Attacks

The "Amerithrax" attacks of 2001 sickened 22 people, resulted in 5 deaths, and also necessitated the largest biological remediation effort that the US civilian sector had ever undertaken. In the fall of 2001, an estimated seven envelopes containing *Bacillus anthracis* spores were mailed to public figures and media companies in the US, including NBC and Tom Brokaw in New York City, American Media Inc. (AMI) in Boca Raton, Florida, and US Senators Patrick Leahy and Thomas Daschle in Washington, DC [3]. The letters were processed through mail facilities in Washington, DC, New Jersey, and New York City, and were delivered to their targets where at least four of the envelopes were opened. Recipients who opened the letters discovered anthrax powder inside, along with threatening notes identifying the substance as "anthrax" (*Bacillus anthracis*) [3, 4].

Environmental contamination and human exposures occurred in a number of ways in this attack: targeted recipients or staff opened the anthrax letters and inadvertently released spores into the surrounding environment; anthrax spores were aerosolized from the letters during processing at mail facilities when they went through the mail sorting machines, exposing postal employees; and cross contaminated mail was delivered to homes and businesses where individuals were exposed to small amounts of the *Bacillus anthracis*.

Of the 22 illnesses resulting from the letters, 11 were inhalation cases caused by breathing in aerosolized spores. Five of those inhalation cases eventually succumbed to their infections, and 9 of the 11 inhalation cases were postal workers. One inhalation anthrax victim, a 94-year-old woman from Connecticut, died from her infection, but the source of her exposure was never definitively determined: No *B. anthracis* spores were recovered from her home, car, or any other belongings despite extensive sampling and testing. It is hypothesized that the woman was exposed via a letter that became contaminated as it passed through a postal facility with other cross-contaminated letters [22]. Because no other routes of exposure were identified and no spores were recovered, the 94-year-old woman—who had chronic pulmonary diseases and was otherwise immunocompromised—may have inhaled a very small number of spores, perhaps as few as one or two. This hypothesis is further supported by the incubation period in this case, which was 3 weeks longer than the 10th patient to be diagnosed with inhalation anthrax. Animal studies show that smaller doses of spores result in longer incubation periods [23].

The amount of *B. anthracis* material involved in the 2001 attacks was small, but the decontamination response was expansive, including at least 42 buildings in 9 states and Washington, DC. Some postal facilities in New Jersey and Washington, DC, and the AMI building in Florida, were decontaminated in their entirety, whereas in other contaminated locations, parts of buildings were fumigated or otherwise decontaminated [4]. Estimates of the cost of remediation from these seven letters range from $320 million [4] when accounting strictly for the decontamination costs, to $2 billion or more when secondary costs of facility closures and lost revenues are accounted for [24]. The US Postal Service (USPS) buildings, in particular the Brentwood postal facility in Washington, DC, and the Trenton, New Jersey, postal facility, were the most contaminated due to mechanical aerosolization of *B. anthracis* spores as the letters passed through high speed mail sorting machines. Complete remediation of Brentwood and Trenton buildings took over 2 years each, with a combined cost of over $200 million [4], and the AMI building never reopened.

Because of both natural variability in human susceptibility to *B. anthracis*—as illustrated by the CT victim example above—and because of uncertainty about *B. anthracis* biology, including its re-aerosolization potential and persistence in the environment, remediation of the Amerithrax attacks was conducted with the clearance goal of zero viable spores detected. While zero risk is obviously most desirable, it is not very practical—reducing contamination to zero spores is nearly impossible to achieve, particularly in a large contamination event. Although this was the official goal of remediation in 2001, and is the still the current policy today [25], it is likely that residual viable spores remained in contaminated buildings even after remediation was completed. No additional cases of anthrax stemming from residual contamination occurred. However, this event prompted discussion among the biodefense community about "how clean is safe" in a *B. anthracis* attack—a discussion that continues more than 15 years later.

6.2.2 *Bacillus anthracis* Drum Remediation

Periodically, human *B. anthracis* exposures occur unintentionally via contact with contaminated soil or animal products, including animal hides and hair. One way these exposures have occurred in the past is through drum making using untreated animal hides. Contaminated hides are stretched, dried, scraped, and otherwise processed for use as drumheads. This process can aerosolize *B. anthracis* spores on the hides, and cause human exposure and environmental contamination, often indoors where these drums are made and used. When these events occur, they are typically small in scale, causing a limited number of illnesses and minimal contamination. However, they do require remediation, and thus provide examples and practice in responding to *B. anthracis* contamination.

One such *B. anthracis* drum incident occurred in 2007 when a man and his son were infected with cutaneous anthrax after the man made drums using goat hides in a shed outside of his house in Connecticut. While the drum making did not occur in the house and the son had not visited the shed, swabs from surfaces in multiple parts of the house and the man's car, as well as from the materials, tools, and environment in the shed all were culture positive for *B. anthracis* [26].

The remediation response to the 2007 *B. anthracis* drum incident shows how difficult and expensive even a small decontamination/remediation response in a contained environment can be. Multiple US state agencies, including the Connecticut Department of Health, Connecticut Department of Environmental Protection, and the Connecticut Police responded, in addition to federal agencies, which included the FBI, the CDC, and EPA. Input from all of these stakeholders was integrated to develop a sampling strategy and to make decontamination decisions. However, individuals and responding agencies disagreed about the approach to both sampling and decontamination. Because of Connecticut's recent experience with the last case from the Amerithrax attacks, there was a greater public perception of risk surrounding anthrax exposures, and it was decided by the unified command that a zero-growth standard for decontamination would be adopted like it had been for Amerithrax. This decision necessitated the implementation of more stringent and expensive decontamination methods.

Complex surfaces in the home environment, including wood in the shed and carpet in the house, posed problems for non-fumigation approaches to decontamination because other methods could not penetrate all contaminated materials and inactivate all existing spores. Given these limitations, chlorine dioxide (ClO_2) fumigation was chosen for decontamination of the entire home, while the shed was decontaminated using bleach, in order to avoid damaging potentially valuable tools (which might oxidize), and because the shed was not air tight, thus making fumigation difficult. Other approaches to decontamination were explored but ultimately ruled out because they would create large amounts of waste, with no safe way to dispose of that waste. For example, removal of the carpets was contemplated, but ultimately dismissed.

The ClO_2 fumigation caused some oxidization of personal property within the house, but the house was eventually found to be free of viable spores, and the family was able to return to their home once the remediation process was complete. In total, remediation of this incident displaced the family for 4 months and cost US$900,000 (US$612,000 of which was for fumigation) [27].

6.2.3 Ebola 2014: Remediation in NYC and Dallas

During the 2013–2016 Western African Ebola virus disease (EVD) epidemic, questions arose about the need for and approach to decontamination in healthcare and other settings. When infected individuals are most ill with EVD, they produce large amounts of highly infectious body fluids and waste, which often contaminate surrounding environments. At the outset of the 2013–2016 EVD response, it was expected that Ebola virus (EBOV), like influenza A virus and many other viruses, would not remain viable for long on most surfaces, but empirical support for this assumption was based on relatively few EVD outbreaks and limited data. With EBOV's rapid spread during the 2013–2016 outbreak, some began to speculate that EBOV was hardier than expected. Additional research was conducted to characterize the stability of EBOV on a variety of surfaces and under a variety of environmental conditions [28, 29]. However, significant uncertainty surrounding environmental persistence remained. In some cases, this uncertainty was used to justify political decisions about remediation policy, which were not recommended or supported by scientists and public health officials.

In the US, EBOV decontamination entered the wider public consciousness first in September and October 2014, when Thomas Eric Duncan traveled from Liberia to Dallas, Texas and fell ill with EVD, resulting in his death and transmission of the virus to two nurses who provided care for him when he was at his sickest in the hospital. When it was determined that Mr. Duncan had EVD, a private remediation company in Dallas was hired to decontaminate the patient's living quarters [30]. A great deal of fear surrounded this process, and individuals living in the apartment complex where Mr. Duncan stayed, faced stigmatization even after remediation. However, a comprehensive plan of action was established in collaboration with the Texas Commission on Environmental Quality, decontamination was completed effectively, and no infections occurred related to these spaces following remediation [31]. While complex and uncertain, this is an example of a fairly well considered decontamination response.

In New York City just a few weeks later, Dr. Craig Spencer became sick with EVD upon his return from treating patients in Western Africa. In the wake of Mr. Duncan's death and the infection of the two nurses in Dallas, the New York City government was faced with difficult decisions about whether to decontaminate the spaces where Dr. Spencer visited during his incubation period. Available scientific evidence suggested that remediation was probably unnecessary because EBOV is not contagious during the incubation period before symptoms appear. Dr. Spencer isolated himself and sought treatment as soon as he recognized his symptoms, but he started feeling feverish during an evening out in NYC, and so it was impossible to say that there was zero risk of contamination. Given this uncertainty, the city government felt political pressure to act, and the decision was made to retrace Dr. Spencer's steps in the days before he became ill and to inspect and decontaminate the places he visited, including a bowling alley where he bowled a few games before he felt feverish, the meatball shop where he ate dinner, the vehicle that he traveled in on the way home, and his apartment in Manhattan [32].

The City hired a contractor—Bio-Recovery Corp—to undertake the decontamination process. Unfortunately, this company had recently been sold to a new owner, who turned out to be fraudulent. The company performed "decontamination" in a number of locations, but the effectiveness of its work was later called into question when it was discovered that the company was unlicensed and its owner had committed fraud on a number of other occasions [33]. No individuals were infected with EBOV in relation to Dr. Spencer, either before or after the decontamination was completed, but it is unknowable whether this was because Bio-Recovery remediated the sites effectively or that it wasn't necessary to remediate them in the first place. This example highlights the need for risk assessment and evidence-based decision-making to inform decontamination in a bio-event, and also the need to identify remediation contractors, verify their credentials, and put operational plans in place prior to an emergency.

6.2.4 Wide Area Attack Remediation

The examples above illustrate relatively circumscribed biological contamination events, involving small amounts of contamination, and delimited contaminated zones. Remediation of each of these events was complicated and costly, but would look minor in comparison to the human and economic costs of a wide area biological weapons attack in an urban area, particularly involving *B. anthracis*.

Concerns about a wide area attack are multifactorial. In addition to the substantial illness and loss of life that would result directly from an *B. anthracis* attack, residual contamination and the risks of re-aerosolization could threaten the viability of a city. In response to an attack, people and businesses would likely be urged or required to evacuate, or would leave of their own volition. Large areas would need to be remediated, and would include not only buildings and indoor spaces, but outdoor areas such as streets and parks. With the policies and capacities in place currently, it has been estimated that remediation of a US city might take 2 years or more. However, after about 6 months businesses and residents might relocate permanently, leaving a city abandoned indefinitely [34]. The following sections outline the current policies and programs in place for remediation of a wide area biological attack, and the remaining gaps in our knowledge and ability to respond.

6.3 Government Policies and Programs for Biological Decontamination/Remediation

6.3.1 Current and Past Decontamination Policies

The policy landscape of environmental biological decontamination is complex, and the ability to apply science to policy formulation is limited due to gaps in scientific knowledge about biological agents and their environmental persistence. Decontamination decisions should be based on conservative risk concepts that are protective of public health. However, conservative decisions can result in labor-intensive—and sometimes unattainable—decontamination options. Scientific uncertainty related to decontamination leads to time-consuming, expensive operations that may leave people displaced from contaminated areas for extended periods of time.

It is important to note that much of the research, and many policies, related to biological decontamination focus on *B. anthracis*, due to its environmental persistence as a spore and the lethality of anthrax disease. Researchers have focused their efforts on *B. anthracis*, with the general understanding that if you can remediate areas contaminated with *B. anthracis* spores, you can also remediate areas contaminated with other, less hardy, agents.

Biological remediation policy and practice are plagued by one constant question: How clean is safe? Thus far, decontamination approaches have revolved around the expectation that contaminated areas will be remediated and re-opened in their pre-incident state (i.e., zero contamination)—a near impossibility. While there has been advancement in biological decontamination methods, the policies governing decontamination remain largely unchanged. For example, one of the principal US government decontamination policies around *B. anthracis* is the concept of "zero viable spores." This concept dictates that decontamination will be completed to ensure that zero viable *B. anthracis* spores remain. The cleanup goal of "zero viable spores" is based on the idea that there is no safe level of exposure to a *B. anthracis*—essentially, one spore can kill you [25]. As discussed in case examples above, there are several reasons why this policy has been put in place. As of 2017, there has not been a agreed-upon minimal acceptable risk for exposure to *B. anthracis* spores. While there has been work to try to develop dose-response models for *B. anthracis*, a cleanup goal has not been identified based on existing research. In addition, there is no reliable way of translating environmental sampling data to identification of an acceptable risk level [25]. While the number of spores identified in a culture of an environmental surface sample can provide evidence of levels of environmental contamination, it cannot provide information as to the likelihood of human exposure to the spores. In the absence of dose-response information and an ability to interpret environmental sampling data, decision-makers have defaulted to a "zero viable spores" cleanup goal for decontamination [25].

So who is responsible for implementing the "zero viable spore" cleanup goal? Various authorities at the local, state, and federal level have roles to play in implementing the cleanup goal, further complicating decision-making during a biological incident. Local and state authorities may address small incidents that do not require specialized expertise or resources, with technical support from federal officials. Larger contamination events and intentional attacks require action from the federal government, with CDC providing guidance from a public health perspective, and EPA leading the decontamination operations upon request. Regardless of the size of the incident, local and state public health officials, or property owners for indoor contamination, retain the final authority for determining if a clearance goal has been achieved [25]. Local, state, and federal officials make decisions within the Unified Command structure, ensuring all levels of government are informed and can make recommendations on actions moving forward.

Achieving a cleanup goal requires significant coordination. Those responders conducting decontamination need to clearly understand the goals and reasoning of public health officials, in order to reach collective cleanup goals. Without this understanding, it can be easy to overreact to small incidents by applying the most stringent standards and methods, and then not scale operations appropriately for larger responses. Moreover, experience levels of responders will heavily influence the success or failure of remediation.

Current federal guidance for response to biological incidents provides a roadmap for how remediation should be approached. Documents like the Federal Emergency Management Agency's (FEMA) Biological Incident Annex detail roles and responsibilities for a federal response. In addition, two advisory committees may be established to aid in answering difficult technical aspects of a decontamination response. One committee, known as the Technical Working Group (TWG), consists of experts from all levels of government, the private sector, and universities, and provides technical advice to Incident Command. The TWG can provide recommendations related to technical challenges such as sampling and analysis, decontamination, and waste management issues. The second group, and arguably the more influential of the two, is the Environmental Clearance Committee (ECC). The difference between the TWG and the ECC is that the ECC is made up of an entirely independent group of experts, none of whom are directly involved in the response. The ECC can assess whether clearance goals have been achieved—providing critical, objective advice to the Incident Command during a time when there will be great political and public pressure to bring the response to a close. While the TWG and ECC input can be influential in a decontamination response, they operate in an advisory capacity, and government is not obligated to follow their recommendations [25].

Determining who is in charge of the federal government's biological remediation response can be challenging—with different scenarios resulting in different federal roles. At the time this chapter was written, the White House National Security Council was composing a new National Biodefense Strategy (NBS), which may have an impact on remediation roles and responsibilities in the future. The NBS, modeled after a report by the Blue Ribbon Study Panel on Biodefense, is the government's attempt to develop a comprehensive strategy to address everything from biodefense research, to prevention, public health preparedness, international health security, and environmental remediation, while reducing redundancy across agencies [35]. The NBS was due to Congress in September 2017, and was released by the White House in September 2018.

As it is currently organized, numerous federal agencies play roles in providing guidance and research on biological remediation. The Department of Homeland Security (DHS) plays a significant role in shaping federal guidance and the government's research agenda. DHS has funded several projects over the years to address gaps in biological decontamination and remediation science. The Wide-Area Recovery and Resiliency Program (WARRP), completed in 2010, identified difficult recovery issues and aided in development of guidance following a chemical, biological, or radiological incident in wide urban areas or military installations [36]. DHS also collaborated with the Defense Threat Reduction Agency (DTRA) to develop the Integrated Biological Restoration Demonstration (IBRD) Program. IBRD took WARRP's concepts and developed a recovery framework based on a scenario involving the release of a biological aerosol over a large, urban area [36]. More recently, DHS funded the Underground Transportation Restoration (UTR) Project, focused on responding to and recovering from a biological attack on a subway station. UTR consisted of multiple studies, including a release of a non-harmful (non-pathogenic) surrogate that behaves much like anthrax spores in the New York City subway during business hours to observe how it would disperse in the transit system, and an Operational Technology Demonstration (OTD) to exercise methods of rapidly decontaminating a subway system [37]. It is through these projects that DHS is able to bring multiple stakeholders together to address critical gaps in environmental decontamination and remediation.

In addition to the work conducted by DHS, EPA has a robust biological remediation research program that covers various facets of a potential response. Led by the agency's National Homeland Security Research Center (NHSRC), EPA conducts research on environmental sampling and analysis methods, decontamination, waste management, and water system security. The NHSRC was established shortly after the 2001 Amerithrax attacks, when it became evident that significant gaps remained in environmental decontamination [38]. NHSRC's work on waste disposal and environmental persistence are areas that have brought greater confidence to how responders will approach a potential decontamination and remediation operation [39].

6.3.2 Current State of Federal Guidance and Research Agenda

As previously stated, the federal government is in the midst of developing the NBS—a broad strategy with an implementation plan to be issued in 2018. With regard to environmental decontamination and remediation, the NBS can help provide clarity concerning policy and research responsibilities. While there is a general understanding in federal government that EPA is the lead agency for environmental decontamination, the funding allocation process for those activities is complex and uncertain. For example, during a wide area biological incident, it is anticipated that there would be a Stafford Act declaration (commonly known as a federal emergency declaration), allowing FEMA to fund EPA's decontamination activities, and enabling EPA to lead the response. However, absent a Stafford Act declaration, funding for response would become more complicated. While without a Stafford Act declaration EPA would maintain its authority to respond to a biological incident through the Comprehensive Environmental Response, Compensation, Liability Act (CERCLA) (better known as the Superfund Law), funding for the response would come out of the already stretched Superfund budget, which routinely funds EPA's core mission of cleaning up hazardous waste sites. Given the restricted funding available for a very expensive decontamination operation following a biological attack, the EPA may urge state and local responders to take on a larger portion of the operation than would be expected in a routine hazardous waste situation. Additional clarity is needed regarding plans to fund and implement biological emergency remediation responses, including frank conversations with state and local officials about their roles and responsibilities.

One area in need of additional research and improved policy, is waste management. Environmental remediation produces large amounts of waste, including used personal protective equipment, liquid waste from decontamination efforts, and contaminated items removed from areas of concern (rugs, furniture, office equipment, etc.). This waste adds up very quickly, and it all needs to go somewhere for final disposition. Laws and practices related to biohazardous waste can vary greatly by state and landfill operator. In addition, peoples' risk perceptions about waste can impact where it will be accepted, thus potentially increasing the cost for waste management. For instance, Ebola virus-contaminated waste associated with Craig Spencer's case in New York City was given a "special Category A" designation, requiring greater levels of destruction and packaging before being transported to the only landfill in the United States willing to accept it. In the end, this one patient generated 352 drums of waste costing tax payers US$1.12 million to dispose [40]. The federal government needs to clearly plan for how to address special waste issues, either by creating a federal waste facility or by providing guidance urging all US jurisdictions to standardize how waste is managed.

Research initiatives also need to be better coordinated and managed by the federal government. There does not appear to be a cohesive approach to funding decontamination research, nor is there an updated research agenda that takes into account persistent knowledge gaps for biological remediation. EPA and DHS, along with other federal partners, have made great advances in closing research gaps. However, the state of the science is at a point where low hanging questions have largely been answered, and more difficult gaps remain unattended. With recent budget cuts to critical biodefense programs, significant progress has been stymied. Future federal research on remediation needs to be conducted with the goal of faster reoccupation of contaminated areas, and greater certainty of the risks involved with re-occupancy.

6.3.3 State of the Science

Research to address persistent gaps in decontamination and remediation have yielded mixed results. While new tools and methods have been developed for decontamination, there are still gaps in approaches to sampling and analysis that would hamper a response to a wide-area incident. Additionally, risk assessment research for biological agents has not yielded the desired result—to move away from the "zero viable spore" clean up goal established by EPA and CDC.

Many of the research successes in remediation have come from the development of decontamination methods. Perhaps due to the ease with which successes can be documented (simply, either a biological agent is inactivated, or it is not), government and industry have developed a toolbox of decontamination options that can be used given various contamination scenarios. EPA has been a leader in decontamination research, developing methods for both large-scale and small-scale contamination scenarios. One challenge EPA has successfully addressed is the decontamination of electronic or sensitive materials that may be destroyed by aggressive methods such as fumigation with chlorine dioxide. Given how ubiquitous electronics are in buildings and critical infrastructure, it would be impossible to thoroughly decontaminate an area by working around areas with electronics. EPA has shown that fumigants such as methyl bromide and ethylene oxide are effective against *B. anthracis* surrogates while leaving electronics such as computers fully operational [41].

Challenges in sampling and analysis continue to be the bottleneck in environmental decontamination and remediation. Collecting environmental samples is still cumbersome, and the process can be slow. Field samples are generally collected in teams of three: a Sampler, an Assistant, and a Team Leader. The setup is primarily driven by the need to prevent cross-contamination between samples. The Sampler, many times in the field known as the "Dirty Person," collects the sample in the field. The Assistant, sometimes known as the "Clean Person," provides the clean sampling materials to the Sampler, thus preventing the Sampler from contaminating other sampling kits during their time in the contamination zone. The third person, the Team Leader, serves several roles, including navigation to the sample location; ensuring sample collection is conducted in accordance to the standard operating procedure; communication with teams outside of the contamination zone; and proper documentation of the sampling event.

There has been limited research to optimize sample collection, but none has had a significant impact on reducing sample collection times. One example of a difficult sampling method is micro-vacuum sampling. The purpose of micro-vacuum sampling is to collect samples on porous surfaces, which includes wood, concrete, and carpeting. Many other methods exist for collecting porous surface samples, but they are all difficult to process in the laboratory. The micro-vacuum sample collects particles onto a 37-millimeter case with a filter inside, trapping any particles collected onto the filter. The method provides good recovery efficiency, and creates a sample that can be safely processed in a laboratory with less risk of contaminating the laboratory itself [42]. While an important development for ensuring a quality sampling event and safe laboratory processing, micro-vacuum sampling is an extremely slow process in the field. The validated method of collecting the sample is to pass the pencil-sized vacuum nozzle across a contaminated surface in an overlapping S-pattern for a 12-in. by 12-in. area at a pace of 3–5 s across—a process that is repeated hundreds of times even in a limited indoor contaminated area. This example illustrates the need to take field practices into account when developing sample collection methodologies, because improved methods in the laboratory can add burden to field personnel.

On the analytical side, clearance of a contaminated area is dependent upon the "gold standard" of culture analysis. The reason for the reliance on culture analysis is the need to determine viability. PCR analysis is able to quickly identify DNA of an agent of concern, but will not give you any information about whether that DNA is from a viable organism or from an inactivated organism in the sample location. Each culture sample requires several culture plates and approximately 36 h to grow, taking up lab space and time. The ability to make quick decisions on re-occupancy is very difficult given this current approach.

A promising method of viability analysis is the Rapid Viability-PCR (RV-PCR) method, which has been used successful with *B. anthracis*. RV-PCR combines the sensitivity of broth culture with the speed of real time-PCR (qPCR) to quickly analyze samples for presence and viability. Samples are washed and extracted onto filter cups, and a growth media is added to promote growth. An initial aliquot is extracted, known as the Time Zero (T0) aliquot. After 9 h of growth, another aliquot from the filter cup is extracted, known as the Time 9 (T9) aliquot, and qPCR analysis is conducted on both aliquots to detect *B. anthracis* DNA. The cycle-threshold (Ct) number for both aliquots are compared, with the delta in the Ct values between 9 and 36 representing an increase in viable *B. anthracis* that germinated over the 9-h period. In total, a sample can be analyzed in 15 h with smaller quantities of growth versus the 36–48 hours required for culture analysis [43, 44]. Unfortunately, validation of this method has not yet been completed, and could not yet be used for response.

A common question asked during environmental remediation is "what's the number?" In other words, how much contamination needs to be removed to achieve acceptable levels of public health risk. That question can be answered with a high level of certainty for most chemical and radioactive hazards. For biological agents, it is a difficult question to answer. In theory, inhalational exposure to one spore of *B. anthracis* can be deadly. However, it is not reasonable to expect that every spore will be inactivated during decontamination. Information regarding low-dose exposures to *B. anthracis* would help make informed decisions on clearance levels beyond what may be an impossible "zero viable spore" standard. EPA took a first step in trying to establish a low dose number by exposing rabbits to wet formulations of the Ames strain *B. anthracis* over several days. Although the researchers were able to come up with an extremely conservative number of 4900 total inhaled colony-forming units as a benchmark dose lower 95% confidence limit for 10% mortality ($BMDL_{10}$), this number cannot be directly used to assess human health risk. First, a biological attack would most likely involve a dry formulation, resulting in perhaps a smaller quantity of spores being able to cause death. Secondly, it is difficult to effectively extrapolate risk from a small study using rabbits. Similar studies using a dry formulation may be able to provide greater confidence on a conservative risk number that can be used during a contamination scenario [45].

The path forward for decontamination and remediation research needs to build upon the work already conducted, and begin to tackle the challenging gaps in scientific knowledge to achieve greater competence in addressing environmental remediation and decontamination. Future work should focus heavily on addressing two issues: outdoor decontamination and remediation, and clearing bottlenecks that prevent quick clearance decisions from being made. Outdoor decontamination may seem to be a matter of scaling up the tactics that would be used in indoor scenarios. However, decontamination of vegetation and soil, streets, storm water drains, and vehicles has not been thoroughly contemplated. A comprehensive effort that aims to control contamination in a dynamic, outdoor environment, while addressing unique decontamination scenarios, are needed to safely and effectively return contaminated outdoor areas to the public.

Sampling continues to be a cumbersome process that could benefit from additional innovation. For example, EPA has conducted initial evaluations of robotic floor cleaners for collecting composite samples, but despite early positive findings, additional research is needed to evaluate whether this technique would be successful in a dynamic environment outside of the laboratory [42]. Anything to reduce the burden on the number of personnel and time it takes to collect a sample would greatly decrease the overall response time.

Additionally, sample analysis continues to be a challenge in remediation response. The key is to develop a method that is both sensitive and also able to determine replication competency. RV-PCR is a good first step, but additional work needs to be done to ensure this method is effective for different biological agents and sample types. A field detection method that can be used to identify an agent and to provide information regarding replication competency with some level of certainty would greatly aid in conducting initial characterization, and perhaps clearance, of a site.

It is easy to neglect environmental decontamination and remediation as part of a national biodefense enterprise because of the urgent need to focus on medical countermeasures (MCM) development and public health surge capacity-building. However, without effective tools for appropriately addressing contaminated areas in a timely manner, there will be no homes or jobs for people to safely return to following a biological attack.

References

1. US Department of Homeland Security. The Biowatch program fact sheet. 2016. https://www.dhs.gov/publication/biowatch-program-factsheet. Accessed 27 Dec 2017.
2. US Environmental Protection Agency. ERLN overview. 2011. https://www.epa.gov/sites/production/files/documents/erln_general_fs.pdf. Accessed 27 Dec 2017.

3. National Research Council. Science and decisions: advancing risk assessment. Washington, DC: The National Academies Press; 2009.
4. US Department of Justice. Amerithrax investigative summary. 2010. https://www.justice.gov/archive/amerithrax/docs/amx-investigative-summary.pdf. Accessed 25 Aug 2017.
5. Schmitt K, Zacchia NA. Total decontamination cost of the anthrax letter attacks. Biosecur Bioterror. 2012;10(1):98–107. https://doi.org/10.1089/bsp.2010.0053
6. Franco C, Bouri N. Environmental decontamination following a large-scale bioterrorism attack: federal progress and remaining gaps. Biosecur Bioterror. 2010;8(2):107–17. https://doi.org/10.1089/bsp.2010.0009
7. Canter DA, Sgroi TJ, O'Connor L, et al. Source reduction in an anthrax-contaminated mail facility. Biosecur Bioterror. 2009;7(4):405–12. https://doi.org/10.1089/bsp.2009.0028
8. Canter DA. Remediating anthrax-contaminated sites: learning from the past to protect the future. Chem Health Saf. 2005;12(4):13–9. https://doi.org/10.1016/j.chs.2004.12.001
9. Lesperance AM, Stein S, Upton JF, et al. Challenges in disposing of anthrax waste. Biosecur Bioterror. 2011;9(3):310–4. https://doi.org/10.1089/bsp.2011.0033
10. NIAID. NIAID emerging infectious diseases/pathogens. 2016. https://www.niaid.nih.gov/research/emerging-infectious-diseases-pathogens. Accessed 27 Dec 2017.
11. Paton S, Thompson KA, Parks SR, et al. Reaerosolization of spores from flooring surfaces to assess the risk of dissemination and transmission of infections. Appl Environ Microbiol. 2015;81(15):4914–9. https://doi.org/10.1128/AEM.00412-15
12. Layshock JA, Pearson B, Crockett K, et al. Reaerosolization of Bacillus spp. in outdoor environments: a review of the experimental literature. Biosecur Bioterror. 2012;10(3):299–303. https://doi.org/10.1089/bsp.2012.0026
13. US Department of Defense. Review committee report: inadvertent shipment of live Bacillus anthracis spores by DoD. US Department of Defense. 2015.
14. US Environmental Protection Agency. Persistence and decontamination testing of Brucella suis. US Environmental Protection Agency, Office of Research and Development; 2011.
15. US Environmental Protection Agency. Persistence of categories A and B select agents in environmental matrices-TB. US Environmental Protection Agency, Office of Research and Development; 2014.
16. Inglis TJJ, Sagripanti J. Environmental factors that affect the survival and persistence of Burkholderia pseudomallei. Appl Environ Microbiol. 2006;72(11):6865–75. https://doi.org/10.1128/AEM.01036-06
17. Sinclair R, Boone SA, Greenberg D, et al. Persistence of category a select agents in the environment. Appl Environ Microbiol. 2008;74(3):555–63. https://doi.org/10.1128/AEM.02167-07
18. National Institutes of Health. Report of the blue ribbon panel to review the 2014 smallpox (Variola) virus incident on the NIH Campus. National Institutes of Health; 2017.
19. Khan AS, Swerdlow DL, Juranek DD. Precautions against biological and chemical terrorism directed at food and water supplies. Public Health Rep. 2001;116(1):3–14. https://doi.org/10.1093/phr/116.1.3
20. Arnon SS, Schechter R, Inglesby TV, et al. Botulinum toxin as a biological weapon: medical and public health management. JAMA. 2001;285(8):1059–70. https://doi.org/10.1001/jama.285.8.1059
21. Audi J, Belson M, Patel M, et al. Ricin poisoning: a comprehensive review. JAMA. 2005;294(18):2342–51. https://doi.org/10.1001/jama.294.18.2342
22. Centers for Disease Control and Prevention (CDC). Update: investigation of bioterrorism-related inhalational anthrax—Connecticut, 2001. MMWR Morb Mortal Wkly Rep. 2001;50(47):1049–51.
23. Barakat LA, Quentzel HL, Jernigan JA, et al. Fatal inhalational anthrax in a 94-year-old Connecticut woman. JAMA. 2002;287(7):863–8. https://doi.org/10.1001/jama.287.7.863
24. Congressional Research Service. The US postal service response to the threat of bioterrorism through the mail. 2002. Accessed 25 Aug 2017.

25. Centers for Disease Control and Prevention – US Environmental Protection Agency (CDC-EPA). Interim clearance strategy for environmental contamination with *Bacillus anthracis*. Centers for Disease Control and Prevention – US Environmental Protection Agency; 2012.
26. Centers for Disease Control and Prevention (CDC). Cutaneous anthrax associated with drum making using goat hides from West Africa—Connecticut, 2007. MMWR Morb Mortal Wkly Rep. 2008;57(23):628–31. mm5723a3 [pii].
27. Guh A, Heyman ML, Barden D, et al. Lessons learned from the investigation of a cluster of cutaneous anthrax cases in Connecticut. J Public Health Manag Pract. 2010;16(3):201–10. https://doi.org/10.1097/PHH.0b013e3181ca650d
28. Joob B, Wiwanitkit V. Ebola virus in sterilized wastewater. Ann Trop Med Public Health. 2016;9(3):209. https://doi.org/10.4103/1755-6783.179122
29. Bibby K, Casson LW, Stachler E, et al. Ebola virus persistence in the environment: state of the knowledge and research needs. Environ Sci Technol Lett. 2015;2(1):2–6. https://doi.org/10.1021/ez5003715
30. Schonfeld Z. This is the cleaning company decontaminating the Dallas Ebola Patient's apartment. Newsweek. 2014. http://www.newsweek.com/cleaning-company-decontaminating-dallas-ebola-patients-apartment-275208. Accessed 30 Sept 2017.
31. Cameron RA. Ebola virus response procedures regarding the remediation of dwellings for the 2014 Dallas Ebola outbreak. Texas Commission on Environmental Quality. 2014. http://www.protectusa.net/wp-content/uploads/Ebola-Virus-Response-Protect-Environmental-Services.pdf
32. Boddie C. Cleaning up is hard to do. 2014. http://www.bifurcatedneedle.com/new-blog/2014/12/10/cleaning-up-is-hard-to-do
33. Campbell A, Kaczynski A. The con artist hired to clean Ebola. BuzzFeed News. 2014. https://www.buzzfeed.com/alexcampbell/how-new-york-city-hired-a-con-artist-to-clean-up-ebola?utm_term=.ltoeXPa2bl#.wwrqnd1J6O. Accessed 30 Sept 2017.
34. Raber E. The challenge of determining the need for remediation following a wide-area biological release. Biosecur Bioterror. 2011;9(3):257–61.
35. Schlegelmilch J, Carlin EP. Congress needs to act now to prevent another biodisaster like anthrax, Zika. The Hill. 2017. http://thehill.com/blogs/pundits-blog/defense/316897-congress-needs-to-act-now-to-prevent-another-biodisaster-like. Accessed 29 Sept 2017.
36. Pacific Northwest National Laboratory (PNNL). Northwest Regional Technology Center for Homeland Security. 2017. http://nwrtc.pnnl.gov/projects/programs.stm. Accessed 29 Sept 2017.
37. US Environmental Protection Agency (EPA). Chemical, biological, radiological & nuclear consequence management advisory division 2016 annual report. US Environmental Protection Agency, Office of Land and Emergency Management; 2016.
38. US Environmental Protection Agency (EPA). About the National Homeland Security Research Center. US Environmental Protection Agency, Office of Research and Development; 2017. https://www.epa.gov/aboutepa/about-national-homeland-security-research-center-nhsrc. Accessed 29 Sept 2017.
39. US Environmental Protection Agency. Remediation following man-made or natural disasters – Homeland Security Research. US Environmental Protection Agency, Office of Research and Development; 2016. https://www.epa.gov/homeland-security-research/remediation-following-man-made-or-natural-disasters-homeland-security. Accessed 29 Sept 2017.
40. US Environmental Protection Agency Region 2 (EPA Region 2). EPA Region 2 Ebola Experience. US Environmental Protection Agency Region 2; 2016. https://www.nrt.org/site/download.ashx?counter=3098. Accessed 29 Sept 2017.
41. US Environmental Protection Agency (EPA). Assessment of the impact of decontamination fumigants on electronic equipment. US Environmental Protection Agency, Office of Research and Development; 2014.

42. US Environmental Protection Agency (EPA). Evaluation of vacuum-based sampling devices for collection of *Bacillus* spores from environmental surfaces. US Environmental Protection Agency, Office of Research and Development; 2013.
43. US Environmental Protection Agency (EPA). Rapid viability PCR method for detection of live *Bacillus anthracis* spores. US Environmental Protection Agency, Office of Research and Development; 2012.
44. US Environmental Protection Agency (EPA). Evaluation of surface sampling for *Bacillus* spores using commercially-available cleaning robots. US Environmental Protection Agency, Office of Research and Development; 2012.
45. US Environmental Protection Agency (EPA). Multiple daily low-dose *Bacillus anthracis* Ames inhalation exposures in the rabbit. US Environmental Protection Agency, Office of Research and Development; 2012.

Maximum Containment Infectious Disease Laboratories as an Integral Part of Emergency Preparedness and Emergency Response

7

Sarah Carpenter and Nahid Bhadelia

7.1 Introduction

Reliable, accurate, and rapid diagnostic capability is a hallmark of public health emergency preparedness and response to infectious diseases threats. Laboratory surveillance assists in early recognition of a bioweapons attack, or of a sentinel case in either an endemic outbreak or a returning traveler. Maximum containment laboratories contribute to critical research towards the creation of novel diagnostics, vaccines, and therapeutics, particularly related to pathogens on which there is limited experience or knowledge. However, as countries expand their level of research capacity, biosafety (the safe containment of pathogens to reduce or eliminate exposure to laboratory workers and environment), and biosecurity (the protection, control and accountability for biological materials to prevent their unauthorized access, loss, theft, misuse, diversion or intentional release) at maximum containment research laboratories become integral components of national and international biodefense planning [1, 2].

Infectious pathogens are categorized into risk groups (by National Institutes of Health and World Health Organization) which characterize their ability to infect and cause disease in humans or animals, the severity of the resultant disease, and whether there are preventive measures and effective treatments available for the disease. Risk groups range from 1 to 4, with pathogens in Risk Group 3 defined as those infectious agents that can cause severe or lethal human disease but for which there may be treatment available. Pathogens in Risk Group 4 are classified as those that pose high

S. Carpenter
Boston University School of Medicine, Boston, MA, USA
e-mail: SCarpe@bu.edu

N. Bhadelia (✉)
Section of Infectious Diseases, Boston University School of Medicine, Boston, MA, USA
e-mail: nbhadeli@bu.edu

S. K. Singh, J. H. Kuhn (eds.), *Defense Against Biological Attacks*,
https://doi.org/10.1007/978-3-030-03053-7_7

individual and community risk and for which we do not yet have effective or widely available medical countermeasures [1]. Many of the Risk Group 3 and all of the Risk Group 4 pathogens are also part of the Federal Select Agent List, a compilation of infectious agents determined to potential to pose severe threat to human, animal and/or plant health by Health and Human Services and United States Department of Agriculture [3]. Lastly, the U.S. Centers for Disease Control and Prevention (CDC) categorization of infectious agents is organized by the risk they pose to national security or the likelihood that they could be used in a bioterrorism event. High risk Category A agents encompass viral hemorrhagic fever (VHF) agents, anthrax, botulism, plague, smallpox and tularemia; Category B agents are those that are easily disseminated but have lower morbidity and mortality and Category C agents are emerging pathogens that could be engineered in the future for mass dissemination [4].

Biosafety levels, also ranging from 1 to 4, define the increasing level of protection built into a laboratory space and related work flow to ensure safety of staff, environment and community based on the nature of the pathogens studied therein. Maximum containment laboratories [also termed biosafety level 4 (BSL-4) laboratories, and used here interchangeably] are defined as facilities that conduct research on high risk (Risk Group 3 and 4) pathogens. Risk group designation of a pathogen often, but not always, correlates with the biosafety level of the laboratory space where safe research can be conducted on it. For example, research on a number of pathogens in the Tick Borne Encephalitis (TBE) complex are restricted to BSL4 laboratories in United States where the diseases are not endemic, despite the fact that these agents are classified as Risk Group 3 [1].

These institutions not only perform critical research related to emerging pathogens but also form an integral part of the response to naturally occurring outbreaks with Risk Group 3 and 4 pathogens providing both critical scientific knowledge and surveillance resources.

Additionally, they serve as repositories for subject matter expertise and medical countermeasures in cases of imported or laboratory acquired infections. During the 2013–2016 Ebola virus disease (EVD) epidemic, the scale and level of involvement of many of these laboratories and their staff in outbreak response was unprecedented [5]. The lessons learned from this experience can guide public health planning about the role of these institutions, not just as scientific powerhouses, but also as sites for critical physical, scientific, and human resources to handle future threats from novel infectious disease outbreaks. In the aftermath of the EVD epidemic, several reports outlined the importance of integrating research during outbreaks into facile diagnostics and effective medical countermeasures. BSL-4 facilities are integral to this charge when it comes to pathogens that cause viral hemorrhagic fevers. Table 7.1 describes the range of activities maximum containment laboratories play in preparedness and response to infectious pathogens of high concern.

This chapter examines how maximum containment laboratories differ from other research facilities and the types of investigation conducted at these sites that contribute to biosafety/biosecurity preparedness. Additionally, the risks and benefits of proliferating BSL-4 capacity at the global level are briefly examined. The recent

Table 7.1 Role of maximum containment laboratories in preparedness and response to infectious pathogens of high concern

Period	Activities
Between epidemics	Continuation of ongoing relevant BSL-4 research Information-sharing between countries and facilities Contribution of biosafety and biosecurity expertise to developing laboratories Contribution to international surveillance resources Development of relevant guidelines for biosafety and biosecurity
During epidemics	Diagnostic support in the field Service on advisory panels Contribution of medical technologies and products Use of experts to train local and national staff Contribution of medical programs created to support BSL-4 research as clinical resource during outbreak response
Future recommendations	
• Invest in research and development of critical products for high-concern pathogens in advance of outbreaks • Develop more long-standing research partnerships with institutions in areas at high risk for outbreaks caused by Risk Group 4 agents • Participate at greater level in public health planning efforts for pandemic preparedness • Participate in more formal global collaboration between BSL-4 facilities	

EVD epidemic is used to highlight the different activities that maximum containment laboratories could undertake between and during outbreaks. Lastly, recommendations for future directions that maximum containment laboratories could undertake to further assist emergency preparedness are presented.

7.2 The Unique Role of BSL-4 Laboratories

As discussed above and in earlier chapters, the biosafety level (BSL) designation attributed to laboratories is a result of not only what pathogens are being studied but also the type of experiments being conducted. For example, BSL-2 laboratories may prepare an acid-fast stain of *M. tuberculosis*, whereas any study of *M. tuberculosis* that generates aerosols must take place in at least a BSL-3 laboratory [1]. Similarly, laboratories characterized as biosafety level 4 are reserved for the study pathogens which if accidentally released pose a high health risk for the laboratory staff (namely many viral hemorrhagic fever-causing viruses, variola virus, agents that have an antigenic or phylogenetic relationship to known BSL-4 agents, or agents that have an unknown risk of transmission), i.e. for viruses that pose the risk of life-threatening disease, potentially by infectious aerosols, and for which no licensed treatment is readily available. However, if the biologic material is separated into its components without capacity to replicate, a pathogen like Ebola virus may be studied in BSL-2 setting [6]. BSL-4 or maximum containment conditions are met by conducting research in an environmentally controlled room in a dedicated laboratory space, either within a biosafety class I or II cabinet while wearing a full-body suit or within a "glovebox" or biosafety class III cabinets [7, 8].

The argument in favor of constructing or maintaining BSL-4 facilities has often cited their use to research agents that may be potentially used in a bioterrorism attack [9]. Apart from research on variola virus, however, [research on which is limited to two centers in accordance with World Health Assembly resolutions (WHA52.10) (WHA 55.15)] much of the work conducted in BSL-4 laboratories focuses on pathogens that are most likely to infect humans through zoonotic spillover (by jumping to humans from another animal) [10]. Lassa fever outbreaks, for example, have historically been started after zoonotic transmission of the virus from an infected rodent to a patient zero [11]. These facilities are critical to identification of and research on emerging infectious diseases, as well as ensuring appropriate responses to outbreaks of infectious diseases. Zoonoses account for two thirds of new human pathogens and pose a rising risk to global public health due to factors such as environmental changes, population growth and agricultural intensification which are bringing humans in greater contact with wild animals [12].

Only within a BSL-4 facility can the pathophysiology of a disease like EVD and the host immune response mounted against the whole virus be studied. Laboratories with biosafety classifications of three or below undoubtedly contribute valuable research to the study of Ebola and other hemorrhagic diseases. Yet these laboratories may only study the virus once it is separated into pieces. The virus as a whole operates differently from the sum of its parts and as a result, BSL-4 research contributes uniquely to the understanding of the disease caused by it. Furthermore, BSL-4 laboratories allow the study of the natural course of the infectious disease in animal model as insight into how disease manifests in humans. Animal disease models have their limitations but they remain the closest proxy available to study human diseases that occur with limited frequency in nature and/or in areas where capacity to conduct clinical research during outbreaks may be hampered by lack of healthcare infrastructure [13]. The research conducted at BSL-4 laboratories has not only helped to elucidate the pathophysiology of EVD, but has also identified potential pharmacologic interventions. When the World Health Organization (WHO) published a report documenting the leading investigational therapies to combat EVD in 2014, many of the therapies on the list—including Biocryst (BCX4430) and ZMapp—had first shown promise during animal experiments conducted in BSL-4 laboratories [14–16]. For example, ZMapp, a mixture of three monoclonal antibodies targeting Ebola virion antigens was developed at two BSL-4 laboratories. Two of the antibodies were developed at Canada's National Microbiology Laboratory and the third was developed at the U.S. Army Medical Research Institute of Infectious Diseases [14, 17]. The contribution experimental therapies developed in maximum containment and other associated resources is discussed at length below.

Additionally, maximum containment laboratories offer the ability to validate novel diagnostic assays against live virus stocks, contributing to identification of high yield candidate diagnostics which can be deployed during outbreak settings. Many of the current diagnostic assays for both Lassa virus and Ebola virus were initially validated in BSL4 settings [18, 19]. There are, however, several challenges and dilemmas to the transfer of Risk Group 4 live virus from countries where they are endemic to a site with a maximum containment laboratory. First there are extensive procedures and regulations applicable to the transport of non-deactivated viruses internationally which cause significant delay in transfer. Aside from Materials Transfer Agreements (MTAs) required for transfer of any materials from one party to another, the transfer of infectious agents is subject to national policies of the country of origin, many of which are based United Nations Model Regulations [which in turn are based on Recommendations made by the Committee of Experts on the Transport of Dangerous Goods (UNCETDG)] and outline extensive packaging, training and shipping requirements [20]. In addition, the transfer of biological samples to the US requires numerous permits pursuant to US Customs and Border Protection regulations and numerous federal regulations (including USPHS 42 CFR—Part 71 Foreign Quarantine) and all select agents must further be registered with the CDC's Select Agent Program in accordance with 42 CFR Part 73) [21].

Secondly, there is growing discussion about the genomic information and biological knowledge that governments and populations are forfeiting when patient and pathogen samples are transferred out of their country to another site [22]. When such transfers are made from resource limited to resource rich settings, it raises further questions about the ownership of samples and equitable returns from research [23]. On the one hand, the work performed at BSL4 laboratories creates common good through increased knowledge about the endemic pathogens affecting communities. On the other hand, as a group of African scientists recently argued, "research collaborations have been largely unfair for African researchers and participants" because they often do not lead to capacity building at the site of research, result in dissemination of data without participation of African populations and often are not in line with African values and cultures. In light of these concerns, H3Africa Working Group on Ethics recently released voluntary guidelines entitled, "Ethics and Governance Framework for Best Practice in Genomic Research and Biobanking in Africa [24]."

7.3 Barriers to BSL-4 Laboratory Operation

In spite of the research that can be performed in maximum containment laboratories between and during epidemics, there are several notable drawbacks to the facilities including cost, community concerns, and infrastructure requirements. Building and maintaining a facility with the capacity to Risk Group 4 pathogens requires an impressive financial investment. Even in resource-rich economies, these laboratories represent a significant financial undertaking. Maximum containment laboratories employ a security force, maintain fences, operate security cameras, and utilize electronic locks that record the identities of every person who attempts to enter the facility. They maintain airlocks for gaseous decontamination of equipment, provide positive-pressure air suits for researchers, and HEPA-filter all exhaust air from rooms in which maximum containment research may be conducted as well as adjacent spaces. The entry and exit out of the laboratory space is arranged in a sequential route through a chemical shower into work space and back out through the chemical shower, into an inner suit room and then a personal shower and outer clean changing room [25, 26]. Effluents must be decontaminated, waste must be sterilized, and emergency power sources must be supplied [27]. The use of these technologies and of highly specialized, security-approved staff members incurs significant cost in addition to the average cost of operating any non-BSL-4 facility of similar size.

Beyond the financial burden, local community opposition to the operation of maximum containment laboratories has prevented several built facilities from conducting research in the past. Such opposition is fueled both by rightful concerns regarding either accidental or intentional release of dangerous pathogens and potential environmental impact but also by public misconception regarding how risk group 4 pathogens can be spread; most of these pathogens require close contact for transmission but are perceived to be airborne [28].

In 1981, the National Institute of Infectious Diseases in Japan developed a BSL-4 glove box facility that was expensive to maintain. Local opposition to the laboratory conducting BSL-4 research and fears about the spread of infectious pathogens forced the facility to operate as a BSL-3 laboratory for decades. It was not until 2015 that the policy concerning the laboratory changed to allow BSL-4 level research—a move inspired in large part by fear of Ebola virus being imported into Japan [29]. In the US, some of the maximum containment facilities faced similar opposition at the time of their opening. Boston University's National Emerging Infectious Diseases Laboratories finished the construction of its BSL-4 suites in 2008 but was not granted a CDC permit for operations until 2016—after settling federal and state lawsuits from community activists [30]. To a lesser degree, the opening of National Institute of Allergy and Infectious Diseases (NIAID)'s maximum containment facility at Rocky Mountain Laboratories saw similar community opposition [31, 32]. Conversely, the opening of the maximum containment laboratories at University of Texas Medical Branch (UTMB) in Galveston, Texas and Public Health Agency of Canada (Canada)'s Canadian Science Centre for Human and Animal Health (CSCHAH) in Winnipeg in Canada were relatively uneventful. The difference in reception appears to be driven in part by the location of the lab and the successful institutional use of early proactive strategies to engage the host community, disburse public education on topics related to maximum containment research and provide extensive transparency to stakeholders during the approval process [31].

In the United States, maximum containment laboratories are heavily regulated by local, state, and national policies. In particular, the Centers for Disease Control and Prevention and USDA are responsible for the registration and oversight of laboratories that possess, use, or Biological Select Agents and Toxins (BSATs) (discussed above) under the Select Agent Program. Select Agent regulations (7 C.F.R. Part 331, 9 C.F.R. Part 121, 42 C.F.R. Part 73) require compliance with operational guidance spanning from training, to transport, storage and handling of agents, security clearance of employees and visitors, occupational health, viral inactivation methods, use of equipment, decontamination procedures, reporting of laboratories accidents and other elements of safety [27, 33]. Hence, coordination is also required with other agencies such as the Department of Justice, and the Department of Transportation, the Occupational Safety and Health Administration.

As the number of maximum containment laboratories has increased and as these laboratories have expanded outside the government into commercial and academic spheres, concerns regarding regulatory oversight have increased. These concerns have risen in the context of a handful of mishaps and near misses at BSL-4 facilities being widely publicized [34–36]. Despite these headlines, at this writing, the existing maximum containment laboratories in the US have operated for decades without even one incident of laboratory-acquired infection or pathogen escape.

7.4 Role of BSL-4 Laboratories in Emergency Preparedness

The research generated at maximum containment facilities advances our understanding of BSL-4 pathogens and moves us closer to efficacious and safe medical countermeasures. Yet the role of these laboratories does not end with their contribution to research. In fact, BSL-4 laboratories serve an important role in building global capacity for epidemic response and planning for outbreaks before they occur.

7.4.1 Contributions to Capacity-Building and Surveillance

The number of BSL-4 laboratories has been increasing in recent years. Prior to 1990, only two BSL-4 laboratories existed in the United States (USAMRIID in Maryland and the CDC in Georgia). A decade later five BSL-4 laboratories were operational. Another 11 years later, 12 total maximum containment laboratories were operating or were in the planning or construction phase [37].

The rapid increase of maximum containment facilities has occurred within the context of increased fear of bioweapons attacks and zoonotic disease threats. As humans alter their environments through deforestation or reforestation, build dams, and expand agriculture, their interaction with animal reservoirs and vectors may increase. At the same time, global travel, and urbanization all increase the likelihood and danger of widespread disease dissemination.

The majority of the naturally occurring disease from BSL-4 agents is prevalent in low- and middle-income areas of the world, yet the countries most affected by many of these diseases are also the very settings in which operating maximum containment facility is often not feasible. As discussed above, the creation of BSL-4 conditions require extensive infrastructure, staffing, and waste management. The establishment of each of the components is cost prohibitive and unrealistic to implement from a regulatory management perspective for health systems in resource-limited settings. Aside from cost and resource issues, political and social instability such as civil wars, refugee crises, and uneven economic growth create security risks that make the operation of a maximum containment laboratory a risky endeavor. Additionally, a recent report outlines the variation of regulation and oversight of existing high containment (BSL-3) laboratories in different regions of the world even before adding the complexity of maximum containment [37].

Existing BSL-4 institutions often find themselves in the role of contributing staff as advisors to nations aiming to shore up their laboratory capacity, providing technical and experiential knowledge regarding requirements for safe and sustainable operations. Dr. Gerald Keusch, the Associate Director of Boston University's National Emerging Infectious Diseases Laboratories, reports that at times the role of the expert consultant is to help elucidate why BSL-4 capacity may not be feasible based on the needs of the country and lack of certain regulations or resources. There is an on-going debate among BSL-4 experts and stakeholders whether global BSL-4 capacity needs to be increased through "a small number of larger, more economical laboratories and a larger number of small laboratories closer to potential endemic sites or existing centers of excellence" [28].

Whatever the strategy, as the work in maximum containment laboratories increases, it requires additional qualified staff to handle the pathogens studied within. Existing BSL-4 facilities hence serve as natural resources for training and mentorship for the next generation of researchers and laboratory personnel in maximum containment. Creation of this type of intellectual capital is part of the mission of most BSL-4 institutions, with their contributions ranging from arranging biosafety seminars and running workshops for new BSL-4 staff to hosting formal training programs.

One of the most extensive training programs for researchers is run by the University of Texas Medical Branch (UTMB). The National Biocontainment Training Center (NBTC) provides experience for laboratories ranging from BSL-2 through BSL-4 in biosafety principles and practices to researchers from around the globe [38].

Given how small the community of BSL-4 researchers is, the leadership and experts at these facilities also contribute to national preparedness through creating consensus on practices and training. For example, in 2008, laboratory directors of most of the existing and proposed BSL-4 laboratories in North America came together to establish a framework for formal training of BSL-4 staff. This framework included guidelines for background preparation, theoretical training in the principles of biocontainment, practical training while wearing personal protective equipment (PPE), and mentored on-the-job training [39]. Notably, as of yet, there exists no standardized certification of training. Instead, depending on one's institution and background, a BSL-4 staff member in North America may be trained through a series of online modules, classroom workshops, field seminars, or through a 2-year fellowship [40].

Capacity-building, however, is not limited to training and assistance with laboratory development. It also includes knowledge-sharing between institutions prior to outbreaks occurring. The Emerging and Dangerous Pathogens Laboratory Network (EDPLN) is a coalition of BSL-3 and BSL-4 laboratories that exist under the umbrella of the WHO Global Outbreak Alert and Response Network (GOARN). The mission of the EDPLN is to enhance the readiness and response of countries to emerging outbreaks, both in terms of detection and of management. To this end, they facilitate the transfer of diagnostic technologies, practices, and training to laboratories around the world [41].

The EDPLN was initially founded in 2008 as a response to the growing need in a globalized world to be able to identify pathogens and contain outbreaks, even when these pathogens emerged in countries without strong surveillance capacity. As of 2012, the network included 25 member laboratories in 18 countries [28]. The coalition solidified a worldwide network of maximum containment facilities to share knowledge and materials amongst participating organizations. The coalition has supported international and regional workshops and coordinated the production and distribution of needed laboratory reagents [41].

BSL-4 laboratories serve as a cornerstone in global surveillance for emerging outbreaks. The utilization of BSL-4 resources enable countries to not only respond to outbreaks once they occur, but to identify their emergence early in the epidemic. This early identification allows for mobilization of appropriate resources for outbreak response and potentially limit the size of an outbreak. In 2017, at the start of an EVD outbreak in the Democratic Republic of the Congo (COD), several Congolese scientists sent samples from their patients to a BSL-3 laboratory in Kinshasa. There, testing identified the etiological agent as "probably" Ebola virus—an identification that was confirmed after the samples were shipped and tested at the Centre International de Recherche Médicale de Franceville (CIRMF), a BSL-4 laboratory in Gabon [42]. The network of BSL-4 laboratories, working in collaboration with the WHO, offers confirmatory diagnostic testing for countries without the capacity to safely isolate Ebola virus on their own. Aside from identification of index cases in outbreaks of viral hemorrhagic fevers, CIRMF has also contributed to discovery of diagnostic modalities, surveillance of animal reservoirs (including evaluating unexplained animal deaths during outbreaks), and sequencing of outbreak virus strains [43, 44].

However, the ability of BSL-4 laboratories to serve as tools for surveillance is dependent on samples being sent to them for diagnostic and confirmatory testing. This dependency requires that clinicians and scientists on the ground have a high degree of suspicion that a Risk Group 4 agent is involved in an ongoing infectious disease outbreak. Such clinical vigilance is particularly difficult as pathogens spread into new geographical territories with health care staff that have not encountered them before. For example, clinicians in countries like COD, which have a long history of EVD epidemics, are likely to identify or suspect Ebola virus as a causative agent in a patient with a particular viral hemorrhagic fever presentation. In contrast, the identification of Ebola virus as the causative agent for the Western African EVD epidemic was delayed by almost 3 months after the index case, partially because there was little clinical suspicion of an EVD epidemic in this geographical area [45].

7.5 Role of Maximum Containment Laboratories During Emergency Response

During the 2013–2016 EVD epidemic, BSL-4 laboratories worldwide were involved at multiple levels of the response (Table 7.2). For some of the laboratories, the commitment to the response was secondary to existing academic or research relationships in the area. Some BSL-4 laboratories operated diagnostic mobile laboratories. Others contributed staff as subject matter experts or as mobile laboratory volunteers. Others formed working public health groups to respond to the outbreak. Some BSL-4 laboratories contributed medical countermeasures to returning responders for investigational new drug (IND) use.

7.5.1 Subject Matter Expertise and Investigational Therapies

BSL-4 researchers, biosafety personnel, and associated clinical support staff provided subject matter expertise to the public health sector or to clinical enterprises. Before the 2013–2016 EVD epidemic, the clinical experience with EVD patients was limited. In addition to information from prior case series, the experience from animal model data both about the pathophysiology of the disease as well as use of IND medical countermeasures guided experimental treatment approaches and care of patients in both Western Africa and expatriated returning responders to Europe and United States. BSL-4 researchers found themselves being pulled into technical working groups for responding agencies such has World Health Organization and U.S. Centers for Disease Control and Prevention [46]. Others provided insight into choice and use of experimental therapies to medical centers caring for returning responders with EVD. An extensive series on care of EVD patients in US and Europe showed that 85% received at least one investigational therapy and 70% received at least two. The therapies ranged from blood products such as convalescent plasma or whole blood, monoclonal antibody cocktails such as ZMapp, Zmab, or MIL77 and antiviral agents with presumed activity against Ebola virus such as TKM-Ebola, favipiravir, brincidofovir, or amiodarone [47].

As discussed earlier, Zmapp was the first medical countermeasure that prevented death in Ebola virus-infected nonhuman primates after the development of clinical signs [48]. Zmapp was given as an experimental treatment to a few patients with Ebola Virus Disease (EVD) and the group showed lower than expected mortality rate (only 2 out of 10 patients died) but uncertainty remained whether the improved survival was due to better supportive care or the investigational therapy [49]. In the US, compassionate use or expanded access to investigational products outside of clinical trials necessitates certain pre-conditions established by the US Food and Drug Administration, including a willingness on the part of the physician and the patient to participate, lack of other treatment options, and sufficient evidence of a candidate medical countermeasure regarding safety and efficacy [50]. Significant discussion and controversy exists on whether compassionate use of treatments with limited data regarding efficacy should have been used during the epidemic and

Table 7.2 Maximum containment laboratories and their response activities during the 2013–2016 EVD epidemic

Laboratory	Country of involvement	Activities
Germany's Bernhard-Nocht-Institute for Tropical Medicine	Multiple	First use of favipiravir for Ebola virus-infected animal models Service on WHO advisory panels Diagnostic support in mobile lab Development and utilization of genomic surveillance system
Canada's National Microbiology Laboratory	Multiple	Development of two antibodies of Zmapp antibody cocktail Consulting for WHO
US Centers for Disease Control and Prevention (in conjunction with NIAID)	Liberia Mali	Operation of mobile diagnostic laboratory Confirmation of Ebola virus-positive samples from SEREFO laboratory
Germany's Institute of Virology, Philipps University of Marburg	Multiple	Consulting for WHO
US National Institute of Allergy and Infectious Diseases Integrated Research Facility at Fort Detrick	Liberia	Diagnostic laboratory support Service as WHO consultants
US National Institute of Allergy and Infectious Diseases Integrated Research Facility at Rocky Mountain Laboratories	Liberia Mali Multiple	Operation of diagnostic laboratory Genetic and phylogenetic analysis of all Ebola virus-positive samples at SEREFO laboratory Consulting for WHO
South Africa's National Institute for Communicable Diseases Special Pathogens Unit	Sierra Leone	Operation of diagnostic mobile lab
Japan's National Institute for Infectious Diseases	Multiple	Service on WHO advisory panels
Italy's National Institute for Infectious Diseases	Liberia Sierra Leone	Deployment of diagnostic team to existing mobile laboratory unit Establishment of mobile diagnostic laboratory Infection prevention and control training
USAMRIID	Multiple	Development of one antibody contained in Zmapp antibody cocktail Development of remdesivir (GS-5734)
US's UTMB	Sierra Leone Multiple	Diagnosis in mobile laboratories Epidemiologic leadership Service on WHO advisory panels
Wuhan National Biosafety Laboratory Chinese Academy of Sciences	Sierra Leone Multiple	Establishment and operation of mobile diagnostic laboratory Phylogenetic analysis of EBOV outbreak strains from Sierra Leone and Guinea

(continued)

Table 7.2 (continued)

Laboratory	Country of involvement	Activities
Central Research Institute and State Research Center of Virology and Biotechnology (VECTOR), Russian Ministry of Defense	Guinea	Development and testing of GamEvac-Combi (a heterologous VSV- and Ad5-vectored) EBOV vaccine Establishment and operation of mobile diagnostic laboratory in Conakry

The activities listed above are illustrative and not exhaustive accounting of contributions made by programs above as well as other partners

whether their use through expanded access actually added to the knowledge about their efficacy in humans as they were given outside of a controlled clinical trial setting and without appropriate controls [51, 52]. The use of ZMAPP during the EVD epidemic also raised controversy about how scarce investigational therapies should be distributed justly and safely during a large public health crises. Moving forward, it remains unclear if the same level of use of investigational therapies for compassionate use will be applied during a future outbreak. This example raises one of the limitations of adopting medical countermeasures developed against Risk Group 4 agents in maximum containment laboratories: to prove efficacy in humans, clinical trials still need to be conducted with these therapies, often in the middle of outbreak response and in settings without the pre-existing appropriate infrastructure for clinical research [53]. ZMAPP was eventually tested in a randomized control trial, namely, as part of the Partnership for Research on Ebola Virus in Liberia (PREVAIL) II study and provided evidence of small mortality benefit which was not statistically significant enough to prove efficacy [54]. The rVSV vectored EBOV vaccine was another countermeasure developed in maximum containment which was tested in a ring vaccination of close contacts post exposure in Guinea and showed promising efficacy and potential for long term immune protection (up to 2 years) [55, 56]. The development and testing of non human animal vaccines for Risk Group 4 agents has had a greater track record of success. For example, the live attenuated Clone 13 and Smithburn vaccines for Rift Valley Fever virus have been used extensively in animal vaccination in endemic countries and Hendra virus vaccine (containing recombinant HeVsG glycoprotein) serves as part of the One Health strategy for control of that virus in Australia [57, 58].

Aside from contribution of potential therapeutics during outbreaks, maximum containment laboratories also provide additional support to clinical research focused on viral hemorrhagic fevers. For example, the Liberian National Reference Laboratory at Liberian Institute for Biomedical Research (LIBR) Ebola Diagnostic capability was established in partnership and with expertise from NIH/NIAID-Integrated Research Facility (IRF) and US Army Medical Research Institute of Infectious Diseases, two maximum containment laboratories operated by the US government. During the West African EVD epidemic, the reference lab provided containment and

PCR diagnostics capability and since then has expanded research capabilities in country [59]. The laboratory updates at LIBR also provided laboratory support for the PREVAIL II trial (mentioned above) and is now supporting subsequent PREVAIL trials (including PREVAIL III (natural history of Ebola) and PREVAIL IV (persistence of Ebola virus in semen of male survivors). Additionally, NIH/NIAID-IRF facility in Fort Detrick, Maryland serves as a repository for sample processing and storage for live virus samples collected in these trials on behalf of the Liberian government [60].

7.5.2 Physical and Human Resources to Promote Ad-Hoc Laboratory Capacity

The importance of diagnostic capacity cannot be overstated during an EVD epidemic. A delay in laboratory-confirmed diagnosis can lead to two troubling outcomes. A patient whose disease fits the EVD case definition but actually has a different disease similar in clinical presentation may stay longer than necessary in an Ebola Treatment Unit among other, potentially Ebola virus-infected patients and therefore be at increased risk of Ebola virus infection. At the end of illness, delays in laboratory confirmation caused EVD patient who had cleared viremia to stay within Ebola treatment units pending results, and occupying a bed which may otherwise have been given to a patient awaiting admission [61]. Laboratory-confirmed diagnosis also helps provide the true scope of an epidemic and identify areas with the greatest need for public health support. As the Western African EVD epidemic grew, the need for access to quick results throughout the affected countries resulted in deployment of more extensive international support and mobile laboratories [18, 62].

Mobile field laboratories from around the globe were sent to Guinea, Sierra Leone, and Liberia. The first was set up in Guinea in March 2014 [63]. Thereafter, mobile units continued to be constructed in Western African towns affected by the virus. As the incidence of EVD waned in some areas and rose in others, mobile units were broken down and transported to new sites where the diagnostic capabilities they offered were more needed. By September 2015, 41 separate sites throughout Guinea, Liberia, and Sierra Leone had, at one point during the outbreak, a mobile field laboratory providing diagnostic support to health care workers [64].

These laboratories were capable of detecting Ebola virus through PCR testing of body fluids, but were also often equipped with the ability to test for specific antigens via ELISA and immunofluorescence assays [63]. They cut the time for Ebola virus detection from previously 3–5 days to a few hours [65].

BSL-4 laboratories contributed staff as part of the on-the-ground effort to mobile laboratories. These staff members served as diagnosticians in some locations and as trainers for national laboratory employees in others (Table 7.2). BSL-4 researchers also volunteered as individuals to assist with laboratory diagnosis in existing field laboratories.

Judy Yen, the BSL-4 Operations Manager for the NEIDL, worked in a mobile laboratory in Sierra Leone during the outbreak, testing samples from patients suspected of being infected with Ebola virus. She noted that the personnel within the units needed to be able to safely handle tissues infected with a BSL-4 pathogen and consistently and accurately diagnose EVD. They needed to be able to do so in settings without power, with limited water supplies, and during inclement weather. Yen notes that among the most difficult aspects of the work for many laboratory technicians was gaining the dexterity required to handle samples through the glove box set-up of a mobile laboratory while preventing cross-contamination. She cites her prior experience as a BSL-4 researcher as having prepared her for her work in Sierra Leone [66]. Dr. Ian Goodfellow, who operated a mobile diagnostic laboratory sponsored by Public Health England in Sierra Leone, was cited reflecting about the moral responsibility with those with the appropriate training to contribute during crises, "The academic virology community had a responsibility to offer support. We couldn't just sit back and watch this massive epidemic explode in front of our eyes with the knowledge that we have skills that could be useful [67]."

7.5.3 Contribution of Medical Programs Supporting BSL-4 Institutions

The clinical care of patients with highly communicable diseases is discussed in Chap. 31. It is, however, worth mentioning that each of the BSL-4 programs in the United States and Europe has an assigned medical back-up site at a local hospital where researchers or staff with suspected or known exposure to or illness caused by a BSL-4 pathogen can be admitted. These programs were created to meet requirements for post-exposure management as specified under the Federal Select Agent Program [68].

The clinical programs contracted for these activities each have trained staff, assigned space, and well-developed protocols and policies both for infection control and for the clinical care of patients with highly communicable diseases [69]. Many of these units—including the NIH Special Clinical Studies Unit (that provides support to NIH's and USAMRIID's BSL-4 programs) and Emory University's Serious Communicable Disease Unit (providing medical backup to CDC's BSL-4 research program) volunteered to receive patients with EVD during the 2013–2016 EVD outbreak [70]. Similarly, in Europe patients were cared for at the Unit for Treatment of Highly Contagious Infections (UTHCI) at the University Medical Center Hamburg-Eppendorf—a hospital which provides back up to the BSL-4 Bernhard Nocht Institute for Tropical Medicine [71].

Additionally, trained healthcare workers from other back-up medical programs such as St. Patrick's Hospital in Missoula (providing back-up to Rocky Mountain Laboratories) and Boston Medical Center (supporting NEIDL) volunteered as individual medical responders in Ebola Treatment Units in Western Africa through other response organizations operating in the field. The readiness and experience with infection prevention and control protocols related to viral hemorrhagic fevers provided a compatible background for clinical work during the 2013–2016 EVD epidemic.

7.6 Future Recommendations

Despite the unique role of maximum containment laboratories in preparedness and response to infectious diseases of high concern, there are several areas that funding agencies, public health bodies, and BSL-4 institutions could further expand or. First, a recent report from the National Academy of Medicine's "Committee on Clinical Trials During the 2014–2015 Ebola Outbreak" underscores the importance of pre-epidemic investment by funding agencies in the US, UK, and Canada for the creation of products that could actually be deployed during crises [53]. Due to government priorities, investments made in EVD preclinical research, vaccines, and post-exposure countermeasures could actually be provided for clinical trials in Guinea, Liberia, and Sierra Leone. Products for many other pathogens of high concern are not at the same level of readiness and speak to the importance of international collaboration and investment in targeted BSL-4 research. One recent advancement towards this goal is the creation of the Coalition for Epidemic Preparedness Innovation, a philanthropic and intergovernmental fund aimed at the development of vaccines for emerging infectious diseases that are likely to pose an immense public health threat. The effort is currently focused on vaccines against Nipah encephalitis, Lassa fever, and Middle Eastern respiratory syndrome [72].

Second, BSL-4 facilities need to be better integrated into efforts to build research capacity prior to outbreak of emergencies in resource-limited countries where BSL-4 pathogens are endemic. Several experts have pointed to the importance of long-standing research collaborations between research enterprises in resource rich and resource limited countries. These collaborations produce participatory science that is rooted in mutual trust and lead to successful and timely collaboration when outbreaks occur [73]. Yet international collaboration can extend beyond these bilateral partnerships. It is conceivable that maximum containment laboratories will continue to contribute to the development and utilization of new diagnostic and medical countermeasure products during future outbreaks of Risk Group 4 pathogens. Even if the involvement of these laboratories is limited to just preclinical research and development, involving maximum containment researchers in public health policy planning, and linking them with stakeholders in low-resource countries, can be mutually beneficial for both outbreak preparedness and advancing science. One obvious arena for involvement of maximum containment laboratories is the Global Health Security Agenda, a multilateral partnership between countries, international organizations, and academic institutions, to strengthen global and national capacities to prevent, detect, and respond to emerging infectious diseases [74].

Lastly, there is room for greater and more formal collaboration in both research and public health response between existing BSL-4 facilities globally. Currently, European BSL-4 facilities are united by their mission to serve as an essential diagnostic and surveillance component of their public health systems [5]. North American BSL-4 facilities reside in governmental, academic or commercial spaces, with different mandates, making it harder to find common space for collaboration. Similarly, communication and collaboration between not just European and North

American laboratories but also those in Asia and Africa can lead to improved sharing of best practices and collaboration during emergencies.

Maximum containment laboratories have proven to be not just sources for critical science in the battle against infectious diseases threats but also as institutions that can contribute subject matter expertise and critical resources during active outbreaks. Although their missions vary wildly depending on their geographic and institutional location, it is certain that their roles in emergency preparedness and response to infectious diseases outbreaks is only likely to increase as the number and frequency of these threats rise.

References

1. Chosewood LC, Wilson DE, Centers for Disease Control and Prevention (U.S.), National Institutes of Health (U.S.). Biosafety in microbiological and biomedical laboratories. 5th ed. Washington, DC: U.S. Dept. of Health and Human Services, Public Health Service, Centers for Disease Control and Prevention, National Institutes of Health; 2009. xxii, p. 415.
2. World Health Organization. Biorisk management: laboratory biosecurity guidance. 2006. Contract No.: WHO/CDS/EPR/2006.6. http://www.who.int/csr/resources/publications/biosafety/WHO_CDS_EPR_2006_6.pdf. Accessed 26 Sept 2017.
3. National Research Council (US). Committee on scientific milestones for the development of a gene sequence-based classification system for the oversight of select agents and National Academies Press (US). Sequence-based classification of select agents: a brighter line. Washington, DC: National Academies Press; 2010. xv, p. 220.
4. Centers for Disease Control and Prevention. Bioterrorism agents/diseases. 2017. https://emergency.cdc.gov/agent/agentlist-category.asp. Accessed 17 Apr 2017.
5. Nisii C, Castilletti C, Di Caro A, Capobianchi MR, Brown D, Lloyd G, et al. The European network of biosafety-Level-4 laboratories: enhancing European preparedness for new health threats. Clin Microbiol Infect. 2009;15(8):720–6. https://doi.org/10.1111/j.1469-0691.2009.02946.x
6. Halfmann P, Kim JH, Ebihara H, Noda T, Neumann G, Feldmann H, et al. Generation of biologically contained Ebola viruses. Proc Natl Acad Sci USA. 2008;105(4):1129–33. https://doi.org/10.1073/pnas.0708057105
7. Bressler D, Hawley R. Safety considerations in the biosafety level 4 maximum-containment laboratory. In: Wooley DP, Byers KB, editors. Biological safety: principles and practices. 5th ed. Washington, DC: ASM Press; 2017. p. 695–717. https://doi.org/10.1128/9781555819637.ch38
8. Smith PL, Edwards S. Working at biosafety level 4—contain the operator or contain the bug? In: Richmond JY, editor. Anthology of biosafety V: BSL-4 laboratories. Mundelein: American Biological Safety Association ABSA; 2015. p. 209–12.
9. National Institute of Allergy and Infectious Diseases. The need for biosafety laboratories: more about biodefense. 2010. https://www.niaid.nih.gov/research/biosafety-laboratories-needed. Accessed 12 Aug 2017.
10. World Health Organization. Programme to facilitate Variola virus research. 2018. http://www.who.int/csr/disease/smallpox/research/en/. Accessed 16 Apr 2018.
11. Mylne AQ, Pigott DM, Longbottom J, Shearer F, Duda KA, Messina JP, et al. Mapping the zoonotic niche of Lassa fever in Africa. Trans R Soc Trop Med Hyg. 2015;109(8):483–92. https://doi.org/10.1093/trstmh/trv047
12. Jones BA, Grace D, Kock R, Alonso S, Rushton J, Said MY, et al. Zoonosis emergence linked to agricultural intensification and environmental change. Proc Natl Acad Sci USA. 2013;110(21):8399–404.

13. Falzaran D, Bente DA. Animal models for viral haemorrhagic fever. Clin Microbiol Infect. 2015. https://doi.org/10.1111/1469-0691.12630
14. Pettitt J, Zeitlin L, Kim D, Working C, Johnson JC, Bohorov O, et al. Therapeutic intervention of Ebola virus infection in rhesus macaques with the MB-003 monoclonal antibody cocktail. Sci Transl Med. 2013;5(199):199ra113. https://doi.org/10.1126/scitranslmed.3006608
15. Warren TK, Wells J, Panchal RG, Stuthman KS, Garza NL, Van Tongeren SA, et al. Protection against filovirus diseases by a novel broad-spectrum nucleoside analogue BCX4430. Nature. 2014;508(7496):402–5.
16. World Health Organization. Background document potential Ebola therapies and vaccines. 2014. http://www.who.int/csr/disease/ebola/ebola-new-interventions-02-sep-2014.pdf. Accessed 3 Aug 2017.
17. Qiu X, Wong G, Fernando L, Audet J, Bello A, Strong J, et al. mAbs and ad-vectored IFN-α therapy rescue Ebola-infected nonhuman primates when administered after the detection of viremia and symptoms. Sci Transl Med. 2013;5(207):207ra143. https://doi.org/10.1126/scitranslmed.3006605
18. Broadhurst MJ, Brooks TJ, Pollock NR. Diagnosis of Ebola virus disease: past, present, and future. Clin Microbiol Rev. 2016;29(4):773–93. https://doi.org/10.1128/CMR.00003-16
19. Raabe V, Koehler J. Laboratory diagnosis of Lassa fever. J Clin Microbiol. 2017;55(6):1629–37. https://doi.org/10.1128/JCM.00170-17
20. World Health Organization. Guidance on regulations for the transport of infectious substances 2015–2016. WHO/HSE/GCR/2015.2. 2015. http://apps.who.int/iris/bitstream/handle/10665/149288/WHO_HSE_GCR_2015.2_eng.pdf;jsessionid=05E6CCB02D28D1A234527E76009F8EEA?sequence=1. Accessed 18 Apr 2018.
21. Boston University. Appendix B: Import and export of infectious biological agents. Institutional Biosafety Committee Biosafety Manual; 2018.
22. Mungwira RG, Nyangulu W, Misiri J, Iphani S, Ng'ong'ola R, Chirambo CM, et al. Is it ethical to prevent secondary use of stored biological samples and data derived from consenting research participants? The case of Malawi. BMC Med Ethics. 2015;16(1):83. https://doi.org/10.1186/s12910-015-0077-x
23. Mahomed S, Behrens K, Slabbert M, et al. Managing human tissue transfer across national boundaries – an approach from an institution in South Africa. Dev World Bioeth. 2016;16(1):29–35. https://doi.org/10.1111/dewb.12080
24. Wellcome Trust. Ethics and governance framework for best practice in genomic research and biobanking in Africa. H3Africa Working Group on Ethics; 2018. https://h3africa.org/images/Ethics_folders/Oct2017/Final%20Framework%20for%20African%20genomics%20and%20biobanking_SC%20.pdf. Accessed 18 Apr 2018.
25. Janosko K, Holbrook MR, Adams R, Barr J, Bollinger L, Newton JT, et al. Safety precautions and operating procedures in an (A)BSL-4 laboratory: 1. Biosafety level 4 suit laboratory suite entry and exit procedures. J Vis Exp. 2016;116. https://doi.org/10.3791/52317
26. Kraus A, Mirazami A. Laboratory biosafety in containment laboratories. In: Weidmann M, Silmann N, Butaye P, editors. Working in biosafety level 3 and 4 laboratories: a practical introduction. New York: Wiley; 2013. p. 5–12 (Incorporated).
27. Shurtleff A, Garza N, Lackemeyer M, Carrion R Jr, Griffiths A, Patterson J, et al. The impact of regulations, safety considerations and physical limitations on research Progress at maximum biocontainment. Viruses. 2012;4(12):3932–51. https://doi.org/10.3390/v4123932
28. Le Duc J. Requirements for and challenges associated with BSL-4 laboratories (plenary session). In: Biosecurity challenges of the global expansion of high-containment biological laboratories: summary of a workshop. Washington, DC: The National Academies Press; 2012. p. 81–8. https://doi.org/10.17226/13315
29. Shen H. Ebola spurs creation of Japan's first maximum-security biolab. Nature. 2015;524(7565):274–5. https://doi.org/10.1038/nature.2015.18179
30. Allen E, Gans F. BU biolab gets approval for research on deadliest pathogens. The Boston Globe; 2016. https://www.bostonglobe.com/metro/2016/12/24/biolab-gets-cdc-green-light-for-

level-research-but-still-needs-city-approval/xgYk6jljuTOyohcWRySTNI/story.html. Accessed 26 Sept 2017.
31. Dickmann P, Keith K, Comer C, Abraham G, Gopal R, Marui E. Report of the International Conference on Risk Communication Strategies for BSL-4 laboratories, Tokyo, October 3–5, 2007. Biosecur Bioterror. 2009;7(2):227–33.
32. Race MS. Evaluation of the public review process and risk communication at high-level biocontainment laboratories. Appl Biosaf. 2008;13(1):45–56.
33. Federal Select Agent Program. Select agent regulations. 2017. https://www.selectagents.gov/regulations.html. Accessed 27 Sept 2017.
34. Penzenstadler N. Report: CDC could do more to keep laboratories safe. USA Today; 2017. https://www.usatoday.com/story/news/2017/05/25/report-cdc-could-do-more-keep-laboratories-safe/102109390/. Accessed 27 Sept 2017.
35. Government Accountability Office. High-containment biosafety laboratories; Preliminary observations on the oversight of the proliferation of BSL-3 and BSL-4 laboratories in the United States. (GAO Publication No. GAO-08-108T). 2007. http://www.gao.gov/new.items/d08108t.pdf
36. U.S. Department of Health and Human Services. CDC generally met its inspection goals for the federal select agent program. However, opportunities exist to strengthen oversight. (Report no. OEI-04-15-00430). 2017. http://library.bryanhealthcollege.edu/c.php?g=3265&p=10833
37. National Academy of Sciences and National Research Council. Biosecurity challenges of the global expansion of high-containment biological laboratories: summary of a workshop. Washington, DC: The National Academies Press; 2012. https://doi.org/10.17226/13315
38. Ksiazek T. National Biocontainment Training Center. 2015. http://www.dtic.mil/dtic/tr/fulltext/u2/1002976.pdf. Accessed 27 Aug 2017.
39. Le Duc J, Anderson K, Bloom M, Estep JE, Feldmann H, Geisbert JB, et al. Framework for leadership and training of biosafety level 4 laboratory workers. Emerg Infect Dis. 2008;14(11):1685–8. https://doi.org/10.3201/eid1411.080741
40. American Association for the Advancement of Science. Biological safety training programs as a component of personnel reliability. 2009. https://absa.org/wp-content/uploads/2017/01/AAASBiosafetyTraining.pdf. Accessed 12 Aug 2017.
41. World Health Organization. Emerging and dangerous pathogens laboratory network (EDPLN). 2015. http://www.who.int/csr/disease/OP_EDPLN_FINAL.pdf?ua=1. Accessed 3 Aug 2017.
42. World Health Organization. New technology allows for rapid diagnosis of Ebola in the Democratic Republic of the Congo. 2017. http://www.who.int/emergencies/ebola-DRC-2017/articles/rapid-diagnosis/en/. Accessed 3 Aug 2017.
43. Leroy E, Gonzalez JP. Filovirus research in Gabon and equatorial Africa: the experience of a research center in the heart of Africa. Viruses. 2012;4(9):1592–604. https://doi.org/10.3390/v4091592
44. Rouquet P, Froment JM, Bermejo M, Kilbourn A, Karesh W, Reed P, et al. Wild animal mortality monitoring and human Ebola outbreaks, Gabon and republic of Congo, 2001–2003. Emerg Infect Dis. 2005;11(2):283–90. https://doi.org/10.3201/eid1102.040533
45. World Health Organization. Factors that contributed to undetected spread of the Ebola virus and impeded rapid containment. One year into the Ebola epidemic. 2015. http://www.who.int/csr/disease/ebola/one-year-report/factors/en/. Accessed 20 Apr 2018.
46. World Health Organization. Key messages from the WHO meeting on clinical aspects of Ebola virus disease. Advancing standards of clinical care. 2015. http://www.who.int/csr/disease/ebola/clinical-care-summary.pdf. Accessed 26 Sept 2017.
47. Uyeki T, Mehta A, Davey R Jr, Liddell AM, Wolf T, Vetter P, et al. Clinical management of Ebola virus disease in the United States and Europe. N Engl J Med. 2016;374(7):636–46. https://doi.org/10.1056/NEJMoa1504874
48. Qiu X, Wong G, Audet J, Bello A, Fernando L, Alimonti JB, et al. Reversion of advanced Ebola virus disease in nonhuman primates with ZMapp. Nature. 2014;514(7520):47–53. https://doi.org/10.1038/nature13777

49. Dodd LE, Proschan MA, Neuhaus J, Koopmeiners JS, Neaton J, Beigel JD, et al. Design of a randomized controlled trial for Ebola virus disease medical countermeasures: PREVAIL II, the Ebola MCM study. J Infect Dis. 2016;213(12):1906–13. https://doi.org/10.1093/infdis/jiw061
50. U.S. Food and Drug Administration. Expanded access (compassionate use). 2017. https://www.fda.gov/NewsEvents/PublicHealthFocus/ExpandedAccessCompassionateUse/default.htm. Accessed 26 Sept 2017.
51. Hantel A, Olopade C. Drug and vaccine access in the Ebola epidemic: advising caution in compassionate use. Ann Intern Med. 2015;162(2):141–2. https://doi.org/10.7326/M14-2002
52. Rid A, Emanuel E. Ethical considerations of experimental interventions in the Ebola outbreak. Lancet. 2014;384:1896–9. https://doi.org/10.1016/S0140-6736(14)61315-5
53. National Academies of Sciences, Engineering, and Medicine. Integrating clinical research into epidemic response: the Ebola experience. Washington, DC: The National Academies Press; 2017. Report No.: 9780309457767, 0309457769. https://doi.org/10.17226/24739
54. Prevail II Writing Group, Multi-National PREVAIL II Study Team, Davey RT Jr, Dodd L, Proschan MA, Neaton J, et al. A randomized, controlled trial of ZMapp for Ebola virus infection. N Engl J Med. 2016;375(15):1448–56. https://doi.org/10.1056/NEJMoa1604330
55. Henao-Restrepo AM, Camacho A, Longini IM, Watson CH, Edmunds WJ, Egger M, et al. Efficacy and effectiveness of an rVSV-vectored vaccine in preventing Ebola virus disease: final results from the Guinea ring vaccination, open-label, cluster-randomised trial (Ebola Ca Suffit!). Lancet. 2017;389(10068):505–18. https://doi.org/10.1016/S0140-6736(16)32621-6
56. Huttner A, Agnandji ST, Combescure C, Fernandes JF, Bache EB, Kabwende L, et al. Determinants of antibody persistence across doses and continents after single-dose rVSV-ZEBOV vaccination for Ebola virus disease: an observational cohort study. Lancet Infect Dis. 2018. Epub. https://doi.org/10.1016/S1473-3099(18)30165-8
57. Dungu B, Lubisi BA, Ikegami T. Rift Valley fever vaccines: current and future needs. Curr Opin Virol. 2018;29:8–15. https://doi.org/10.1016/j.coviro.2018.02.001
58. Middleton D, Pallister J, Klein R, Feng YR, Haining J, Arkinstall R, et al. Hendra virus vaccine, a one health approach to protecting horse, human, and environmental health. Emerg Infect Dis. 2014;20(3):372–9. https://doi.org/10.3201/eid2003.131159
59. Liberian Institute of Biomedical Research. Programs and capabilities of LIBR. 2017. http://libresearch.org/programs.html. Accessed 18 April 2018.
60. Olinger G. Personal communication with Nahid Bhadelia. NIH/NIAID-IRF, MRI Global; 2017.
61. O'Dempsey T, Khan SH, Bausch DG, et al. Rethinking the discharge policy for Ebola convalescents in an accelerating epidemic. Am J Trop Med Hyg. 2015;92(2):238–9. https://doi.org/10.4269/ajtmh.14-0719
62. Sealy T, Erickson B, Taboy C, Stroher U, Towner JS, Andrews SE, et al. Laboratory response to Ebola – West Africa and the United States. MMWR Morb Mortal Wkly Rep. 2016;65(3):44–9. https://doi.org/10.15585/mmwr.su6503a7
63. Wölfel R, Stoecker K, Fleischmann E, Gramsamer B, Wagner M, Molkenthin P, et al. Mobile diagnostics in outbreak response, not only for Ebola: a blueprint for a modular and robust field laboratory. Euro Surveill. 2015;20(44):pii 30055. https://doi.org/10.2807/1560-7917.ES.2015.20.44.30055
64. World Health Organization. Ebola situation report, September 16, 2015: Map of Mobile Laboratories. 2015. http://www.who.int/csr/disease/ebola/situation-reports/16-september-2015-map5.png?ua=1. Accessed 3 Aug 2017.
65. Zhang Y, Gong Y, Wang C, Liu W, Wang Z, Xia Z, et al. Rapid deployment of a mobile biosafety level-3 laboratory in Sierra Leone during the 2014 Ebola virus epidemic. PLoS Negl Trop Dis. 2017;11(5):e0005622. https://doi.org/10.1371/journal.pntd.0005622
66. Yen J. Personal communication with Sarah carpenter. National Emerging Infectious Diseases Laboratories; 2017.
67. University of Cambridge. Call of duty: fighting Ebola in Sierra Leone. 2017. http://www.cam.ac.uk/research/features/call-of-duty-fighting-ebola-in-sierra-leone. Accessed 13 Apr 2018.

68. Federal Select Agent Program. Occupational health program. 2017. https://www.selectagents.gov/ohp-app2.html. Accessed 27 Sept 2017.
69. Best M. Medical emergency planning for BSL-4 containment facilities. In: Richmond JY, editor. Anthology of biosafety V: BSL-4 Laboratories. Mundelein, IL: American Biological Safety Association ABSA; 2015. p. 98–295.
70. Courage K. Inside the 4 U.S. Biocontainment Hospitals that are stopping Ebola. Scientific American; 2014. https://www.scientificamerican.com/article/inside-the-4-u-s-biocontainment-hospitals-that-are-stopping-ebola-video/. Accessed 26 Sept 2017.
71. Kreuels B, Wichmann D, Emmerich P, Schmidt-Chanasit J, de Heer G, Kluge S, et al. A case of severe Ebola virus infection complicated by gram-negative septicemia. N Engl J Med. 2014;371(25):2394–401. https://doi.org/10.1056/NEJMoa1411677
72. Coalition for Epidemic Preparedness Innovations. Mission. 2017. http://cepi.net/mission. Accessed 26 Sept 2017.
73. Yozwiak N, Happi C, Grant D, et al. Roots, not parachutes: research collaborations combat outbreaks. Cell. 2016;166(1):5–8. https://doi.org/10.1016/j.cell.2016.06.029
74. Global Health Security Agenda. Global health security agenda. 2017. https://www.ghsagenda.org/. Accessed 27 Sept 2017.

8 Infection Prevention and Control: A Biodefense Measure

Saskia Popescu

8.1 Infection Prevention and Control

Infection prevention and control is a series of practices, processes, and programs within healthcare to reduce the spread of infection among patients and staff. Infection prevention and control (IPC) programs within hospitals and healthcare facilities typically consist of infection preventionists (IPs) who are commonly nurses, epidemiologists, and microbiologists. The duties of the IP and IPC program includes a variety of responsibilities; disease surveillance, case investigation for hospital-associated infections, education, training of staff, policy review, risk assessments for construction in patient care areas, emergency preparedness, reporting of communicable diseases to local health departments, reporting of healthcare-associated infections for reimbursement and federal requirements, etc. Traditionally, the role of the infection prevention program is to focus on reducing healthcare-associated infections (HAIs) however, communicable diseases play a large role in their duties. Properly training and educating staff on the correct use of personal protective equipment (PPE) and ensuring isolation precautions are correctly maintained is a daily responsibility for the IPC department. Aside from ensuring staff compliance to PPE and isolation precautions, IPC works to educate and facilitate compliance among visitors, as they can be a source for disease spread within a healthcare facility. The role of the IPC program and infection preventionists is also as the connective tissue within healthcare facilities to ensure that infectious diseases have the proper response across the healthcare sector, which includes departments like environmental services, security, emergency medicine, facilities, administration, interventional radiology, sterile processing departments, etc. Infection and control programs are the

S. Popescu, MPH, MA, CIC, PhD(c) (✉)
George Mason University, Fairfax, VA, USA

Phoenix Children's Hospital, Phoenix, AZ, USA
e-mail: spopesc2@gmu.edu

S. K. Singh, J. H. Kuhn (eds.), *Defense Against Biological Attacks*,
https://doi.org/10.1007/978-3-030-03053-7_8

sentinels for underlying issues within the health of the community and healthcare facility.

8.1.1 Practitioners

Infection prevention and control departments are comprised of infection preventionists, who have a variety of backgrounds. Traditionally the IP role has been primarily nurse-driven, but recent years have seen an influx of epidemiologists as the role increasingly involves hospital epidemiology, outbreak investigation, and public health response. The IP, even at the most novice level, must have a rudimentary understanding of microbiology, disease transmission, and public health interventions. IP competencies are built up according the Association for Professionals in Infection Prevention and Epidemiology (APIC) model, which is considered the gold standard. The APIC competency model is based upon a text that includes topics that highlight the range of IP duties—occupational health, sterilization and disinfection, bioterrorism, emergency preparedness, outbreak response, infection prevention in a variety of settings (e.g., behavioral health, correctional facilities, dental services), aseptic technique, infection prevention for specialty care populations (e.g., burns, pediatrics, organ transplants), as well as the environmental components to infection prevention work (e.g., waste management, heating and ventilation, healthcare textiles). The duties and capabilities of the IP are required to be extensive and diverse. Infection preventionists are expected to sit for the Certification in Infection Control (CIC) after 2 years. The CIC is a complex certification which aims to ensure the IP is well-versed in the variety of issues that plague healthcare systems. Since the 2002 severe acute respiratory syndrome (SARS) coronavirus and 2013–2016 Ebola virus disease outbreaks, infection preventionists' duties have increasingly involved high-consequence pathogens and emergency preparedness. Overall, infection preventionists are responsible for all manner of infectious disease threats within a healthcare facility, regardless of origin.

8.1.2 Surveillance and Coordination with Health Departments

Perhaps one of the most crucial responsibilities of the infection prevention and control program is to perform surveillance and report communicable diseases to local health departments. While IPC programs are frequently responsible for reporting hospital-associated infections within their facility, it is the responsibility for reporting communicable diseases that is relevant to defense against biological attacks. Communicable disease reporting requirements (also known as Infectious Disease Reporting) vary within each state and even county, but overall, there are state laws requiring medical providers to report notifiable disease to local public health departments [1]. While there is variation in the lists of notifiable diseases that states require healthcare providers to report, there is a general consistency of common and concerning microorganisms that includes all Category A agents and a majority of those listed within the Center for Disease Control and Prevention's (CDC) bioterrorism agents page [2–5]. Within hospitals and healthcare facilities, it is generally the IPC program that is responsible for reporting communicable diseases to local public health authorities. While laboratory reporting can and does happen, it is mostly incomplete information and commonly limited to the positive laboratory result and the name of the patient. The legal responsibility of reporting the communicable diseases ultimately tends to fall upon the infection prevention and control program, which must not only report such diseases with complete patient and laboratory information, but also within the required time frame. State reporting laws dictate the list of notifiable diseases and their required reporting time frame. For example, in California, the California Code of Regulations requires a report for West Nile virus infection be submitted by electronic transmission, telephone, or mail within one working day of identification, while a human or animal case of plague requires immediate reporting by phone [6]. These lists mirror that of the CDC's National Notifiable Diseases Surveillance System (NNDSS). In essence, the reporting structure commonly begins with the IP performing surveillance based off laboratory reports and daily reports of admitting diagnoses for newly admitted patients. From this surveillance, reporting is performed to the local health department, which will then report to the state health department, which then reports it to the CDC. In short, national diseases surveillance and reporting frequently starts with hospital IPC programs.

IPC programs have a critical role in communicable disease reporting, which aids in public health response. The strength of the infection prevention and control program will be seen in their ability to perform more specific surveillance and real-time disease reporting. Poorly supported or minimally staffed IPC programs will likely rely solely on microbiology reports and might be delayed in their surveillance and reporting efforts. Some diseases, like Kawasaki syndrome, cannot be confirmed by laboratory tests, which can be challenging, but is overcome through daily chart review and a strong relationship with physicians and other clinical staff. Moreover, once a case is identified as a communicable disease that must be reported, the skilled IP will not only report the required information (e.g., laboratory results, demographics, contact info) but also relevant medical record information for the condition, which will include a timeline of illness, travel history, physician notes, etc. Such attention to detail facilitates a more rapid and efficient response from public health authorities. Overall, the infection prevention and control program within healthcare facilities is frequently the primary source of communicable disease reporting, which is a vital part of public health response during biological events.

8.1.3 Coordination Among Internal Departments

IPC programs are a connective tissue within the healthcare setting, working across professions and job roles to ensure the response to infectious disease threats is appropriate. It is the responsibility of the IPC program to ensure that clinical staff follow infection control guidelines, but also ancillary staff, which includes but is not limited to environmental services, facilities, sterile processing and disinfection, laboratory, security, or food services. IPC coordinates with these departments to ensure the proper education and compliance is ongoing as interaction with patients is not solely limited to healthcare providers. Aside from ensuring infection control practices are followed, the IPC program must also coordinate with different departments within the healthcare facility/system to ensure the proper expertise are being utilized. Consider the role of construction or facilities maintenance. A relatively simple task, like installing cabling or renovating a new space can bring with it substantial patient safety concerns due to dust, mold, etc. IPC must consider the full spectrum of affected people and departments during biological incidents, for example; during a public health emergency or treatment of a patient in enhanced precautions, IPC must coordinate with environmental services to ensure proper waste disposal is being performed. In the event of a pediatric patient with Ebola virus disease, IPC must ensure that Child Life Services are able to provide recommendations and guidance in regards to the unique needs of pediatric patients. Such unique situations require IPC response and coordination of expert opinions to ensure the most appropriate actions are taken. During an influx of patients with an infectious disease or in the wake of an attack, IPC will be an integral component to hospital response and coordination of efforts. In emergent situations, the Hospital Incident Command System (HICS) may be initiated, of which IPC plays a vital role in providing instructions and expert recommendations.

8.2 Isolation Precautions and Personal Protective Equipment

Isolation precautions are singularly one of the most critical components to preventing the spread of infectious diseases during patient care. The rapid isolation and adherence to personal protective equipment guidelines are imperative in daily patient care, but will become even more pertinent during infectious disease crises. Adherence to standard precautions and hand hygiene is fundamental for all patient interactions. However, additional transmission-based precautions may be needed during the medical care of patients with infectious diseases. Transmission-based precautions (isolation precautions) are often required *in addition* to standard precautions. Full information on microorganism-specific isolation precautions and detailed PPE directions can be found within several CDC and World Health Organization (WHO) publications. IPC recommendations follow CDC guidelines for isolation precautions and corresponding PPE based upon the identified microorganism (e.g., bacterial, viral, fungal). The recommendations within this chapter serve as broad guidance for utilization and compliance. Reference is made to the patient room, but isolation precautions should be applied to the area in which the patient is present (e.g., treatment room, post-procedural recovery bay).

8.2.1 Personal Protective Equipment

Personal protective equipment (PPE) in the healthcare setting includes gloves, gowns, masks, shoe covers, and eye-protection. These items are the most common and frequently used PPE during patient care, but additional PPE may be recommended in emergent situations. PPE provides a mutual benefit to both patient and healthcare worker as it reduces the spread of diseases between the two parties. Correct utilization of PPE decreases the microbial burden in the environment as well, preventing the spread of disease through cross-contamination. PPE is mutually beneficial within the care process, which should be taught in training to encourage compliance.

Donning and doffing PPE is a critical component to efficacy. Improper use of PPE is common among healthcare workers, as the time to properly don can be taxing or because it is thought to be easier to continue wearing PPE if only leaving a patient's room momentarily (e.g., keeping a facemask on while running to the medication room). For PPE to be effective, it must be donned prior to entering the patient room and doffed prior to leaving or upon exiting (depending upon the layout of the care area) to avoid contamination of the hospital unit. PPE should always be changed between patients and should never leave the patient's room, regardless of how short a simple task may appear. If a patient is in isolation, anyone entering the room should don PPE, regardless of whether the planned task does not involve patient interaction or expected contact with the environment. Once in the patient's room, one cannot predict what will be needed or if an emergency will occur, meaning that PPE should be worn at all times while caring for a patient in isolation.

Isolation precautions and the required PPE range from standard precautions to airborne isolation, all of which have different and distinct requirements. Adherence to CDC isolation precaution guidelines and manufacturer recommendations for the PPE are imperative for the treatment of patients with infectious diseases. Regardless of origin, vigilant adherence to proper isolation and PPE guidelines will aid in infection prevention and control efforts, which can halt an outbreak in its tracks. PPE can and should be worn by visitors of the patient however, this is limited to PPE that does not require fit-testing (i.e., respiratory protection for patients with airborne microorganisms, like N-95 masks, should not be worn by visitors or patients as they require special fit-testing to ensure proper sizing). During hospitalization, the room of an contagious patient should have the proper signage on the door indicating which transmission-based precautions are being used so that all staff or visitors entering may wear the necessary PPE.

8.2.2 Standard Precautions

Standard precautions (also referred to Universal Precautions) are basic precautions that should be utilized on all patients as a means of reducing the transmission risk of, for instance, bloodborne pathogens [37]. The most basic premise is to use hand hygiene and assess the risk of a potential exposure to blood or other potentially contaminated material (e.g., urine, feces, breast milk) and use the corresponding PPE. Standard precautions encourage the healthcare worker to consider the tasks being performed and apply critical thinking to the risk of potential exposure and then select PPE based upon this assessment. For example, drawing blood typically only calls for gloves, but if the patient is thrashing or aggressive, the healthcare worker should consider eye protection. Another component to standard precautions is respiratory hygiene and cough etiquette. This lesson was developed and became essential during the SARS outbreak and has now become a staple within universal precautions. Respiratory hygiene and cough etiquette highlights the importance of education for healthcare workers, patients, and visitors, to ensure they cover their mouth and nose when coughing or sneezing, perform hand hygiene following exposure to respiratory secretions, and practice spatial separation of febrile persons with respiratory symptoms. The goal of standard precautions is to ensure that, at the most basic level, healthcare workers are protected against blood and other potentially contaminated material through hand hygiene and risk-based PPE utilization. Additional transmission-based precautions, such as contact isolation precautions, build upon standard precautions, meaning that standard precautions will always be used.

8.2.3 Hand Hygiene

The most basic and effective tool in the infection prevention arsenal is hand hygiene. Competent hand hygiene practices can prevent the transmission of diseases, making it the foundation for infection prevention [32]. Efforts to maintain compliance can be challenging as healthcare workers are expected to perform hand hygiene before and after patient interaction [30, 31]. The new WHO model for hand hygiene goes beyond the "before and after patient care" practice and emphasizes moments during the patient care process. The Five Moments of Hand Hygiene (frequently called "5 moments") model emphasizes a broader approach to hand hygiene that encompasses the intricacies of patient care. The WHO's 5 Moments model is increasingly becoming the standard practice and includes five moments in which hand hygiene should be performed: before patient contact, before aseptic tasks, after body fluid exposure risk, after patient contact, and after contact with patient surroundings [7]. Hand hygiene should also be performed after removing gloves. Gloves can easily be seen as a cure-all protective barrier however, hand hygiene should be performed following their removal as leakage during care and cross-contamination during removal is possible [8].

Hand hygiene is performed with either soap and water or alcohol-based hand sanitizer. According to the CDC, washing with soap and water for 15–20 seconds should be performed when hands are visibly soiled, after known exposures to *Clostridium difficile*, after known or suspected exposures to patients with diarrhea during a norovirus outbreak, when exposure to *Bacillus anthracis* is suspected or proven, before eating, and after using the restroom. If none of those situations are applicable, alcohol-based hand sanitizer is recommended and considered the most effective product. Alcohol-based hand sanitizer (composed of 60–95% alcohol to maintain efficacy) is an easy and effective manner for hand hygiene that takes roughly 20 seconds to dry. Depending upon the situation, alcohol-based hand sanitizer can easily provide staff with an effective and timely method for cleaning their hands, which significantly improves compliance.

8.2.4 Contact Precautions

Contact isolation precautions are used for microorganisms that are transmissible through touch or contact with the patient or a contaminated item (fomite). Contact precautions include wearing a gown and gloves during interaction with the patient or their environment. The most common examples of infections in which contact isolation is used include diarrheal illnesses, methicillin-resistant *Staphylococcus aureus* (MRSA), and draining abscesses. Contact precautions should also be taken during the care for patients infected with known or suspected multi-drug resistant microorganisms, Human Parainfluenza virus (HPIV) infection in pediatric patients, poliovirus, vaccinia virus (adverse events following vaccination, like generalized vaccinia or conjunctivitis), etc. Broadly speaking, contact precautions, should be used for patients with diarrhea or draining wounds, and should be considered for children who have respiratory illnesses. The necessary PPE for contact precautions, gown and gloves, should be donned appropriately before entering the room, and removed at the threshold, prior to exiting. Depending on the make of the gown, it is important to follow manufacturers' recommendations for removal (like pulling away from the body to break any ties in the back.) and wear (i.e., wearing the gown over the shoulders, not hanging down around the elbows). Contact precautions are an effective practice for protecting the healthcare worker against infectious diseases that are spread through touch and fomites.

8.2.5 Droplet Precautions

Droplet isolation precautions are used to help stop the spread of microorganisms that are spread through tiny droplets that are disseminated through sneezing and coughing. The most common causes of droplet precautions are pneumonia, seasonal influenza, pertussis, and bacterial meningitis. Droplet precautions can frequently be confused with airborne precautions, but these precautions focus on those microorganisms that are transmitted via droplets during close proximity interactions [9]. To protect against droplet-spread microorganisms, a surgical mask should be worn while in the patient's room. The mask should be put on prior to entering the room and removed upon exiting the room. While most PPE should not be worn outside of the room, a surgical mask should be discarded in the trash after leaving the room [10]. Hand hygiene should be performed after the mask is removed upon exiting of the room. Masks should always be disposed of after leaving the room and should not be re-used for multiple patients as the mask is considered contamination after being touched. In emergent conditions, there have been situations that call for re-using masks if supplies are limited however, these should be evaluated during the event and not a common practice. The Association of Perioperative Registered Nurses (AORN), Food and Drug Administration (FDA), WHO, and CDC all recommend disposal of FDA-approved medical masks after single use and that healthcare workers should don a new mask for each contact with a new patient [11]. AORN highlights the hazards of reusing surgical masks by saving them or hanging them around the neck. Droplet precautions are an effective infection control strategy for preventing the spread of infectious diseases that are spread via droplets and close contact.

8.2.6 Airborne Precautions

Airborne isolation precautions are required during the care of patients who have or are suspected of having diseases like tuberculosis, measles, or chickenpox, which are caused by small organisms (<5 μm in diameter) that easily flow on air currents. Patients infected with such pathogens are capable of transmitting disease through respiratory secretions that are aerosolized during coughing, sneezing, suctioning procedures, etc. Airborne precautions are perhaps the most challenging as they require healthcare workers to wear appropriate respirators that have been fit-tested. Adherence to airborne isolation requires healthcare workers to be fit-tested to ensure the National Institute for Occupational Safety and Health (NIOSH)-approved N-95 or higher level respirators maintain a proper seal around the face. These masks should be donned prior to entering the room, removed after exiting the room, and then disposed of.

One of the most crucial aspects of airborne precautions is that the patient must be in an appropriate negative air pressure room. Negative pressure rooms (also called airborne isolation rooms or AIIR) are designated rooms with the proper engineering controls to ensure that aerosol particles containing microorganisms <5 μm in diameter are kept inside the room. It is vital to maintain negative air pressure, as it ensures that the air flow is from outside into the room, which prevents such microorganisms from being carried into the hallway and adjacent spaces (more information on negative air pressure rooms can be found in the section on environmental infection control). If a patient is suspected of having a disease that requires airborne isolation, it is critical to have them wear a surgical mask until a negative pressure room is available. One of the most common sources for exposures is delay in placing the patient into airborne isolation. Healthcare workers should don N-95 respirators while in the patient's room and the patient should wear a regular surgical mask during transportation or in an area that is not negative pressure. The use of a surgical mask for patients acts as a protection for those in the rest of the healthcare space.

Airborne isolation is perhaps the most prudent during bioweapons-related or high-consequence biological events. Patients with high-consequence pathogens, such as monkeypox, SARS, smallpox, viral hemorrhagic fevers, MERS, and influenza, require airborne isolation with some additional isolation precautions. Airborne isolation is the most challenging aspect of dealing with such biological threats as it requires the unique environmental controls that can pose challenging during an influx of patients with infectious diseases or in resource poor areas. A recent WHO report highlighted the role of healthcare facilities and infection control failures in MERS outbreaks, which underscores the consequences of PPE and infection control failures [12, 13]. It is a common practice in healthcare that a patient with a cough/fever has to wear a mask until infectious etiologies can be ruled out. By simply having a potentially sick patient wear a surgical mask during the triage process, healthcare workers and other patients can be protected, regardless whether the patient has an infectious disease that is spread by droplet or airborne mechanisms. Although airborne isolation can be the most taxing in terms of ensuring properly fitted masks are used and negative pressure rooms are available, these are vital components to infection control for high-consequence and highly transmissible diseases.

8.2.7 Combination or Enhanced Precautions

In many cases, combining isolation precautions is most beneficial. As medical care grow is in complexity and more people are involved in patient care, there is a higher risk for spreading microorganisms via aerosol-generating procedures or contaminated surfaces/objects. Additional PPE, like boot/shoe covers, caps, etc., may be utilized in such isolation practices. However, proper PPE donning and doffing practices should always be followed [33]. Some microorganisms, like *Human herpesvirus* 3 (HHV-3, varicella-zoster virus), require a combination of airborne and contact isolation due to the nature of the illness they cause and transmission mechanisms. It is also common in pediatric care to combine contact and droplet isolation during the care of patients with respiratory viruses (e.g., human metapneumonvirus, respiratory syncytial virus, rhinovirus) as children frequently spread such microorganisms through touch. For example, according to CDC guidelines, respiratory infectious diseases (except for influenza, pertussis, etc.) only require standard precautions in adults. However, in infants and young children, standard plus contact isolation is encouraged, which emphasizes the need to assess the patient population when considering isolation precautions during care. Combination precautions are especially beneficial if microorganisms are isolated from multiple sources, such as a patient with Carbapenem-resistant Enterobacteriaceae isolated from both a draining wound and an endotracheal secretion. Typically, contact precautions are sufficient. However, if the microorganism is isolated in respiratory secretions, it is suggested that droplet precautions also be used.

There is an increasing practice of combining and enhancing isolation precautions for severe, high-consequence, or novel infections as research is ongoing regarding their transmission mechanisms. Enhanced isolation goes beyond just combining transmission based precautions, but adding additional measures and processes [33, 34]. Ebola virus disease, SARS, MERS, and many other diseases require a combination of isolation precautions that include airborne and contact prevention measures, but also enhanced isolation practices that include additional PPE and precautions. Novel or emerging infectious diseases may require an amalgamation of isolation precautions, especially in advanced medical settings. The 2013–2016 Ebola virus disease outbreak highlighted the challenges of managing patients in American hospitals and required substantial revisions to existing PPE and isolation guidelines. At the most rudimentary level, isolation precautions should be designed around the transmission mechanisms of the infectious disease. SARS and MERS coronaviruses are primarily spread through respiratory droplets, which caused significant issues in the early stages of outbreaks due to delayed airborne isolation. On the other hand, Ebola virus disease cases within the United States revealed gaps in isolation guidance and preparedness when responding to patients with emerging infectious diseases. Although aerosol-generating procedures were a topic of concern for SARS and MERS, respiratory precautions were already considered for isolation precautions. The limited experience and knowledge among healthcare workers surrounding Ebola virus disease also posed unique challenges for training on enhanced isolation precautions and PPE during the 2013–2016 outbreak. Overall, enhancing isolation precautions is a helpful strategy for handling a novel or emerging infectious disease, especially if newer or more invasive medical procedures or pediatric patients are involved. One of the most common lessons infection preventionists highlight for clinical staff is that it is always better to quickly initiate isolation precautions and remove them later than find out hours or days into a patient's care that they were contagious. Patients may voice frustration with being placed in isolation, but delay in isolating a patient who potentially has an infectious disease, runs the risk of furthering disease transmission.

8.2.8 Isolation and PPE Compliance in Visitors

Visitors play a significant role in the well-being of hospitalized patients however, they also can facilitate the transmission of infectious diseases. Healthcare facilities will ultimately need to grapple with balancing patient satisfaction and the potential for visitors to act as amplifiers of disease, especially during outbreaks. Adherence to isolation precautions and hand hygiene is vital among healthcare workers, but the role of visitors has also become a topic of concern. There is substantial evidence for the horizontal transmission of even multi-drug resistant microorganisms involving the hands and attire of healthcare workers. However, consideration must also be given regarding the role of visitors in infection prevention and control [14]. While there is no universal recommendation for handling visitor compliance to standard or more advanced precautions, it is encouraged that those visiting a contagious patient should use PPE for their own safety. The Society for Healthcare Epidemiology of America (SHEA) has provided recommendations that highlight the challenges in enforcing visitor compliance, but also notes that it is crucial to consider the infectious microorganism when dealing with visitors who are unwilling to follow precautions. Proper education and guidance on PPE donning and doffing should be given to visitors. However, hand hygiene is perhaps the most important and likely practice to be followed. SHEA notes that "hospital visitors may play a significant role in the ecology of healthcare-associated infections" but few visit more than one patient, which makes the transmission between visitors and multiple patients less likely [14]. There is a consistent challenge in enforcing compliance with visitors and little research that has focused solely on their compliance and role within healthcare-associated disease transmission. Several things should be considered to address visitors of patients in isolation for an infection. If the visitor is a household contact and likely to already be exposed the concern then becomes if they are already infected and potentially contagious. If the visitor is a breast-feeding mother who is involved in the care of the patient, it is unlikely isolation precautions can be enforced. Many hospitals within the U.S. also make it a practice to screen visitors during respiratory virus season to prevent symptomatic visitors from exposing patients or staff [35]. This practice should be considered for visitors of such patients, especially if they are house-hold or close contacts and may be a source for further transmission.

Patients with high-consequence diseases, like Ebola virus disease or SARS, may pose a unique challenge for healthcare workers in terms of visitors and isolation precautions and should addressed on a case by case basis. CDC recommendations for visitors of Ebola virus disease patients include that healthcare workers should "avoid entry of visitors into the patient's room" and that visits should be scheduled and controlled [15]. Visitors played a significant role in furthering the spread of SARS and MERS during outbreaks and consideration should be given to restricting the number of visitors during such events. A recent WHO report on MERS found that of the cases between December 2016 and July 2017, 31% were associated with transmission in a healthcare facility, including healthcare workers, patients sharing rooms/wards with MERS patients, and visitors [12, 13]. CDC guidance for infection control during treatment of patients with SARS also encourages limiting visits to those who are necessary for the patient's emotional well being. Ensuring public health during outbreaks inherently trumps visitation [16]. Visitors should be educated and encouraged to wear the necessary PPE when visiting an isolated patient. At the bare minimum, it is prudent to educate and encourage visitors to practice hand hygiene upon entering and exiting the patient's room.

Airborne isolations pose a challenge for visitor compliance. Since hospitals can not support fit-testing visitors for N-95 masks or other respirators, each situation should be evaluated on an individual basis. Surgical masks can provide some protection but will not provide a true seal. It is a common practice for many healthcare institutions to allow house-hold contacts to visit a patient in airborne isolation without a mask (or wearing a surgical mask) since exposure has likely already occurred. In this situation, consideration should be given for exposure from that visitor to healthcare workers and others after they have left the negative pressure room. It may be prudent to have visitors in such situations wear a surgical mask upon exiting the patient's room and wear it until they exit the hospital.

Visitor adherence to isolation precautions is challenging for even the most common hospitalizations. Staff compliance and vigilance can be a strong role-model and reminder for visitors, especially in regards to hand hygiene. In the event of a bioweapons-related attack or outbreak of a highly infectious disease (i.e., Ebola virus disease), visitor protocols should be established. Unique events and microorganisms can pose a challenge for isolation precaution compliance and recommendations may need to be modified for novel or specific situations (i.e., caring for a pediatric patient and maintaining parent/guardian visitor isolations). Changes to IPC guidelines for such unique situations should be done in conjunction with CDC (and other expert) guidance and discussed in a multidisciplinary setting to ensure support is facility-wide. In these situations, allowance of visitors and PPE recommendations should be made on a case by case basis. Visitors do play a significant role in infection prevention and control within the healthcare environment, which makes their adherence to standard precautions and isolation precautions important and challenging.

8.2.9 Discontinuation of Isolation Precautions

Transmission-based precautions should be maintained until the patient is no longer considered capable of transmitting the disease. The CDC's Guidelines for Isolation Precautions is an excellent resource for duration of isolation. Consideration should be given for immune status of the patient as immunocompromised patients may need to be isolated for longer periods of time due to lengthier microorganism shedding.

8.3 Patient Placement and Cohorting

Patient placement is a vital component to infection control measures, especially during an influx of ill patients. Isolation practices and cohorting (grouping like patients in a shared hospital room or designated unit) plays a significant role in reducing the microbial burden within the healthcare environment. Patient cohorting is a common practice during outbreaks or limited availability of negative pressure rooms however, it is important to consider transmission-based precautions during the assignment of patient rooms and establishing cohorted units.

8.3.1 Placement of Patients

Isolating patients with the appropriate transmission-based precautions is often done during admission. However, if patients are transferred to the healthcare facility and there is already suspicion of an infectious disease, this may also determine where the patient is placed. There is a myriad of components that impact patient placement and such efforts should be performed with IPC in mind.

Although many healthcare facilities have private rooms, there is always the possibility that patients may have to share rooms. Efforts should be made to place patients with infectious diseases that are transmitted via the contact, droplet, or airborne route, in private rooms. Consideration for route of transmission should always be made during patient admission, which would determine both isolation status and patient placement. Patients needing airborne isolation require rooms that maintain negative pressure, which are typically only a handful of rooms per care area. These patients should be place within private negative pressure rooms. There is an increasingly common practice of requiring private rooms for patients infected with highly resistant microorganisms, such as carbapenem-resistant Enterobacteriaceae (CRE) or multidrug-resistant TB (MDR-TB). Patient placement should be determined with transmission-based precautions in mind and efforts should be made to ensure patients with highly contagious or high consequence pathogens (e.g., Ebola virus, SARS and MERS coronaviruses) are in private isolation rooms per CDC guidelines.

8.3.2 Cohorting of Patients and Staff Assignments

Despite our best efforts, during peak respiratory virus season or an influx of patients, private room availability may be constrained. Cohorting is a common practice during times in which room availability may be scarce. There are two kinds of cohorting; sharing of private rooms or grouping patients within a defined entire hospital unit. If private rooms are not available, patients infected or colonized with the same microorganism (e.g., influenza A virus, MRSA) can be cohorted together to share a room. During respiratory virus season, cohorting based off clinical symptoms is often utilized. While it was necessary to cohort Ebola virus disease patients during the Western African 2013–2016 Ebola virus disease outbreak, CDC guidelines for U.S. hospitals is that patients should be placed in a private room. If there should be an influx of patients with a high consequence pathogen, consideration for cohorting within the same room should be made on a case-by-case basis with infection prevention and control coordinating with the appropriate public health authorities.

During an outbreak or influx of patients in which hospital bed capacity is strained, cohorting can be utilized for an entire hospital unit or well-defined space. Designated hospital units can be cohorted for infected patients. These units should either be dedicated to patients affected by an outbreak (and the same microorganism) or of all patients in isolation who are cohorted within their rooms based off the previously mentioned criteria. This is an effective practice for segregating the patients and their visitors from healthy/non-infected patients. Cohorting is also an effective strategy for reducing the risk of cross-contamination by staff. The CDC notes that "cohorting healthcare personnel to care only for patients infected or colonized with a single target pathogen limits further transmission to uninfected patients but is difficult to achieve in the face of staff shortages in hospitals and residential healthcare sites. However, if continued transmission is occurring after implementing routine infection control measures and creating patient cohorts, cohorting of healthcare personnel may be beneficial." [17]. Designated staff can be assigned to cohorted patients or units. It is highly encouraged that healthcare workers (especially nurses) caring for cohorted patients, should not also be assigned to patients who are immunocompromised. Cohorting in single rooms or in well-defined units can be an effective mechanism for reducing the spread of infectious diseases during patient care.

8.3.2.1 Designated Staff For Specific Microorganisms

The Ebola virus disease outbreak in 2013–2016 highlighted the role of healthcare worker transmission, not only between themselves, but also among patients. CDC guidelines encouraged healthcare facilities to minimize staff with direct patient care. It was encouraged that those staff caring for a person under investigation (PUI) or infected with Ebola virus, be dedicated to that patient as to minimize the potential for additional exposure. During the 2011 outbreak of carbapenem-resistant *Klebsiella pneumonia* at the National Institute of Health's Clinical Center, IPC staff adopted a series of control measures that included cohorting, limiting activities of hospital staff exposed to infected patients, and hiring monitors to ensure healthcare providers were wearing appropriate PPE and using hand hygiene [18]. Facilities should have policies developed and in place for those staff who may be exposed during their duties. Incidents involving bioweapons, novel or emerging infectious diseases, or high consequence pathogens, will surely require unique guidelines that could test the capacity for hospitals. It behooves healthcare facilities and preparedness efforts to consider staffing and cohorting needs during such challenging situations.

8.4 Patient Screening and Movement

The flow of patients from triage to discharge also poses vulnerabilities for infectious disease transmission. The initial screening of patients can be an effective infection control measure for rapid isolation and PPE utilization. Proper communication and triage questions can aid in the rapid identification of patients with increased risk for infectious diseases. Hasty identification of a patient with a potential infection within their first moments within the healthcare facility can substantially decrease the risk for additional exposure. Once admitted to a healthcare facility, patients are frequently transported for diagnostic tests and care. The transportation of a patient can also act as a facilitator for disease transmission if infection control practices are not maintained.

8.4.1 Emergency Department Screening and Triage

The initial interaction a patient within a healthcare facility or urgent care unit is the first opportunity to identify risk for infection. The triage process determines the degree of urgency of the patient's condition, which then decides the order they are seen. During this process, there are many opportunities to not only identify patients with potentially contagions infections, but also rapidly isolate them to prevent further exposure to staff and other patients. It is encouraged for triage staff to provide patients presenting with a cough, (i.e., a potential respiratory infection) a mask to wear during their time in a waiting room. This simple act can be hugely preventative, whether the patient has influenza, MERS, or TB.

While the triage process primarily focuses on medical questions, this is also an opportunity for questions regarding international travel and fever. Since the SARS outbreak in 2002, it has become a relatively common practice to incorporate a question within the patient screening/triage process regarding international travel and fever. Travel history is always an important component to the patient's medical history but we should not forget its significance during triage and patient screening. Presence of fever is a standard part of the medical triage and intake during initial screening and if the patient is febrile, questions regarding travel history become that much more important. There are a mixture of recommendations regarding processes, but it is consistently thought that if a patient with an acute illness (e.g., fever within 2 weeks, diarrhea) has traveled outside the U.S. within the last 21 days, the patient should be isolated. Isolation should be based upon the presenting signs and symptoms but providers should also consider the area of travel and if there are ongoing epidemiologically-significant outbreaks. During outbreaks, like those of SARS and MERS, screening and triage can include questions regarding not just travel, but also exposure or close contact with a sick person who recently traveled to those areas or was infected. The screening and triage process is especially beneficial for early detection and isolation of potentially infectious patients because it can easily be modified. Screening questions can be added or removed based upon healthcare needs. While the standard questions remain the same, questions about specific travel or exposures can be added during times of public health crises. Questions related to travel history should be maintained during all times though and it is encouraged that patients with an acute illness and international travel within 21 days, be placed in a private room until further evaluation by a healthcare provider. Measles and TB are frequent suspected in patients with respiratory symptoms and a recent international travel history. However, the inclusion of travel history in medical triage and screening can also potentially capture cases of avian influenza, MERS, viral hemorrhagic fever, etc.

8.4.2 Transportation of Patient Within Hospital

Patients are frequently transported throughout the hospital during their care, which can make maintaining isolation precautions challenging. It is encouraged for patient transportation to be limited, but there are several recommendations to further reduce the spread of disease during patient movement. Using appropriate barriers is crucial during the transportation of an infectious patient.

Patients in contact isolation are encouraged to change into a new hospital gown or attire and should be transported on a wheelchair or bed covered with a clean sheet (or other barrier). Infectious skin lesions should be covered. The transporter should wear the appropriate PPE while assisting the patient into or out of the covered wheelchair or bed, which should then be removed upon exiting the room (gown and gloves are not required during the transport of patients in contact precautions). The CDC notes that "when transport or movement in any healthcare setting is necessary, ensure that infected or colonized areas of the patient's body are contained and covered." [17]. Those patients in droplet or airborne isolation should wear a mask during the transportation process. It is important that notification of the patient's isolation status be done to the receiving department or healthcare personnel to facilitate the continuation of isolation precautions.

8.4.3 Exposure Response

Despite best efforts, patients are not always isolated or transported appropriately. If a patient with an infectious disease has potentially exposed a considerable number of staff and other patients, infection prevention and control programs may be used to the contain further spread. All exposures are not equal, meaning that MRSA is a common microorganism within the community and an exposure response is often not warranted, whereas exposure to mycobacteria or a more severe illness may require follow-up. These responses should be determined by the IPC program, Occupational Health, and the medical directors associated with these programs in conjunction with hospital administration. Post-exposure prophylaxis may be necessary for staff and patients, but medication distribution should be determined by the appropriate hospital experts. Hospitals and IPC programs should have a functioning process for determining patients with exposure to specific staff or in other patients (i.e., those in the waiting lobby of an emergency department). Commonly, exposure monitoring and notifications are used for patients, visitors, and staff exposed to measles, pertussis, etc., as post-exposure prophylaxis is recommended. IPC programs might monitor admitted patients who were exposed or potentially isolate them during the incubation period. Exposures may also be the result of infectious healthcare staff, which may require patient notification. Overall, these events should be addressed on an individual basis and adequate response measures should be applied to reduce the time between exposure and follow-up.

8.5 Environmental Infection Control

Although isolation and PPE utilization are a significant portion of preventing the spread of disease during hospitalization or medical treatment, the role of environmental transmission should not be forgotten. Many diseases can be spread through fomites and poor consideration for the environmental aspects of infection control only fuel transmission. There is considerable literature on the recommendations and guidelines for environmental infection control, specifically by the CDC, OSHA, WHO, and APIC. This section will provide a very high-level overview of these publications. The goal of this section is to raise awareness for environmental mechanisms of transmission and encourage consideration during the development of preparedness and response plans. Healthcare environmental components, such as building design, waste management, disinfection and sterilization, and air monitoring, are critical to reducing the spread of microorganisms, but the ability to make changes may be hindered by high costs and other restraints.

8.5.1 Cleaning, Disinfection, and Sterilization

Cleaning and disinfection of non-critical surfaces is a basic premise of infection control practices, especially standard precautions [17]. These practices become even more significant when caring for patients with infectious diseases. Cleaning is the process of removing pathogens, whereas disinfecting kills germs on the surface, and sanitizing lowers the number of pathogens on an object to a safe level. Rutala and Weber provide an extensive outline of such methods for cleaning, which include:

- Sterilization: destroys all microorganisms, including bacterial spores
- High-level disinfection: destroys all microorganisms except high numbers of bacterial spores
- Intermediate level disinfection: destroys vegetative bacterial, mycobacteria, most viruses, most fungi, but not bacterial spores
- Low-level disinfection: destroys vegetative bacteria, some fungi and viruses, but not mycobacteria or spores.

Such practices should be closely followed, especially in the event of novel or severe illnesses requiring enhanced precautions. Disinfection recommendations may be revised based upon recommendations from OSHA and the CDC, as was seen during treatment of patients with Ebola virus disease in U.S. hospitals. Environmental contamination can be a significant mechanism for disease transmission during patient care and invasive procedures can increase this substantially. Poorly disinfected or sterilized equipment has been the culprit of many healthcare-associated infections and outbreaks. High-touch surfaces or objects, like sinks and bedrails, require frequent cleaning and disinfection. Frequency and intensity of cleaning and disinfection may also need to be increased depending up on the "degree of environmental contamination" [17]. Environmental Protection Agency (EPA)-registered disinfectant should be used to combat the environmental contamination that occurs during patient care. It is important to note that certain pathogens may be resistant to routine disinfectants (e.g., rotaviruses, noroviruses, *Clostridium difficile*, *Bacillus anthracis*), meaning that healthcare workers should ensure the cleaning product they are using is effective against the specific microorganism. Disinfection and sterilization of medical equipment should always follow manufacturer recommendations and be handled with care.

There are frequently forgotten objects and surfaces that are common culprits for contamination and should be highlighted during routine cleaning. Computer keyboards, stethoscopes, and clinical work stations are all areas that routinely have a high microbial burden and are frequently neglected during cleaning. There is extensive literature on practices for disinfection and sterilization, but at the most basic level, preparedness and response efforts should work to ensure routine cleaning and disinfection is done appropriately and correctly [36]. In emergency situations, a quick lesson is always to work "from clean to dirty" and consider all surfaces and objects. Environmental contamination is a continued source for disease transmission in healthcare. These practices should be strengthened during routine care and also refined to meet special needs during public health emergencies and care of patients in enhanced precautions.

8.5.2 Ventilation Systems

Hospital ventilation systems are a critical component to patient safety. These systems maintain negative pressure within airborne isolation rooms and positive pressure in operating/procedure rooms, as well as protective precaution rooms (for severely immunocompromised patients). Negative pressure rooms will become a critical need in outbreaks of microorganisms that require airborne isolation. These rooms should have high-efficiency particulate air (HEPA) and should also follow CDC guidelines for environmental infection control regarding air exchanges per hour (ACH) (i.e. $\geq$12 ACH for new construction as of 2001; $\geq$6 ACH for construction before 2001) and be monitored in accordance with engineers' and manufacturers' recommendations [19]. The negative pressure rooms should maintain air pressure by being properly sealed and are encouraged to have self-sealing doors and ante-rooms. The heating, ventilation, air conditioning (HVAC) systems should also have humidity monitors to ensure the moisture levels are adequate as to discourage microbial growth. The CDC has specific recommendations for air-handling systems in healthcare facilities, which are vital to patient safety and reduction of disease transmission. In the event of a biological attack or outbreak, ensuring that these systems are operational and can maintain proper negative pressure in designated rooms will be of the utmost importance. Although negative pressure rooms are limited on each hospital unit, there is potential to convert entire units to negative pressure through adjustments to the ventilation system and sealing of the unit [20].

8.5.3 Environmental Services

Environmental cleaning, waste management, and handling of textiles are all components of environmental service work in healthcare. These duties are especially important upon discharge of the patient when a deep clean of the room is required (commonly called "terminal cleaning"). Within a single day of admission, the composition of a hospital room's microbes begin to change and mirror that of the patient [21]. Combatting the "bacterial communities" that develop within the healthcare environment is an ongoing struggle that emphasizes the important of adequate and consistent environment services. Environmental services staff frequently manage these tasks and should be aware of the patient's isolation status so that staff utilizes proper PPE and ensures use of proper Environmental Protection Agency (EPA)-registered disinfectants (example: *Clostridium difficile* requires the use of bleach products to kill spores). Typically, hospitals have contracts with linen services and waste management companies (for both regular and medical waste). However, practices may change during the care of patients with high consequence pathogens that require enhanced isolation. CDC guidelines regarding medical waste, dishes and eating utensils and textiles all note that these objects have not been implicated in SARS coronavirus transmission and that routine procedures should be utilized [22]. In the case of Ebola virus, recommendations were significantly different. For example, despite Ebola virus forming enveloped virions (which are susceptible to a wide range of hospital disinfectants), no EPA-registered disinfectants had labels that specifically cited efficacy against the virus. Consequently, CDC recommendations were changed to encourage use of EPA-registered disinfectants that were effective against viruses that form non-enveloped virions, which are more resistant to hospital germicidal disinfectant cleaning products. CDC recommendations also included discarding of all textiles to avoid staff exposure during the laundering process [15]. Many hospitals developed Ebola virus preparedness plans that placed environmental service duties more on the clinical staff caring for a PUI instead of potential exposing additional staff. Environmental services practices should be closely followed during routine patient care but may require modification during the care of patients in enhanced precautions or during outbreaks.

In the pursuit of more effective and efficient cleaning and disinfecting processes, newer technologies have been utilized, like that of ultraviolet disinfection technology and copper-infused linen. Environmental monitoring to check room cleaning quality is also a frequently used measure to test true cleaning of the space. Overall, environmental infection control by means of environmental services staff, textile care, and waste management are critical components to breaking the chain of infection. The efficacy of these duties and the adherence to transmission-based precautions when working throughout a healthcare facility is a monumental task and becomes even more critical during outbreaks.

8.5.4 Water

There is a considerable amount of literature and regulatory guidance on environmental infection control as it relates to water however, this section will focus on only a handful. Firstly, facilities should consider the mechanisms for preventing the spread of waterborne microorganisms through proper hand hygiene, cleaning of sinks, avoidance of decorative fountains and fish tanks in patient-care areas, and maintaining proper temperatures for hot and cold water [19]. Should water disruptions occur, corrective decontamination of the hot water system may be necessary, which can include flushing the water lines and outlets with chlorine and/or superheated water. Hospitals should have contingency plans for water in the event that water disruptions are prolonged due to intrusions or contamination.

8.5.5 Construction and Renovation

Construction, renovation, remediation, repair, and demolition within a healthcare environment can be sources for infection. All work must be reviewed by infection prevention and control to coordinate prevention methods that include containment barriers and post-construction cleaning. Such efforts are necessary as demolition can lead to airborne infections associated with fungal spores in immunocompromised patients. Moreover, if the proper planning is not done, engineering outages can impact negative pressure rooms, which decreases their efficacy. Conversion of spaces to negative pressure wards or specialty isolation units should be planned and tested ahead of time to ensure proper ventilation is possible.

8.6 Challenges

8.6.1 Poor Utilization and Support

In the day-to-day world of medical care, infection prevention and control practices, like hand hygiene, frequently have poor compliance [23]. While there are a multitude of issues that IPC programs face, perhaps the most damning one is internally. Typically, IPC programs are seen as cost centers and not revenue generators, which frequently translates to lackluster support and prioritization [24]. Medicine and healthcare are competitive markets, meaning that those programs seen from this light are likely to receive cuts. While there have been studies evaluating the cost of healthcare-associated infections in relation to that of an IPC program, there is a considerable gap within the research regarding IPC programs and response during outbreaks or cases of high-consequence pathogens. Given that a single healthcare-associated infection (HAI), like a surgical-site infection, can cost upwards of $50,000 and preventing just a handful of these infections can pay for the entire budget of an IPC program, it is not unreasonable to think that such cost-saving could be applied to biothreat response within hospitals [25]. The cases of Ebola virus disease in Dallas, Texas, in 2014 exposed the lack of preparedness within U.S. hospitals, poor compliance with infection control, and the cost of treating such a severe illness with enhanced precautions. It was estimated that the cost of treating a hospitalized patient with Ebola virus disease in the United States is $30,000 per day [26]. Despite spending $60 billion on biodefense since 2001 to prepare for biothreats, a single imported case of Ebola virus disease revealed deep breaks within U.S. preparedness efforts [27]. During the 2013–2016 Ebola virus disease outbreak, it frequently fell on the IPC programs to coordinate hospital-wide preparedness and response efforts to ensure facilities were prepared to identify, isolate, and potentially treat a patient. While such events highlight the importance of IPC programs and the diversity of their responsibilities, for many in healthcare the looming threat of Ebola has passed and so has the support. The focus for many IPC programs has gone back to preventing HAIs and Ebola virus disease response plans have been neglected. A 2015 APIC survey found that 92% of IPC program leaders believe they were more prepared than the previously year to receive a patient with a highly infectious disease [28]. While this is encouraging, would hospitals and IPC programs feel the same today, several years after the 2013–2016 Ebola virus disease outbreak? There is a substantial need within healthcare to truly support IPC programs for not only day-to-day HAI prevention, but also as a means of establishing an all-hazards approach. Infection prevention and control, as shown within this chapter, involves collaboration with many internal and external partners to ensure patient safety. IPC programs are sentinels within hospitals, alerting internal and external partners of issues, so that they may form a collective response effort. It behooves bio-preparedness efforts to utilize IPC programs and consider IPC practices as a vital resource for disease control, medical countermeasures, and surveillance.

8.6.2 Invasive Medical Interventions

While advances in medicine have given us the capabilities to increasingly counter disease, the invasiveness of certain medical procedures and treatments can pose substantial infection control risks. The treatment of Thomas Duncan, the first Ebola virus-diagnosed patient in the United States, in a seemingly "regular" hospital without specialized training or facilities to handle a patient under enhanced precautions, points to this very dilemma. In the weeks following his hospitalization, there were conversations in hospital around the U.S. as to the safety and ethical considerations for more invasive and riskier medical interventions during the treatment of a patient with such an infectious disease. Medical interventions like invasive surgery, dialysis, insertion of central lines, ventilation, extracorporeal membrane oxygenation (ECMO), or even labor and delivery, all carry inherent risks for the staff performing them. The outbreaks of SARS and MERS coronaviruses, and Ebola virus disease all challenged PPE and isolation recommendations, requiring changes to protect healthcare workers and visitors during treatment, especially those interventions that generate aerosols. These outbreaks have taught vital lessons regarding the life-saving capabilities of modern medicine and technology, but also reveal a lack of preparedness consideration for medical interventions in patients with such severe and uncommon diseases. Future efforts should establish treatment and procedure plans with these diseases in mind, drawing on lessons learned and creating new protocols so that afflicted patients may receive the full range of medical care and healthcare workers can perform such work with minimal risk.

8.6.3 Conflicting Guidelines

Perhaps one of the most challenging aspects of infection prevention during the Ebola virus disease response in American hospitals was the continuous evolution of guidelines for isolation, PPE, disinfection, handling of medical waste, etc. The Blue Ribbon Study Panel on Biodefense highlighted this very issue in its recommendations, pointing to the "need for a standardized process to develop and issue clinical infection control guidance for biological events." [29]. The recommendations from the CDC regarding isolation precautions, specifically the use of negative pressure rooms/airborne isolation and the necessary PPE, were an evolving process in the days following Thomas Duncan's hospitalization. As an infection preventionist, these days were taxing due to the constant scramble to not only train staff on the appropriate donning and doffing procedures, but even acquire the correct PPE. Novel diseases like SARS and MERS also tax the public health and healthcare response systems in terms of providing adequate recommendations for isolation. While these fluctuations are understandable, in the event of a biological attack, fear and hysteria will likely play a larger role than in naturally occurring outbreaks, which makes the stability of IPC recommendations that much more important. Healthcare workers must feel confident in not only their training, but also the efficacy of their PPE. It is crucial that future efforts include stronger processes to standardize infection control recommendations and guidelines during these critical times.

8.7 Infection Control in Biodefense Efforts: Minimal Focus Despite Maximum Return

Biodefense and global health security efforts work to strengthen capacity to respond to infectious disease threats. These efforts are imperative to preventing, detecting, and responding to biological events, whether they are intentional, accidental, or natural. Within these plans, there is often a considerable gap regarding the role of infection prevention and control and the programs within healthcare facilities. There have been several reports and recommendations in recent years to strengthen biodefense efforts. The Health Sector Resilience Checklist for High-Consequence Infectious Diseases from the Johns Hopkins Center for Biosecurity points to the importance of infection control practices and their significance, but there is little overall mention of the programs within hospitals. This checklist focuses on the adherence to infection control practices, like that of the Blue Ribbon Study Panel on Biodefense report, but tends to neglect the role of these programs. The same can be said of more recent pandemic influenza and hospital preparedness plans. Unfortunately, there tends to be a gap in the literature regarding IPC within biodefense efforts and although most focus on larger components (disease surveillance, PPE, medical countermeasures), the utilization of IPC programs should not go ignored. Hospitals already have this tool available and it behooves preparedness efforts to employ it.

There is a continuous trend of pointing to infection prevention and control processes and neglecting the programs responsible for ensuring they are followed. The systemic gap in preparedness efforts citing infection prevention and control practices but failing to account for the program and role of infection preventionists reveals a fundamental lack of truly understanding the role of these programs and internal hospital operations. Biodefense efforts inherently include infection control practices as they are the foundation for preventing the spread of disease, but this gap regarding the inclusion of IPC programs does them a disservice. Hospitals already struggle with following IPC practices to reduce the spread of infection. Biodefense efforts should strengthen the existing IPC infrastructure and ensure processes are effectively followed. If hospitals are already struggling with hand hygiene and PPE compliance, how will they manage bioterrorism acts or pandemics?

The problem is twofold: current compliance to infection control processes is often substandard, and the IPC programs are frequently under-utilized and/or -prioritized. Biodefense efforts should seek to increase infection prevention and control efforts and strengthen processes while building upon the existing IPC program infrastructure that can be an effective tool. Infection prevention and control is a significant resource in the biodefense arsenal and should be seen as an independent component, like surveillance and medical countermeasures. These programs are already in place and while they focus on the day-to-day problems of HAIs, they can be utilized to better serve and strengthen the U.S. healthcare infrastructure and thus preparedness. Strengthening IPC programs and processes benefits health security whether it is a local outbreak, an act of bioterrorism, or a pandemic.

Hospitals are the frontline for biological events and their capacity to adequately detect and respond to such threats can either halt or help the spread of disease. Infection prevention and control practices are existing guidelines and recommendations that seek to combat disease transmission during medical care. IPC programs within hospitals work to ensure staff are educated on and compliant with such practices to protect patients, visitors, and healthcare workers. These programs are also an integral source for infectious disease surveillance and a connective tissue within hospitals, making them a significant contributor to public health surveillance. Outbreaks of SARS, MERS, Ebola virus disease, and multi-drug resistant microorganisms, have all revealed the importance of IPC and the devastating consequences of poor compliance. Continued failure to prioritize IPC programs and practices, as well as the role they serve within outbreaks and growing biological threats, will only serve to weaken preparedness and response efforts. Defense against biological threats requires strong infection control practices, programs, and the ability to adequately employ these efforts prior to and during public health emergencies. Healthcare is a critical infrastructure, meaning that if hospitals are amplifiers for disease transmission and sources for negative patient outcomes, the impact can be devastating. Infection prevention and control is an essential resource against all biological threats, regardless of their origin.

References

1. Doyle T, Glynn K, Groseclose S. Completeness of notifiable infectious disease reporting in the United States: an analytical literature review. Am J Epidemiol. 2002;155(9):866–74.
2. Centers for Disease Control and Prevention. Bioterrorism agents/diseases. 2017. https://emergency.cdc.gov/agent/agentlist.asp. Accessed 7 Apr 2017.
3. Centers for Disease Control and Prevention. Clean hands count for healthcare providers. 2017. https://www.cdc.gov/handhygiene/providers/index.html. Accessed 19 May 2017.
4. Centers for Disease Control and Prevention. Standard precautions for all patient care. 2017. https://www.cdc.gov/infectioncontrol/basics/standard-precautions.html. Accessed 19 May 2017.
5. Centers for Disease Control and Prevention. Handwashing: clean hands save lives. 2017. https://www.cdc.gov/handwashing/show-me-the-science-hand-sanitizer.html. Accessed 16 June 2018.

6. California Department of Public Health. Reportable diseases and conditions. 2017. https://archive.cdph.ca.gov/HealthInfo/Pages/ReportableDiseases.aspx. Accessed 10 July 2017.
7. World Health Organization. Five moments for hand hygiene. 2006. http://www.who.int/gpsc/tools/Five_moments/en/. Accessed 19 May 2017.
8. Korniewicz D, Kirwin M, Cresci K, Larson E. Leakage of late and vinyl exam gloves in high and low clinical settings. Am Ind Hyg Assoc J. 1993;54(1):22–6.
9. Salath M, et al. A high-resolution human contact network for infectious disease transmission. Proc Natl Acad Sci U S A. 2010;107(51):22020–5.
10. Association for Professionals in Infection Control and Epidemiology – APIC. Rules for standard and isolation precautions. 2017. http://professionals.site.apic.org/10-ways-to-protect-patients/follow-the-rules-for-isolation-precautions/. Accessed 3 April 2017.
11. Committee on the Development of Reusable Facemasks During an Influenza Pandemic, the National Academies. Reusability of facemasks during an influenza pandemic. Washington, DC: The National Academies Press; 2006.
12. World Health Organization. Evidence of hand hygiene to reduce transmission and infections by multi-drug resistant microorganisms in healthcare settings. http://www.who.int/gpsc/tools/Five_moments/en/. Accessed 19 May 2017.
13. World Health Organization. WHO MERS-CoV global summary and assessment of risk. 2017. http://www.who.int/emergencies/mers-cov/risk-assessment-july-2017.pdf?ua=1. Accessed 28 July 2017.
14. Munoz-Price I, et al. SHEA expert guidance: isolation precautions for visitors. Infection control & hospital epidemiology. 2015. https://shea-online.org/index.php/practice-resources/41-current-guidelines/422-expert-guidance-isolation-precautions-for-visitors. Accessed 2 Aug 2017.
15. Centers for Disease Control and Prevention. Infection prevention and control recommendations for hospitalized patients under investigation (PUIs) for Ebola virus disease. 2015. https://www.cdc.gov/vhf/ebola/healthcare-us/hospitals/infection-control.html. Accessed 4 Feb 2017.
16. Rogers S. Why can't I visit? The ethics of visitation restrictions – lessons learned from SARS. Crit Care. 2004;8(5):300–2.
17. Centers for Disease Control and Prevention. Guidelines for isolation precautions: preventing transmission of infectious agents in healthcare settings. 2007. https://www.cdc.gov/infectioncontrol/guidelines/isolation/prevention.html. Accessed 2 May 2017.
18. Vastag B. Superbug stalked NIH hospital last year, killing six. The Washington Post. 2012. https://www.washingtonpost.com/national/health-science/superbug-stalked-nih-hospital-last-year-killing-six/2012/08/22/5be18b1a-ec66-11e1-9ddc-340d5efb1e9c_story.html. Accessed 28 July 2017.
19. Centers for Disease Control and Prevention and the Healthcare Infection Control Practices Advisory Committee (HICPAC). Guidelines for environmental infection control in health-care facilities. Morb Mortal Wkly Rep. 2003;52(RR10):1–42.
20. Miller S, et al. Implementing a negative-pressure isolation ward for a surge in airborne infectious patients. Am J Infect Control. 2017;45(6):652–9.
21. Lax S, et al. Bacterial colonization and succession in a newly opened hospital. Sci Transl Med. 2017;9(391) https://doi.org/10.1126/scitranslmed.aah6500.
22. Centers for Disease Control and Prevention. SARS, infection control in healthcare facilities. 2005. https://www.cdc.gov/sars/guidance/i-infection/healthcare.html. Accessed 14 July 2017.
23. Erasmus V, et al. A qualitative exploration of reasons for poor hand hygiene among hospital workers lack of positive role models and of convincing evidence that hand hygiene prevents cross-infection. Infect Control Hosp Epidemiol. 2009;30(5):415–519.
24. Reed D, Kemmerly S. Infection control and prevention: a review of hospital-acquired infections and the economic implications. Ochsner J. 2009;9(1):27–31.
25. Dick A, et al. A decade of investment in infection prevention: a cost-effectiveness analysis. Am J Infect Control. 2015;43(1):4–9.
26. Sun L. Cost to treat Ebola in the U.S.: $1.16 million for two patients. The Washington Post. 2014, November 18. https://www.washingtonpost.com/news/post-nation/wp/2014/11/18/cost-

to-treat-ebola-in-the-u-s-1-16-million-for-2-patients/?utm_term=.e6a9091e93d8. Accessed 29 Jan 2017.

27. Hayden E. Biodefense since 9/11: the price of protection. Nature. 2011; 477:150–152. http://www.nature.com/news/2011/110907/full/477150a.html. Accessed 30 July 2017.
28. Association for Professionals in Infection Control and Epidemiology – APIC. Ebola preparedness one year later: a poll of APIC members. 2015. https://apic.org/Resource_/TinyMceFileManager/Topic-specific/APIC_Ebola_Survey_Results_November_2015.pdf. Accessed 15 Aug 2017.
29. Blue Ribbon Study Panel on Biodefense. Recommending and promoting changes to U.S. policy and law to strengthen national biodefense while optimizing resource investments. 2015. http://www.biodefensestudy.org/. Accessed 8 July 2017.
30. Bloomfield SF, et al. The effectiveness of hand hygiene procedures in reducing the risks of infections in home and community settings including handwashing and alcohol-based hand sanitizers. Am J Infect Control. 2007;35(10):S27–64.
31. Boyce J, Pittet D. Guidelines for hand hygiene in health-care settings: recommendations of the healthcare infection control practices advisory committee and the HICPAC/SHEA/APIC/IDSA hand hygiene task force. Am J Infect Control. 2002;30(8):S1–46.
32. Grayson ML, Jarvie LJ, Martin R, Johnson PD, Jodoin ME, McMullan C, et al. Significant reductions in methicillin-resistant Staphylococcus aureus bacteraemia and clinical isolates associated with a multisite, hand hygiene culture-change program and subsequent successful statewide roll-out. Med J Aust. 2008;188(11):633–40.
33. Johns Hopkins University Center for Health Security. Health sector resilience checklist for high-consequence infectious diseases – informed by the domestic US Ebola response. 2017. http://www.centerforhealthsecurity.org/our-work/pubs_archive/pubs-pdfs/2017/HCID_Final_Report_05.23.2017.pdf. Accessed 22 June 2017.
34. Occupational Safety and Health Administration. Ebola: control and prevention. https://www.osha.gov/SLTC/ebola/control_prevention.html. Accessed 8 July 2017.
35. Quinlan B, Loughrey S, Nicklin W, Roth V. Restrictive visitor policies: feedback from healthcare workers, patients, and families. Hosp Q. 2003;7(1):33–7.
36. Rutala W, Weber D. Disinfection and sterilization: an overview. AJIC. 2013;41(5):S2–5.
37. World Health Organization. Standard precautions in healthcare. 2007. http://www.who.int/csr/resources/publications/EPR_AM2_E7.pdf?ua=1. Accessed 19 May 2017.

9 Clinical Management of Patients Infected with Highly Pathogenic Microorganisms

Thomas Kratz, Bruria Adini, August Stich, and René Gottschalk

9.1 Introduction

Providing clinical care for patients suffering from high-consequence infectious disease (HCID) is challenging. Medical therapy for patients suffering from HCID may frequently be sophisticated and complex but lacking in evidence base and in extensive experience. Despite maximal intensive care treatment that is administered, including the most advanced resources, mortality among such patients is mostly very high. On the other hand, the natural impulse as a medical practitioner to provide immediate and efficient medical care to the patient may be compromised by the need for health care workers' (HCW) protection with personal protective equipment (PPE). Biological weapons agents are frequently rare, highly virulent microorganisms causative for HCID (Table 9.1).

This chapter focuses primarily on individual patient care with a high level of both patient care and infection prevention and control. Mass casualty management strategies for HCID are discussed as well.

9.2 Initial Approach Towards an HCID Patient Potentially Requiring High Containment Isolation

The initial approach towards an HCID patient differs. Scenarios might be:

(a) *A patient turns up at a medical facility on her/his own initiative as a suspect case*

T. Kratz (✉) · B. Adini · A. Stich · R. Gottschalk
Federal Information Centre for Biological Threats and Special Pathogens, Robert Koch-Institute, Berlin, Germany
e-mail: KratzT@rki.de; august.stich@medmissio.de; rene.gottschalk@stadt-frankfurt.de

S. K. Singh, J. H. Kuhn (eds.), *Defense Against Biological Attacks*,
https://doi.org/10.1007/978-3-030-03053-7_9

Table 9.1 Causative agents of HCID potentially requiring high level containment

Agent	Disease(s)	US Biosafety Level (BSL) requirement[a]
Middle East respiratory syndrome coronavirus (MERS-CoV)	Middle East respiratory syndrome (MERS)	3
Severe acute respiratory syndrome coronavirus (SARS-CoV)	Severe acute respiratory syndrome (SARS)	3
Yersinia pestis	Pneumonic plague Bubonic plague Plague meningitis Plague sepsis Other forms of plague	3
Variola virus (VARV)	Smallpox	4
Monkeypox virus (MPXV)	Monkeypox	4
Crimean-Congo hemorrhagic fever virus (CCHFV)	Crimean-Congo hemorrhagic fever (CCHF)	4
Ebola Virus (EBOV) Sudan Virus (SUDV) Taï Forest Virus (TAFV) Bundibugyo Virus (BDBV)	Ebola virus disease (EVD)	4
Marburg Virus (MARV) Ravn Virus (RAVV)	Marburg virus disease (MVD)	4
Lassa virus (LASV)	Lassa fever	4
Other viruses for which US BSL requirement is 4 and which are causative of viral hemorrhagic fevers (e.g., Junín Virus, Machupo Virus)		

[a]Laboratory biosafety levels according to: https://www.cdc.gov/biosafety/publications/bmbl5/bmbl5_sect_iv.pdf

This might include the so-called "worried well" who did not encounter a significant source of infection. Particularly in the case of a mass casualty scenario, the "worried well" may slow down clinical management of genuinely affected patients [1]. Furthermore, a significant role is played by patient and staff flow as well as structural capacities of the hospital.

(b) *A patient is brought to a medical facility by ambulance as a suspected case*

This scenario is probably managed most easily. Patient and staff flow, area management, and protective measures (such as donning PPE) are regulated quite restrictively in Standard Operating Procedures (SOPs) of high-level isolation units as well as of other tertiary care hospitals.

(c) *A patient is admitted to hospital for a differential diagnosis (e.g., malaria, influenza). An HCID is diagnosed later on in the hospital or after discharge*

Secondary transmissions of HCID in Africa and the Western World frequently follow this scenario. For example,

- In May 2015, a patient travelled from Liberia via Morocco to the USA, where he was admitted to hospital due to a sore throat, fever, and tiredness. He reportedly did not indicate travel to Western Africa and was discharged home the same day. Three days later, he was re-admitted due to clinical deterioration and tested positive for Lassa virus (LASV) infection. He succumbed to Lassa fever [2].
- A patient who was initially diagnosed with malaria was medically evacuated from Togo to Germany in March 2016. After his death, his body was retrospectively tested and found to be infected with LASV. The mortician who handled the corpse contracted the virus, developed Lassa fever, was administered experimental treatment with ribavirin and favipiravir and survived [3, 4].
- In August 2016, a man presented to a university hospital in Madrid, Spain, with a history of high fever, abdominal pain, malaise, nausea, and diarrhea. His clinical condition deteriorated, he developed multi-organ failure, and died in spite of intensive care treatment. All tests for routine infections were negative. Retrospectively, he was diagnosed with Crimean-Congo hemorrhagic fever (CCHF). A nurse who had assisted with the endotracheal intubation of the index patient contracted CCHFV, was treated with ribavirin, and survived [5].

Patients might present to a health care facility of any level. It is probable that the initial physician confronted with an HCID patient sees an HCID clinical picture for the first time in her or his life. Hence it is crucial to have immediate technical support by medical professionals experienced in infectious diseases as well as reference hospitals close by. Numbers and bed capacities of high-level isolation units (HLIU) differ largely between countries [6]. During the 2013–2016 Ebola virus disease (EVD) outbreak in Western Africa, medical evacuation of non-EU/non-US citizens was an issue due to the scarcity of suitable isolation facilities in their home countries [7].

Infection prevention and control measures are crucial. Please refer to Chap. 8 for further instructions on universal precautions. The particular approach differs according to the suspected pathogen. Causative pathogens of viral hemorrhagic fevers (VHFs) such as Ebola virus (EBOV) and LASV have a relatively low stability, and fluid transmission plays the key role [8]; hence national and international guidelines on EVD recommend a safety distance of 1.5–2 m to avoid transmission via macro droplets in case PPE is not worn—such as in patient screening [9, 10]. If physical contact to the patient and/or her/his bodily fluids has to occur, wearing PPE is of utmost importance. For PPE recommendations please refer to the Sect. 9.7.

Health care facilities of any level should include the initial approach towards suspected and confirmed HCID cases in their emergency preparedness plans and SOPs. These procedures might differ largely between hospitals, depending on factors such as human resources and the architectural setup of the medical facilities.

9.3 Area Management

The location where an HCID patient receives medical care is potentially highly contaminated. PPE should be worn at any time. Areas in which there is a low risk of infection and which serve as a supply base (e.g., storage room, staff room) versus sites that are high-risk zones (places where patient care takes place) should be defined very clearly. The 2013–2016 EVD outbreak has shown that poor area management can be a significant source of infection [11].

There are different concepts of area management. For example,

- In EVD treatment centers in field settings, usually a "low-risk zone" and "high-risk zone" are defined.
- A differentiation between a red zone (contaminated patient area), a yellow zone (inner chamber), and a green zone (outer chamber/non-contaminated area) is made in the context of occupational health dealing with HCID in Germany [12].

The inner chamber ("yellow zone") allows space for PPE decontamination and doffing (Fig. 9.1).

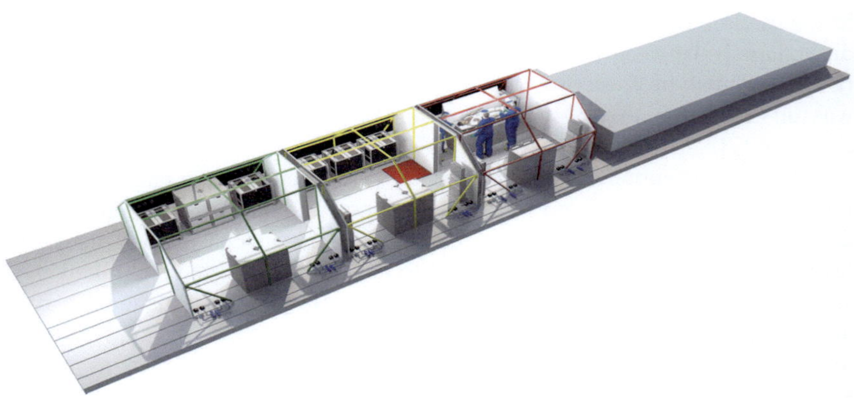

Fig. 9.1 Example of Area Management: Medical evacuation aircraft "Robert Koch". The patient is treated in the red zone by staff wearing personal protective equipment (PPE). The yellow zone is designated for PPE decontamination and doffing. Staff can don PPE and store supplies in the green zone

9.4 Supportive Treatment

Supportive treatment should be based on clear plans and guidelines, such as those for intensive care and sepsis management (http://www.survivingsepsis.org). Furthermore, national and international (WHO) specific guidelines and therapeutic recommendations for the particular disease should be consulted. Intensive care largely depends on available resources in the limited nursing setting. There are indications that intensive care may improve clinical outcomes: Out of 20 patients who were medically evacuated during the 2013–2016 EVD outbreak, 16 survived, yielding a case–fatality ratio (CFR) of 20% [13]. On the other hand, the overall case–fatality ratio for all confirmed cases in Guinea, Sierra Leone, and Liberia with recorded clinical outcomes was 63% [14]. Even more, case fatality ratios for Marburg virus disease (MVD) fluctuate between ~22% for an outbreak which occurred in West Germany in 1967 (supportive treatment and frequent blood samplings for clinical surveillance were performed at that time) and ~88% for an outbreak in Uíge/Angola in 2005 [15, 16].

One means of intensive care not listed in Table 9.2 is extracorporeal membrane oxygenation, the use of which has been documented for one case of Middle East

Table 9.2 Supportive care measures reportedly delivered at EVD treatment facilities during the 2013–2016 Western African outbreak [13, 17, 18]

Facility	Supportive care measures available
Goderich Treatment Centre (Sierra Leone)	Constant bedside nursing Continuous blood pressure/heart rate/respiratory rate monitoring Pulse oximetry Arterial and venous cannulation Nasogastric tube feeding Invasive ventilation Continuous renal replacement therapy Diagnostic biochemistry and hematology Ultrasonography Plain radiography
Mathaska Ebola Treatment Unit (Sierra Leone)	Nasogastric tube feeding Bedside ultrasound Intraosseous cannulation for intravenous fluid resuscitation
Frankfurt High-Level Isolation Unit (Germany)	Constant bedside nursing Continuous blood pressure/heart rate/respiratory rate monitoring Pulse oximetry Peripheral arterial and venous cannulation Central venous cannulation/placement of Sheldon Catheter Invasive ventilation Renal replacement therapy Diagnostic biochemistry and hematology Ultrasonography Plain radiography

respiratory syndrome (MERS) treated in France [19]. Severe acute respiratory syndrome (SARS), MERS, and pneumonic plague mainly affect the respiratory system. A long-term follow-up study indicates that patients affected by adult respiratory distress syndrome (ARDS) can benefit from extracorporeal membrane oxygenation [20]. Equipment and qualified staff are available in designated extracorporeal membrane oxygenation centers. A directory of these centers is available at https://www.elso.org/.

9.5 Specific Treatment

Specific treatment for HCID is often still experimental and/or only administered as compassionate use. This lack of progress is partly due to a lack of patients and costly large-scale clinical trials, hence precluding regular drug approval by national drug authorities. For example, the use of ribavirin as a treatment of Lassa fever is mainly based on a small clinical study without a control group performed in the 1980s [21]. However, none of the HCID dealt with in this chapter are listed as neglected tropical diseases, according to WHO [22]. Examples of specific treatments for HCID are indicated in Table 9.3.

National and international guidelines which mention specific treatments against HCID are scarce. Some of the above-mentioned specific treatments are more frequently used for treating more 'common' diseases, making available more safety data and facilitating easier drug procurement, for example:

- Antibiotics such as quinolones/tetracyclines/cotrimoxazole against bacterial infections
- Oral ribavirin against hepatitis C and
- Favipiravir against influenza.

On the other hand, drug procurement and stockpiling of rarely used drugs or those commercially undeveloped (so-called 'orphan drugs') are a challenge. The fact that synergies are not implemented but country and state authorities tend to do 'their own thing' can be detrimental for efficient stockpiling strategies [41]. Drug procurement strategies should take into account established contacts to the drug manufacturer and to the public health/drug/customs authorities and, if applicable, should also consider benefitting from the compassionate use program. For examples of such programs, please refer to [42–44].

Table 9.3 Examples of specific treatments, post-exposure prophylaxis and vaccines for HCID (all experimental)

Disease(s)	Treatment	Post-exposure prophylaxis	Vaccines	International specific treatment guidelines
Pneumonic plague	Quinolones Aminoglycosides Tetracyclines Cotrimoxazole [23, 24]	Quinolones, cotrimoxazole [23, 24]	Sub-unit vaccine based on the F1- and V-antigens [25]	WHO guidelines available [23]
Smallpox	Brincidofovir [26] Tecovirimat (ST-246) [27]	Smallpox vaccine [28]	Smallpox vaccine (WHO Emergency stockpile 'SVES' available) [29]	
Crimean–Congo hemorrhagic fever	Ribavirin [30] Favipiravir [31]			WHO guidelines in development [32]
Ebola virus disease[a]	ZMAPP [33] Favipiravir [34]	VSV-ZEBOV vaccine [35]	VSV-ZEBOV ChAd3-vaccine [36]	WHO guidelines available [37]
Lassa fever	Ribavirin [21] Favipiravir [38]	Ribavirin [39]		

For more details, please consult chapters dealing with the particular disease as well as national and international guidelines

[a]In the context of the 2013–2016 Western African EVD outbreak, research activities were mainly focused on Ebola Virus (EBOV) and did not target other ebolaviruses (SUDV, TADF, BDBV) or marburgviruses (MARV, RAVV). For future epidemics caused by filoviruses other than EBOV, there are thus still significant research gaps concerning specific treatment and vaccination [40]

9.6 Dead Body Management

Handling dead bodies is a high-risk task. Staff managing dead bodies (e.g., morticians) are rarely trained at the same level as medical practitioners, although they may be exposed to the same risks. As for EVD, the highest viral loads can be found in dead bodies [45]. Having well-trained staff and PPE available is crucial. All procedures that involve any contact with the dead body and/or body fluids should be performed while wearing PPE. All avoidable procedures (e.g., embalming) should be avoided. The body should be placed in a body bag. Cultural sensitivities should be respected when handling a dead body as the evacuation of a dead body can be traumatizing for the population. For example, during the Western African EVD outbreak, burial teams were attacked [46]. Exact procedures for decontamination of the dead body, body bag, and surroundings as well as exact burial procedures depend on the causative pathogen. For example, *Yersinia pestis*, though sensitive in aerosols, can survive in carcasses for up to 2 months at 35 °C [47]. ICRC guidelines for dead body management in disaster contexts were updated recently [48]. A specific WHO dead body management/burial guideline for EVD is available [49].

9.7 Personal Protective Equipment (PPE)

For further instructions on universal precautions, please refer to the Chap. 8 in this book. PPE should be safe, user-friendly, and appropriate. Choosing the ‘appropriate’ PPE is complex because of to the following elements:

- National and international PPE recommendations vary. For example, during the Western African EVD outbreak, differences between PPE standards of WHO, CDC, MSF, and other actors were the subject of debate [50].
- The resources in terms of materials and qualified staff are different from place to place.
- National work security regulations differ (e.g., restricting duration of shifts).
- In the event of a possible bioweapons attack, the causative agent might be unknown (unknown pathogen or, even more complex, generally unknown chemical/biological/radiological/nuclear CBRN-spectrum agent) or multi resistant (example: multi-resistant *Y. pestis*). Particle sizes for different pathogens differ significantly, compromising the choice of the appropriate respirator. For example, if a chemical agent cannot be excluded in a scenario of a generally unknown CBRN-spectrum agent, a self-contained breathing apparatus should be considered.

Types of PPE are

- *PPE with gown/overall*, FFP2/3 (or N95/N98) respirator, double gloving, goggles/face shield and other facultative items (e.g., apron, rubber boots): This PPE is inexpensive, widely available, and WHO standards have been set for clinical care for patients infected with filoviruses such as EBOV: http://www.who.int/csr/resources/publications/ebola/ppe-guideline/en/.

  The challenge is that this PPE is composed of singular items, each of which was usually tested according to norms (in the European Union: EN standards). In practice, due to the complexity of donning, doffing, and decontaminating PPE, errors which compromised biosafety have been documented [51]. Furthermore, particular agents for decontamination (e.g., peracetic acid) cannot be used due to lack of filtering of chemical agents by FFP2/3 (or N95) respirators. Finally, because of the lack of ventilation, exhaustion due to heat is frequent and the wearing time thus limited [52].
- *Powered Air Purifying Respirator (PAPR)* is frequently used in European HLIU and, as for the EU, defined by EN standard 12,941 [53]. Depending on the exact type, they are mostly easy to don and doff. As ventilated equipment, PAPR is comfortable to wear, allowing longer working times [54]. Decontamination can be performed by, for instance, taking a shower with a decontaminating agent,

provided that this agent is compatible with the filter chosen. However, PAPR is quite expensive and hence may not be available in resource-poor settings.

- *Self-Contained Breathing Apparatus* is frequently used by fire brigades in Western countries (e.g., for intervening in smoky locations or those contaminated by chemical agents). This apparatus is heavy, complex to use, allows only very limited time of operation while wearing it, and is, according to national work security regulations, only to be used by staff who underwent a medical examination for aptitude. The choice of a self-contained breathing apparatus should hence be reserved to settings in which the presence of an agent leaking through the filters of other PPE (e.g. PAPR) cannot be excluded [54].

The choice of PPE for different procedures in environmental and health care settings should be circumscribed as clearly as possible *before* an HCID-related event takes place.

Case example 1: Choice of PPE in HLIU in Germany

The choice of PPE in Germany is mainly prescribed by regulations (e.g., Ordinance on biological substances or technical regulations for biological substances—TRBA). Annex 1 of TRBA No. 250 describes the infection prevention and control measures that have to be undertaken in an HLIU. There, PAPR according to the norm EN 12941 is set forth as mandatory. Each one of the seven German HLIU has its individual PAPR. Integrated systems as well as a combination of an overall and a ventilated hood are used.

Case example 2: Choice of PPE in mass casualty management in Israel

The type of PPE to be used in Israel by medical first responders and hospital teams in each type of emergency scenario is determined by the Israeli Ministry of Health. The decision is made based on the lethality of each agent, availability or lack of a specific effective treatment, extent of morbidity, and level of infectivity. In case of agents that are not communicable from person to person (e.g., *Bacillus anthracis*), standard droplet protection is required consisting of surgical masks, protective glasses/facial protection, and a gown while performing actions that involve aerosol exposure (e.g. suction of secretions). During pandemics, aerosol protection is required especially if the lethality of the disease is high relative to seasonal influenza. This protection includes surgical masks (to be replaced at least every 4 h); gowns over clothes; gloves to prevent contact with mucous membranes, blood secretions, and fluids; and protective goggles only when performing activities that pose a risk of spraying onto mucous membranes. In cases when aerosolization is likely, N95 masks are required (e.g., cases of airborne infection). Only very severely ill patients are treated by staff donning the strictest PPE against aerosol agents, including negative-pressure protective systems and ventilated hoods.

9.8 Decontamination and Disinfection

This chapter deals with decontamination and disinfection in the context of clinical management. For general advice on decontaminating capabilities and facilities, please refer to Chap. 6 in this book. The decontaminating agent should be chosen depending on the pathogen and site (PPE, surface, medical equipment, skin/mucous membranes or eyes, environmental decontamination). International and national recommendations concerning decontaminating agents should be taken into account, considering the pathogen (bacteria, viruses, or fungi). The 'worst-case pathogen' is *Bacillus anthracis* due to its extreme stability in most environments [55].

Concerning PPE decontamination, detailed documentation is available concerning its application and decontamination at a CBRN site ('hot zone') [56]. On the other hand, the decision of how to decontaminate PPE after having performed clinical management of HCID patients is complex: Concerning EVD, CDC currently recommends disinfection of gloved hands as well as of visibly soiled parts of the PPE. PAPR should be cleaned and disinfected in accordance with the manufacturer's recommendations [57]. As for high-level isolation units in Germany, PAPR is systematically decontaminated after use. For example, when an EVD case was cared for in Hamburg/Germany in 2014, the PPE was decontaminated after use by two shower cycles with 2% perchloric acid for 2 min, a residence time of 7 min, and rinsing off of acid residues by showering with water [17].

Surface decontamination can be complex if the surface is porous and/or if medical equipment is involved. Krauter et al. [58] discuss general decontamination strategies following a bio-contamination event.

The *decontamination of medical devices and the needed measures* are, in detail, discussed by WHO [59] and RKI [60]; differentiating between:

- non-critical equipment which neither invades the skin nor comes in contact with mucous membranes (e.g., ECG electrodes)
- semi-critical equipment which comes in contact with mucous membranes (e.g., speculum, gastroscope), and
- critical equipment which is invasive (e.g., retractors, trocars).

The relevant guidelines should be consulted to decontaminate medical devices adequately.

Decontamination of skin, mucous membranes, or eyes applies mainly in the context of immediate contamination with a bioweapons, such as *B. anthracis* spores. Please refer to Chap. 6 in this book as well as Krauter et al. [58].

Disinfection of skin, mucous membranes, or eyes is subject to universal precautions (see Chap. 8 in this book). A particular challenge can be the disinfection of mucous membranes and eyes due to their vulnerability. For disinfection of mucous membranes, the Framework Ebola Virus Disease document in Germany recommends octenidine dihydrochloride/phenoxyethanol or chlorhexidine-containing drugs or povidone–iodine complexes (7.5%) (e.g. Octenisept®, Skinsept mucosa®, Braunol®). For application on the eyes, 5% povidone–iodine complex can be suitable [61]. A particular scenario might be an accidental exposure to bodily fluids, for instance by a splash of infective bodily fluids on unprotected skin or into the eyes, or by a needle stick injury. Medical facilities should have clear SOPs for such incidents.

For environmental decontamination, please refer to Chap. 6 in this book. This topic is furthermore discussed in detail by Franco and Bouri [55].

9.9 Practical Case Examples: Lassa Fever and Ebola Virus Disease in Germany

1. *Ebola Virus Disease in Germany in 2014*

Health care worker (HCW) infections were an important and very sad effect of the Western African EVD outbreak. As of 31 March 2015, 815 HCW had contracted EVD. Out of 635 health care workers for whom the outcome was available, 418 (65.9%) died [62]. In total, three medically evacuated HCW were treated in HLIU in Germany in 2014, two of whom survived. The patients were administered experimental treatments and likely benefited from intensive care. Infection prevention and control was thoroughly maintained. Staff caring for the patients wore PAPR as PPE. Neither secondary infections nor imported cases occurred in Germany [17, 18]. In spite of this fortunate epidemiologic 'outcome', there was an emotional debate in Germany driven by fear [63]. Public health authorities made a tremendous effort to enhance emergency preparedness towards the disease, and an elaborate document 'Framework Ebola Virus Disease' was released led by Robert Koch Institute, Germany RKI [64]. A sophisticated medical evacuation airplane with resources for performing comprehensive intensive care measures aboard was designed, launched in November 2014 and, because of the end of the humanitarian crisis related to EVD, dismantled in [65, 66].

2. *Lassa Fever in Germany 2015*

In February 2016, a 46-year-old HCW who worked in Togo fell sick with developing fever, malaise, and a sore throat. He was admitted to a local hospital where treatment for suspected malaria was started immediately. Because of clinical deterioration he was medically evacuated from Togo to Germany nearly 2 weeks later. He was admitted to the University Hospital of Cologne, Germany. After arriving at this hospital, the clinical condition of the patient deteriorated dramatically, resulting in the patient's death shortly after hospital admission. The post-mortem examination did not reveal an underlying infectious disease. Simultaneously, after histologic examination of liver specimens of the index case a viral hemorrhagic fever was suspected. Blood and liver samples were sent to the Bernhard Nocht Institute for Tropical Medicine in Hamburg, Germany, for further analysis. The institute confirmed LASV infection by PCR. The body was hence cremated [67].

A mortician who handled the primary case's corpse, and who was reported to have worn gloves and did not recall being exposed to body fluids, contracted the disease. He had overlapping clinical signs of an upper respiratory tract infection at the time of contact. Eight days after exposure, a first PCR for LASV was negative. A subsequent PCR was positive and the patient was admitted to the isolation unit in Frankfurt. He received experimental treatment with ribavirin and favipiravir, clinically improved with resolving fever from the 7th day of disease onwards, and survived [4, 68].

This was the first recorded transmission of LASV infection outside Africa. The case underlines the need for rapid and robust diagnostics of HCIDs and for maintaining proper infection prevention and control measures by using PPE whenever HCIDs cannot be ruled out. The incident also demonstrated that larger outbreaks can be prevented successfully when the Health Protection Authorities (HPA) react quickly and efficiently and install effective infection control measures. Difficulties were related to the unreliableness of contact persons who did not adhere to communicated control measures, thereby provoking an increase in the number of contacts. Also, one explicit lesson learned was that information between different authorities involved should be shared rapidly, facilitated by a central coordination [69].

9.10 Role of the Public Health Sector

The interface between HCID expertise in clinical management and public health sectors should be well established. For example, for referring a patient, a health care facility might need an isolation transport ambulance—and hence the assistance of the public health service for coordination. For contact-tracing activities, the public health service requires information that is gathered by the health care facilities from affected people. HPA are responsible for responding to any problem concerning the infected patient outside of health care facilities, as well as for limiting secondary transmissions. Despite an increasing knowledge about HCID, major challenges still remain for the physicians and HPA who treat these patients.

In Germany, the Permanent Working Group of Competence and Treatment Centres for highly infectious, life-threatening diseases (STAKOB) was established in 2014. It regroups HCID experts from all seven HLIU—mainly infectious disease specialists—and from seven corresponding HPA [70]. The STAKOB members HLIU and HPA are located dispersedly on the state level to assure proximity. Additionally, specialists from the Federal Institute for Vaccines and Biomedicines, the medical services from the Federal Foreign Office, the medical service of the German Federal Armed Forces, and others participate in the activities. The STAKOB office is affiliated to the Robert Koch Institute (RKI). Regular meetings are held twice a year, and conference calls are made supplementarily and *ad hoc* in the event of an occurring crisis. In addition to professional exchange and emergency preparedness activities, STAKOB was involved in the provision of medical care to the three EVD cases evacuated to Germany in 2014 and in the response to Lassa Fever that occurred in Germany in 2015 (see case descriptions above).

9.11 Particularities in HCID Mass Casualty Management: The Israeli Example

9.11.1 Surge Capacity

An HCID outbreak might rapidly develop andrequire management of mass casualties that may present to the hospitals. The influx of patients to the emergency departments (EDs) necessitates preparedness to expand surge capacities of the varied hospital facilities including the ED, intensive care units, inpatient departments, and laboratory capacities [71, 72]. Nevertheless, compared to natural disasters or other types of emergency situations, an HCID outbreak presents different challenges, as the patients do not necessarily present to the emergency departments simultaneously and for a short period of time but rather do so over a prolonged time span [65]. The patients may present a high risk for transmitting the infection and thus pose a risk for spreading the disease to the personnel or other patients that are hospitalized. Therefore, those managing the event need to take into account unique elements of surge capacity such as designated infrastructure and equipment (i.e., isolation facilities), communication measures, and means for infection control [65, 73].

It is recommended that establishment of a designated "biological ED" be considered, separate from the routine EDs, to facilitate safety measures and prevent potential infection of staff and other patients that present to the hospital concurrently. Regardless of the decision to operate a designated "biological ED" or admit patients in the routine ED, expansion of surge capacity of place, resources, infrastructure, and medical personnel should be instated to enable management of the numerous patients that are expected [72, 74].

The overload of inpatients may necessitate application of modifications in the hospitalization policy, compared to routine management, to optimize resource management [75, 76]. For example, the proportion of patients who require ventilation may rise to a level of 3–4 times the average number during routine periods, rendering the Intensive Care Units (ICUs) incapable of treating all the patients. Expansion of ventilation capacities is needed, based on designation of sites that consist of needed infrastructure for ventilation, such as in ambulatory surgical departments or step-down units.

Elective and ambulatory activities may be reduced and, in extreme situations, may be ceased altogether. General hospitalization guidelines may be modified, such as restricting hospitalization periods after childbirth or surgeries, and may adopt altered standards of care [76]. The resources that will be made available resulting from the modified operations can then be redirected to the HCID mass casualty management.

To overcome the shortage of hospital beds, additional facilities within the hospital may need to be involved as admitting or hospitalization sites. Hospital temporary wards may thus be installed in designated sites such as dining rooms, ambulatory clinics, and libraries or other areas that are sufficiently spacious to be converted for this purpose. Regardless, the scope of inpatients may exceed the overall capacities of the hospitals, necessitating operation of auxiliary facilities such as geriatric hospitals, nursing homes, community clinics, or even hotels, which may be temporarily used as extension hospital sites [77, 78].

9.11.2 Triage

Triage of patients is a crucial component of managing an HCID mass casualty event to prevent overload and facilitate an effective management of resources [79]. The main objective of triage processes is to facilitate saving as many salvageable patients as possible by identifying those that will survive as a result of the administered medical attention who would otherwise not live [79, 80]. Triage during HCID events focuses on sorting those severely and critically ill patients who require immediate medical intervention from other casualties whose treatment may be delayed or administered in alternative facilities [81]. Another objective of triage in HCID situations is to use the available resources (i.e., manpower, equipment, or infrastructure) optimally to mitigate chaos and restore order, so as to resume routine operations and levels of medical care in the shortest time possible [82].

The triage process may represent a significant challenge to health care personnel as it is conducted under potentially stressful circumstances when many patients present to the hospital simultaneously. As the staff are required to work under biosafety precautions and regulations, deviations from the routine mode of operation pose a complex challenge. Modifications to the routine mode of patient admittance are instated, and the staff operate while wearing personal protective gear that may be perceived as uncomfortable or daunting [83, 84].

A triage site must be defined to direct the patients to the appropriate ED. A senior internist or pediatrician (according to the designation of the site) should be appointed as Triage Medical Officer, responsible for managing the triage operations and directing each patient to the appropriate treatment site. Maintaining calm, adhering to principles of emergency response, and leading the initial sorting of the patients is a vital component of an effective emergency management of an HCID event [85].

Patients who are referred to the hospital by official agencies, such as the emergency medical services (EMS) or their primary physicians, should be directed immediately to the "biological ED"; patients who present to the hospital independently should be referred to the Biological Site according to the decision of the Triage Medical Officer.

Initial treatment in the ED is targeted to provide urgent (life-saving) treatment and to decide whether the patient requires hospitalization or may be discharged to be cared for in the community [83]. The triage must be administered strictly; patients may be referred from the ED to hospitalization only following the decision of a senior physician. Therefore, hospitals need to ensure the ongoing presence of senior staff in the ED around the clock as long as the situation persists [71].

9.11.3 Deployment of a Hospital Incident Command Center

As the significant influx of patients may overwhelm the hospital's capacities and to achieve optimal management of resources, an Incident Command Center needs to be established in an HCID epidemic [72]. The Incident Command Center is responsible for providing support to the medical teams to enable them to administer optimal medical care, despite the shortage in resources [71]. The main tasks of this Center are to update and direct the hospital's staff concerning the mode of operation, both clinically and organizationally; maintain ongoing contact with internal and external bodies; manage appropriate allocation of resources including manpower and equipment; monitor admission of patients; manage allocation of protective measures for the personnel and inpatients; and coordinate reinforcements of imaging and laboratory capacities.

9.11.4 Isolation Facilities in the Mass Casualty Management Context

Technical requirements to ensure capacity to treat patients with HCID have been defined and include components such as negative-pressure airborne infection isolation facilities and aerosol-tight doors and windows [86, 87]. Although many countries have installed such measures that allow them safely to admit and treat patients infected with biological agents, only few of them have facilities that allow for a massive influx of such patients. It is therefore expected that the surge of patients presenting to the hospitals will significantly surpass the available isolation capacities [6, 88]. Considering the high cost of installing isolation facilities that comply with the strict requirements of suitability to manage HCID events, it stands to reason that the current infrastructure will not change dramatically. Accordingly, a modified mode of operation is needed to facilitate effectively the management of a mass casualty HCID event while maintaining the safety of the staff and other inpatients of the hospital, such as designating areas in the hospital that will service exclusively HCID patients.

Miller [87] have reported that it is possible to design a temporarily designated negative-pressure isolation ward and operate it in a separate pre-determined location within the hospital. In their study, though, this site was activated over a period of only 24 h, and they have expressed some doubts as to whether such a facility may also be functional during a longer span of time.

9.11.5 Reinforcement of Resources

9.11.5.1 Staff

Reinforcement of hospital personnel should be recruited from two main sources: internal and external. Internal reinforcement should be based on a freeze of vacations, training and education programs, and participation in congresses or other activities external to the medical facility; modification/expansion of working shifts (following agreement with the staff and/or unions); and re-direction of personnel who were formerly assigned to elective or ambulatory services to the newly established wards [89]. External reinforcement may include medical or nursing students as well as recruitment of retired personnel who are still proficient in their basic profession. The hospital should identify areas/activities that may be performed by volunteers from the community and identify such potential volunteers who are available in the vicinity of the hospital.

Those managing personnel must consider not only the number of staff needed at any point in time, over a prolonged period of days or weeks, but must also plan to cope with the fears and anxiety that often accompany outbreaks of HCID [65]. The severity of the disease, lack of effective treatments, and the potentially high transmissibility may create uncertainty among the staff concerning their safety and even distrust of the system at large and specifically the hospital administration [90].

9.11.5.2 Equipment

As the routine medical equipment and devices may be insufficient in mass casualty events, each hospital needs to plan ways of reinforcing vital means, including ventilation machines and PPE [74]. As much as possible, electronic ventilators should be used, but because of the extreme numbers of patients which may require ventilation, manual ventilators may need to be used. According to the type of pathogen, appropriate filters will need to be installed in the ventilators. Plans for procurement of relevant medications should also be developed according to the estimated number of patients expected to present to the hospital. Nonetheless, under austere conditions, some procedures may not be provided considering the scarcity of resources; for example, it may be decided that extracorporeal membrane oxygenation will not be provided during an HCID mass casualty event, despite its effective use during the 2009 H1N1 influenza pandemic [71].

Decision making concerning types and amounts of equipment that should be purchased to maintain the capacity to manage an HCID needs to consider the risk, surge capacity, and costs of stockpiling and managing the equipment [74]. A significant challenge is balancing between the need to maintain continuous preparedness while minimizing waste and the need to destroy (and consequently substitute) drugs or other materials that have expired [71, 74].

9.11.5.3 Training

Training and education are crucial for ensuring the ability of medical professionals to manage HCID mass casualty events [91]. The knowledge required of hospital personnel during an HCID epidemic encompasses both familiarity with the approach to a single/individual patient and his/her contacts and the overall organizational preparedness for mass casualties, increase of encounters of medical/nursing staff with patients, and the need to provide medical care in ambulatory and modified hospital conditions [92, 93]. Medical as well as organizational aptitude is required [77, 93]. Three main levels have been recognized as significant for inclusion in training programs for HCID events: (1) information concerning the varied biological agents, their consequences, the courses of the disease, and means to prevent and/or mitigate its infectivity; (2) ways to develop case definitions, criteria for hospitalization, medical protocols for in-house and ambulatory care, and provision of care for patients' contacts (including staff and family members); (3) policy and decision-making processes [91, 94]. All hospital staff should be trained in the basic level of required knowledge, medical personnel need also to be trained in the second level, while managers of the EDs and the various hospital departments need to be trained in all three levels of knowledge and competencies.

HCIDs present a significant challenge to HCW because of potentially extreme working conditions, close encounters with infectious patients which may involve a substantial exposure to a lethal disease, preventive care that may be only partly effective, and high psychological pressure resulting from a dual commitment to the family and the workplace. Hospitals' training programs should accordingly focus not only on knowledge and competencies of the staff but also relate to the personnel's expected perceptions and attitudes [95, 96].

9.11.6 Crisis Communication

Risk communication is a vital component of managing a mass casualty HCID event, targeted to reduce uncertainty, confusion, and fear [87, 97].

Reporting of hospital staff to work during HCID may be challenging, especially when they perceive a threat to their own or their families' well-being. Based on lessons identified during former HCID events, such as SARS, influenza, or EVD outbreaks, elements which impact on staff's willingness to report to work include perceived cohesiveness and sense of belonging to their team/place of work, and a feeling of a "mission" and commitment to patients and co-workers. Hospital administrators should communicate to the staff the criticality of their tasks, the perception of their contribution as vital for saving lives and for the continuous functioning of the medical facility, and its significant impact on medical consequences [98]. Risk communication should also include reference to actions that were taken to preserve their health and safety, such as preventive measures (physical protection, isolation facilities, infection control means, or preventive medications). Communicating messages during an HCID event must be based on relaying accurate and trustworthy information, conveyed calmly to avoid the creation of confusion and panic, and respectfully regarding the diversity prevailing among the population, including persons with special needs [65, 99]. Both the medical teams and the public need to be updated continually concerning the HCID event, its development, and potential consequences, as a dynamic process, transparently and credibly [98, 100].

The risk communication messages should be developed and published differentially according to the target population: the patients, medical personnel, family members of the staff or the patients, managers of interface agencies that operate in the vicinity of the hospital, decision and policy-makers, or the public at large [101]. Proactive risk communication should be adopted to achieve wide coverage. Clear, easily understandable information should be distributed concerning questions such as who should seek medical care, who needs to be vaccinated, who should approach the hospital or alternative facilities, or what means should be used for evacuation to hospitals (e.g., ambulances or private cars) [102]. Use of a tele-information center is recommended to provide the concerned public with answers to their worries. Leaflets delineating answers to most frequently asked questions should be prepared, translated into languages of the major fractions of the local population. This information should be uploaded to the hospitals' websites to be readily available to all interested parties. Proficient spokespersons should be available to relay information and updated data to local media and, through them, to the population. The spokespersons and the administrators of the hospitals should collaborate closely with other entities and authorities that operate in the community to coordinate the messaging and ensure a synchronized risk communication policy [87, 102].

9.11.7 Reporting (Internal and External)

Effective management of an HCID event is dependent on early identification of the first patients before the outbreak reaches uncontainable magnitudes; this can only be achieved by sharing information concerning patients or suspected contacts [103, 104]. Once the outbreak has created mass casualties, the need for early detection and identification evolves into a vital need for sharing information concerning the patients, their differential diagnoses, and efficiency of treatments that were administered, in an attempt to contain the event and/or mitigate its effects [103, 105].

Patient charts should also be maintained during an HCID, similar to routine procedures. Nonetheless, additional reporting is needed to manage the situation including Epidemiological Investigation Forms that delineate the patients' details, areas where they stayed during the event, tentative diagnosis, and treatment that was administered. The hospital needs to record and report the number of patients admitted to the EDs, the number of suspected patients infected with biological agents, the number of patients hospitalized in ICUs (adult and pediatric), and the number of deaths from HCID. Aggregation and analysis of unbiased data will facilitate their interpretation and subsequently their utilization for decision making concerning ways to manage the situation [105]. The analyzed data should be further disseminated and shared transparently to promote scientific collaboration and effective systemic management [103].

9.11.8 Risk Assessment and Risk Reduction

Outbreaks of HCID which cause numerous casualties constitute a continuous global risk; as the consequences of such events may be devastating to humans and infrastructures, risk assessment and risk reduction must be intrinsic components of a holistic risk management program [106, 107]. Managing mass casualty HCID events necessitates a robust organizational and preparedness program, including vigorous risk assessment and decision-making processes [108]. Risk assessment is vital for both planning and response phases to facilitate decision making and implementation of effective interventions [98, 109]. As the health caregivers may be at risk of being infected, either primarily by the HCID agent or from secondary transmission of the pathogen by the patients they treat, each hospital must have the capacity to identify the risk, assess its potential damage, and implement measures to reduce the risk and ensure the safety of its manpower [99].

Risk reduction in HCID events is dependent on hospital personnel's strict adherence to infection control procedures [65]. If such measures are not implemented as needed, the hospitals themselves may contribute to spreading the pathogen rather than containing or controlling its evolvement. Previous studies have shown that during the SARS outbreak in both Canada and Taiwan, over 75% of the cases could be traced back to virus transmission within hospitals [110].

9.11.9 Ethical Dilemmas

The goal of caring for patients is founded on the ethical objective to provide the optimal protection of lives, health, and welfare of the population in the most effective and successful manner. Ethical management considers presence of factors that may not be known or accurately estimated prior to the event [111], such as the biological features of the agent (lethality, epidemiology, infectivity, and sensitivity to antiviral treatments); characteristics of the population (density, scope of risk, and age distribution); and behavioral patterns of the population. Major ethical aspects which need to be considered include allocation of medical resources under austerity, limitation of civil rights, enforcing professional ethics of medical teams during an HCID event, competing commitments of health care providers to their place of work versus responsibility to their families and themselves, and integrating ethical as well as religious aspects in decision-making processes [111, 112]. Bio-ethical professionals should be involved in the decision-making processes.

References

1. Stone. The "worried well" response to CBRN events: analysis and solutions. Defense technical information center. 2007.
2. CDC. Lassa fever confirmed in death of U.S. Traveler returning from Liberia [Online]. 2015. https://www.cdc.gov/vhf/lassa/announcements/death-of-us-traveler-returning-from-liberia.html. Accessed 18 July 2017.
3. Maisa. Federal state level management of an imported Lassa fever case to North Rhine-Westphalia, Germany, March 2016. Int J Infect Dis. 2017.
4. Raabe. Favipiravir and ribavirin treatment of epidemiologically linked cases of Lassa fever. Clin Infect Dis. 2017;65:855–9.
5. Negredo. Autochtonous crimean-congo hemorrhagic fever in Spain. N Engl J Med. 2017;377:154–61.
6. Schilling S. Isolation facilities for highly infectious diseases in Europe – a cross-sectional analysis in 16 countries. PLoS One. 2014;9(10):e100401.

7. McLean. The politics of fear. 2017.
8. Sagripanti. Persistence in darkness of virulent alphaviruses, Ebola virus, and Lassa virus deposited on solid surfaces. Arch Virol. 2015;155:2035–9.
9. RKI. Rahmenkonzept Ebolafieber [Online]. 2014. http://www.rki.de/DE/Content/InfAZ/E/Ebola/Rahmenkonzept_Ebolafieber.html. Accessed 18 July 2017.
10. WHO. Clinical management of patients with viral haemorrhagic fever. A pocket guide for the front-line health worker. 2016.
11. Kratz. Ebola – the politics of fear. 2017.
12. BBK. Handbuch biologische Gefahrenlagen.
13. Leligdowicz. Ebola virus disease and critical illness. Crit Care. 2016;20(1):217.
14. Garske. Heterogeneities in the case fatality ratio in the West African Ebola outbreak 2013–2016. 2017.
15. Jeffs. The Médecins Sans Frontières Intervention in the Marburg Hemorrhagic Fever Epidemic, Uige, Angola, 2005. I. Lessons Learned in the Hospital. 2007.
16. Martini. Über eine bisher unbekannte, von Affen eingeschleppte Infektionskrankheit: Marburg-Virus-Krankheit. DMW. 1968.
17. Kreuels. A case of severe Ebola virus infection complicated by gram-negative septicemia. N Engl J Med. 2014;371:2394–401.
18. Wolf. Severe Ebola virus disease with vascular leakage and multiorgan failure: treatment of a patient in intensive care. Lancet. 2014;385:1428–35.
19. Guery. Clinical features and viral diagnosis of two cases of infection with Middle East Respiratory Syndrome coronavirus: a report of nosocomial transmission. Lancet. 2013.
20. Linden. ECMO in ARDS: a long-term follow-up study regarding pulmonary morphology and function and health-related quality of life. Acta Anesthesiol Scand. 2009;53(4):489–95.
21. McCormick. Lassa fever. Effective therapy with ribavirin. NEJM. 1985;314(1):20–6.
22. WHO. Neglected tropical diseases [Online]. 2017. http://www.who.int/neglected_diseases/diseases/en/. Accessed 19 July 2017.
23. Poland. Treatment of plague. 1999.
24. CDC. Plague: resources for clinicians [Online]. 2015. https://www.cdc.gov/plague/healthcare/clinicians.html.
25. Titball. Vaccination against bubonic and pneumonic plague. Vaccine. 2001;19:4175–84.
26. Olson. In vitro efficacy of brincidofovir against variola virus. 2014.
27. Durrafour. Tecovirimat, a p37 envelope protein inhibitor for the treatment of smallpox infection. IDrugs. 2010;13:181–91.
28. Keckler. The effects of post-exposure smallpox vaccination on clinical disease presentation: addressing the data gaps between historical epidemiology and modern surrogate model data. Vaccine. 2013;31(45):5192–201.
29. WHO. Smallpox vaccines [Online]. 2017. http://www.who.int/csr/disease/smallpox/vaccines/en/. Accessed 19 July 2017.
30. Tignor. Ribavirin efficacy in an in vivo model of Crimean-Congo hemorrhagic fever virus (CCHF) infection. Antivir Res. 1993;22(4):309–25.
31. Oesterreich. Evaluation of Antiviral Efficacy of Ribavirin, Arbidol, and T-705 (Favipiravir) in a Mouse Model for Crimean-Congo Hemorrhagic Fever. PLoS Negl Trop Dis. 2014;8:e2804.
32. WHO. Crimean-Congo haemorrhagic fever (CCHF) [Online]. 2017. http://www.who.int/csr/disease/crimean_congoHF/en/. Accessed 19 July 2017.
33. PREVAIL. A randomized, controlled trial of ZMapp for Ebola virus infection. N Engl J Med. 2016;375:1448–56.
34. Sissoko. Experimental treatment with favipiravir for Ebola virus disease (the JIKI Trial): a historically controlled, single-arm proof-of-concept trial in Guinea. PLoS Med. 2016;13:e1001967.
35. Henao-Restepro. Efficacy and effectiveness of an rVSV-vectored vaccine expressing Ebola surface glycoprotein: interim results from the Guinea ring vaccination cluster-randomised trial. Lancet, 2015. https://doi.org/10.1016/S0140-6736(13)60982-4.

36. Santis. Safety and immunogenicity of a chimpanzee adenovirus-vectored Ebola vaccine in healthy adults: a randomised, double-blind, placebo-controlled, dose-finding, phase 1/2a study. Lancet Infect Dis. 2016;16(3):311–20.
37. WHO. Ebola publications: case management, infection prevention and control [Online]. 2017. http://www.who.int/csr/resources/publications/ebola/infection-prevention/en/. Accessed 19 July 2017.
38. Mendenhall. Effective oral favipiravir (T-705) therapy initiated after the onset of clinical disease in a model of arenavirus hemorrhagic fever. 2011.
39. Bausch. Review of the literature and proposed guidelines for the use of oral ribavirin as post-exposure prophylaxis for Lassa fever. Clin Infect Dis. 2010;51:1435–51.
40. Burk. Neglected filoviruses. FEMS Microbiol Rev. 2016;40:494–519.
41. Klafki. Debatte Schutz vor Pandemien – Tödliche Kleinstaaterei. TAZ. 2017.
42. Bedell. Global access to medicinal products: compassionate use programs. 2010.
43. BFARM. Compassionate use programmes [Online]. 2017. http://www.bfarm.de/EN/Drugs/licensing/clinicalTrials/compUse/_node.html. Accessed 19 July 2017.
44. FDA. Expanded Access (Compassionate Use) [Online]. 2017. https://www.fda.gov/NewsEvents/PublicHealthFocus/ExpandedAccessCompassionateUse/default.htm. Accessed 19 July 2017.
45. Vetter. Ebola virus shedding and transmission: review of current evidence. J Infect Dis. 2016;214:S177–84.
46. Wilson. Sierra Leone Ebola burial team attacked despite lockdown. Daily Telegraph. 2014.
47. Rakin. Yersinis pestis – Eine Bedrohung für die Menschheit. Bundesgesundheitsblatt. 2003.
48. ICRC. 2017.
49. WHO. How to conduct safe and dignified burial of a patient who has died from suspected or confirmed Ebola virus disease [Online]. 2014. http://www.who.int/csr/resources/publications/ebola/safe-burial-protocol/en/. Accessed 24 July 2017.
50. Macintyre 2015. Uncertainty, risk analysis and change for Ebolapersonal protective equipment guidelines. International Journal of Nursing Studies, 52(5), 899–903.
51. Kang. Use of personal protective equipment among health care personnel: Results of clinical observations and simulations. Am J Infect Control. 2016;45(1):17–23.
52. Coca. Physiological and subjective evaluation of PPE using a sweating thermal manikin. Extrem Physiol Med. 2015;4(Suppl 1):A27.
53. de Iaco G, Puro V, Fusco FM, Schilling S, Maltezou HC, Brouqui P, Gottschalk R, Bannister B, Brodt HR, Siikamaki H, Perronne C, Brantsaeter AB, Fjellet AL, Ippolito G. Personal protective equipment management and policies: European network for highly infectious diseases data from 48 isolation facilities in 16 European countries. Infect Control Hosp Epidemiol. 2012;33:1008–16.
54. REMM. PPE classification system from OSHA and EPA [Online]. 2017. https://www.remm.nlm.gov/osha_epa_ppe.htm. Accessed 19 July 2017.
55. Franco. Environmental decontamination following a large-scale bioterrorism attack: federal progress and remaining gaps. Biosecur Bioterror. 2010;8(2):107–17.
56. Bodurtha. Decontamination science and personal protective equipment (PPE) selection for chemical-biological-radiological-nuclear (CBRN) events. 2016.
57. CDC. Guidance on personal protective equipment (PPE) to be used by healthcare workers during management of patients with confirmed Ebola [Online]. 2015. https://www.cdc.gov/vhf/ebola/healthcare-us/ppe/guidance.html. Accessed 21 July 2017.
58. Krauter. A systematic methodology for selecting decontamination strategies following a biocontamination event. Biosecur Bioterror. 2011;9(3):262–70.
59. WHO. Decontamination and reprocessing of medical devices for healthcare facilities. 2016.
60. RKI. Hygiene requirements for the reprocessing of medical devices [Online]. 2001. https://www.rki.de/DE/Content/Infekt/Krankenhaushygiene/Kommission/Downloads/Hygiene_Requirements_Medical_Devices_2012.pdf?__blob=publicationFile. Accessed 21 July 2017.

61. RKI. Measures regarding disinfection and waste management as pertaining to an at least probable case of Ebola virus infection in Germany [Online]. 2015. http://www.rki.de/EN/Content/infections/epidemiology/outbreaks/Ebola_virus_disease/Ebola_disinfection.pdf?__blob=publicationFile. Accessed 21 July 2017.
62. WHO. Health care worker infections in Guinea, Liberia and Sierra Leone. A preliminary report. [Online]. 2015. http://www.who.int/csr/resources/publications/ebola/health-worker-infections/en/. Accessed 9 Apr 2018.
63. Sorge. Panik, Schlagzeilen und Leichenbeschau. Cicero. 2014.
64. Schmiedel. Ebolafieber in Westafrika und in Deutschland. Bundesgesundheitsblatt. 2015.
65. May. Catastrophic events versus infectious disease outbreak: distinct challenges for emergency planning. Reason Pap. 2015;37:54–64.
66. Lufthansa. Lufthansa hands over "Robert Koch" evacuation aircraft [Online]. 2014. https://www.lufthansa-technik.com/press-releases-content/-/asset_publisher/9Mf5/content/id/1563897. Accessed 24 July 2017.
67. Kochanek LF. Transmission and control of Lassa fever outside of Africa Accepted article – to be published. 2017.
68. WHO. Lassa fever – Germany [Online]. 2016. http://www.who.int/csr/don/23-march-2016-lassa-fever-germany/en/. Accessed 21 July 2017.
69. Maisa. Federal state level management of an imported Lassa Fever case to North Rhine-Westphalia, 2016. 2016.
70. Gottschalk. Seltene hochkontagiöse und lebensbedrohliche Erkrankungen. Bundesgesundheitsblatt. 2015;58:655–6.
71. Hick. Allocating scarce resources in disasters: emergency department principles. Ann Emerg Med. 2012;59:177–87.
72. Waseem. Impact of operational staging to improve patient throughput in an inner-city emergency department during the novel H1N1 influenza surge: a descriptive study. Pediatr Emerg Care. 2012;28:39–42.
73. Therrien. Bridging complexity theory and resilience to develop surge capacity in health systems. J Health Organ Manag. 2017;31:96–109.
74. Rebmann. Best practices for healthcare facility and regional stockpile maintenance and sustainment: a literature review. Health Secur. 2017;15(4):409–17.
75. Bar-El. Moral dilemmas faced by hospitals in time of war: the Rambam Medical Center during the Second Lebanon War. Med Health Care Philos. 2014;17:155–60.
76. Timbie. Systematic review of strategies to manage and allocate scarce resources during mass casualty events. Ann Emerg Med. 2013;61:677–89.
77. Alfred. Preparing for disasters: education and management strategies explored. Nurse Educ Pract. 2015;15:82–9.
78. Bruria. Case study–Israel. Dordrecht: Springer; 2013.
79. Culley. A review of the literature on the validity of mass casualty triage systems with a focus on chemical exposures. Am J Disaster Med. 2014;9:137.
80. Albanese. Clinical guidelines for responding to chemical, biological, radiological, nuclear and trauma/burn mass casualty incidents: quick reference guides for emergency department staff. J Bus Contin Emer Plan. 2014;8:122–33.
81. DOH U. Chapter 2: Health and medical care delivery in a mass casualty event. 2011.
82. Curnow. Mass casualty triage performance assessment tool. No. ARI/FB-RP-2015-02. Army Research Institute for the Behavioral and Social Sciences Fort Belvoir, VA, 2015. 2014.
83. Barilan. Triage in disaster medicine: ethical strategies in various scenarios. Dordrecht: Springer; 2014. p. 49–63.
84. Landsdowne. Recent advances in medical device triage technologies for chemical, biological, radiological, and nuclear events. Prehosp Disaster Med. 2015;30(3):320–3.
85. Russo. Mass casualty disasters: who should run the show? J Emerg Med. 2015;48:685–92.
86. Bannister B. Framework for the design and operation of high level isolation units: consensus of the European network of infectious diseases. Lancet Inf Dis. 2009;9(1):45–56.

87. Miller. Implementing a negative-pressure isolation ward for a surge in airborne infectious patients. Am J Infect Control. 2017;45:652–9.
88. Rubinson. Augmentation of hospital critical care capacity after bioterrorist attacks or epidemics: recommendations of the working group on emergency mass critical care. Crit Care Med. 2005;33:2393.
89. Kearns. Hospital bioterrorism planning and burn surge. Biosecurity Bioterrorism Biodefense Strateg Pract Sci. 2014;12:20–8.
90. Gonsalves. Panic, paranoia, and public health—the AIDS epidemic's lessons for Ebola. N Engl J Med. 2014;371:2348–9.
91. Djalali. TIER competency-based training course for the first receivers of CBRN casualties: a European perspective. Eur J Emerg Med. 2016;24:371–6.
92. Lewis. Mass casualty incident management preparedness: a survey of the American College of Surgeons Committee on Trauma. Am Surgeion. 2016;82:1227–31.
93. Siegel. Prioritization of pediatric chemical, biological, radiologic, nuclear, and explosive disaster preparedness education and training needs. Clin Pediatr Emerg Med. 2014;48:685–92.
94. Pesiridis. Development, implementation and evaluation of a disaster training programme for nurses: a switching replications randomized controlled trial. Nurse Educ Pract. 2015;15:63–7.
95. Kim JH. Learning from UK disaster exercises: policy implications for effective emergency preparedness. Disasters. 2014;38:846–57.
96. Quevillon. Helping the helpers: assisting staff and volunteer workers before, during, and after disaster relief operations. J Clin Psychol. 2016;72(12):1348–63.
97. Barry. A literature review on effective risk communication for the prevention and control of communicable diseases in Europe. J Health Commun. 2013:1–19.
98. Kern. Global epidemics, pandemics, terrorism: risk assessment and European responses. ISPSW Strateg Ser. 2016.
99. Penta. Of earthquakes and epidemics: examining the applicability of the all-hazards approach in public health emergencies. Risk Hazards Crisis Public Policy. 2017;8:48–67.
100. Barrelet. Unresolved issues in risk communication research: the case of the H1N1 pandemic (2009–2011). Influenza Other Respir Viruses. 2013;7(s2):114–9.
101. Farrar. The Ebola emergency—immediate action, ongoing strategy. N Engl J Med. 2014;371: 1481–95.
102. Dickmann. Biological risks to public health: lessons from an international conference to inform the development of national risk communication strategies: report of an international conference on risk communication strategies before, during, and after public health emergencies, Rabat, Morocco, October 22–23, 2015. Health Secur. 2016;14:433–40.
103. Fisman. Nuanced risk assessment for emerging infectious diseases. Lancet. 2014;383:189.
104. Ruggiero. CBRN communication scorecard. Behavior and communication in CBRN crisis: findings and recommendations in case of chemical, biological, radiological, and nuclear attacks on society. Lengerich: Pabst Science; 2015.
105. Lipsitch. If a global catastrophic biological risk materializes, at what stage will we recognize it? Health Secur. 2017;15. https://doi.org/10.1089/hs.2017.0037
106. Jansen. Biological warfare, bioterrorism, and biocrime. Clin Microbiol Infect. 2014;20: 488–96.
107. Lioy. Preparedness and response to chemical and biological threats: the role of exposure science. Ann N Y Acad Sci. 2016;1378:108–17.
108. Watson. Expert views on biological threat characterization for the US government: a delphi study. Risk Anal. 2017;37(12):2389–404.
109. Rainisch. Modeling tool for decision support during early days of an anthrax event. Emerg Infect Dis. 2017;23:46.
110. Grow. The challenge of hospital infection control during a response to bioterrorist attacks. Biosecurity Bioterrorism Biodefense Strateg Pract Sci. 2003;1:215–20.

111. Lor. Key ethical issues discussed at CDC-sponsored international, regional meetings to explore cultural perspectives and contexts on pandemic influenza preparedness and response. Int J Health Policy Manag. 2016;5:653.
112. WHO. Guidance for managing ethical issues in infectious disease outbreaks. [Online]. 2016. http://apps.who.int/iris/handle/10665/250580. Accessed 22 Aug 2017.

Swords to Ploughshares and Back: The Continuing Threat of Immunomodulatory Research and Development

10

Joseph P. Cornish and Reed F. Johnson

10.1 Introduction

Immunology is the scientific study of the functions and components of the immune system, a network of organs, tissues, cells and proteins throughout the human body that combats infections and cancers and regulates inflammation through a range of mechanisms. Through its components and their interactions with other cells and tissues of the body, the immune system plays an enormous role in human health by causing or resolving disease. Antibodies and T-cells targeting specific proteins are generated in response to the detection of components specific to tumors or pathogens known as antigens [1]. Soluble factors and cell surface receptors mediate immune and non-immune system cells to coordinate, regulate and target inflammation [2–4]. Dendritic cells scour the body for signs of pathogens or tumors, triggering inflammation and transporting detected antigens for subsequent targeting of responses [4]. Together, these immune system components target and destroy cancerous tissues and invading pathogens. At the same time, the destructive and inflammatory capabilities of the immune system have the potential to cause disease or even death. Inappropriate reaction to healthy tissues by the immune system leads to autoimmune diseases and uncontrolled inflammation during immune responses that can be lethal [5–7]. In some cases, the immune system fails to respond strongly enough, allowing pathogens and tumors to continue to proliferate [8–10].

Application of the insights and understanding of the immune system has led to the development of entire classes of therapeutics to treat diseases. For instance, cancer-targeting or drug-delivering viruses have been modified to evade the humoral

J. P. Cornish (✉) · R. F. Johnson
Emerging Viral Pathogens Section, Laboratory of Immunoregulation, Division of Intramural Research, National Institute of Allergy and Infectious Diseases, National Institutes of Health, Frederick, MD, USA
e-mail: joseph.cornish@nih.gov; johnsonreed@niaid.nih.gov

S. K. Singh, J. H. Kuhn (eds.), *Defense Against Biological Attacks*,
https://doi.org/10.1007/978-3-030-03053-7_10

immune response, thereby increasing the efficacy of drug target tissue destruction or payload delivery [11]. Genes encoding potent immunomodulatory proteins, antibodies, and antigens have been added to the genomes of viruses to develop improved cancer treatments that aid and direct the immune system's attack on a tumor. Immunomodulatory antibodies have been developed to specifically dampen overly potent immune responses or to increase insufficient responses [12]. The common theme among these therapies and therapeutic approaches is control: the ability to modulate the immune system to respond in a desired way. Through activating, subduing, or even avoiding a function or response of the immune system, these therapies can provide advances in treating human disease.

Approaches for activating, suppressing, and avoiding the immune system have been developed in response to a need for new and improved therapies for a range of diseases, but these very same capabilities represent serious dual-use risks. Just as gene-therapy viruses can be modified to evade host antibodies to increase their therapeutic efficacy, highly virulent viruses can also be modified. Methods created to produce potent cancer vaccines could be subverted to create bioweapons (BW) to attack healthy tissues. Virus-based immunotherapy could be used to enhance inflammatory damage caused during infection by a BW agent. At each step, progress in our ability to control the immune system to enhance human health provides the same potential for enhancing BW agents.

In most cases, discussions on abuses of modern biological theory and technology have been limited to hypothetical scenarios rather than past abuses. Thankfully, many of these risky, modern advances have occurred after the end of all publicly known BW programs capable of abusing these advances [13, 14]. However, immunological insight and immunomodulation are a rare exception, as historical evidence demonstrates utilization of immunomodulation to enhance BW. The Soviet BW program of the Cold War era was a sprawling network of research institutions, production plants, and testing facilities run by various government entities within the Soviet Union [14]. Capable of producing enormous amounts of viral and bacterial agents, possessing extensive facilities for BWs testing, and exploring advanced concepts for enhancing BWs, the Soviet BW program was the largest and most extensive program known [14]. Among other efforts to develop enhanced weapons, such as antibiotic-resistant *Bacillus anthracis*, the Soviet BW program explored immunological insights and immunomodulatory approaches to enhance BWs [14, 15]. These scientists explored methods to evade vaccines and detection, augmented the disease caused by a pathogen, and even engineered pathogens to yield novel BW agents. Importantly, the Soviet BW effort relied extensively on understanding the immune system to identify paths to desired effects, which were then pursued with biotechnologies at the system's disposal. Though the extent to which efforts yielded functional weapons is unclear, the Soviet biological weapons program made significant successes in enhancing biological weapons through immunomodulatory approaches [14].

When developing technologies that modulate the immune system, the difference between a BW effect and a therapeutic is simply application. Guided by insight and understanding of the immune system and developed through biotechnology, control of the immune system is the sole component of both immunomodulatory therapies and BW. The end effect, whether BW or therapeutic agents, is a decision of the user and limited only by the user's technological capability. In the study of the immune system and development of immunomodulatory approaches, both the products and the information are dual-use. While both pose threats, the risk provided by information is a far more persistent and challenging to address. All experiments, even those that do not produce any physical products, have the potential to generate information that could serve as inspiration or guidance for a party wishing to develop enhanced biological weapons. The dual-use risks of an experiment's results may go unnoticed as the potential abuses of that information may be masked in the intricacies of the immune system or involved subject area. As separation of use and abuse of information is impossible, our ability to improve human health through immunology will paradoxically increase our ability to harm health.

When considering paths forward, the scenario is not if risks exist, but which scenario generates the greatest risk to human health. Restricting immunological research and development will reduce the dual-use threat but will come at the cost of human health and lives by delaying or preventing therapeutic enhancements. Similarly, allowing research and development to continue with little to no regulation will only increase the likelihood that BWs may be enhanced in their lethality. In the end, the only option is to achieve balance between improving health, and avoiding, or preparing for, abuses of immunological research and development. Balance however, requires that the involved issues be properly defined and understood by all stakeholders. The nature and severity of dual-use risks of the growing understanding and capacity to control the immune system is best understood by exploring the motivations, developments, generalized capabilities, and, where applicable, historical abuses. Defining the nature of dual-use risk in immunological research and development is an immense task: dual-use potential exists at every level of research and development and the risks impact health, society, and international stability. Only after these issues and their risks are defined can achieving balance be approached.

10.2 Antibody Evasion

10.2.1 Therapeutic Background and Motivation

Antibodies are proteins produced by the immune system in response to exposure to an antigen. An antigen is a foreign substance that elicits a response from the immune system and can be anything from a component of food to a protein found in a pathogen. Antibodies will specifically bind the antigen that elicited the response that lead to antibody generation. Antibodies act through a number of mechanisms, such as neutralizing a protein through binding in a way that prevents it from functioning or by marking the pathogen or cell expressing a given antigen for destruction [16–18]. Humans are exposed to countless microorganisms throughout life, including pathogenic viruses [19, 20]. Exposure to a pathogenic virus may result in the development of immunity to the pathogen. If the host re-encounters the pathogen, the host's immune system rapidly detects and counters the microorganism. Long-lasting immune responses often rely on the production of antibodies that directly or indirectly neutralize the invading organism [21]. Many vaccines, such as those for yellow fever and diphtheria, function in part by eliciting an antibody response against antigen(s) found on a pathogen. Upon infection by that pathogen, a host will already have antibodies capable of slowing or halting infection [1]. Whether elicited naturally through infection or artificially through vaccination, antibodies are a critical component of the immune system's ability to combat infection.

Genetic diseases such as inherited immunodeficiencies or metabolic disorders, in which pathology is caused by defective copies of a gene or genes in a person's genome, are immensely difficult to treat [9]. Treatment often centers around supplementing the patient with functional cells or proteins which may have immense side effects or are extremely expensive and must be repeated as the functional cells and proteins degrade or die [9, 22]. Gene therapy offers the chance of long-term or curative treatments by supplying fully functional gene copies to the patient to produce properly functioning proteins. One approach to delivering genes is to use a virus, known here as a vector, that has been altered to carry the desired functional gene copies [23]. Viruses rely on the machinery of the cell they have infected to replicate. At the heart of a virus's replication approach is the delivery of a copy of its genome into the host cell. This step can be utilized to deliver genes into a host cell by altering the genome of the virus (vector) to carry copies of the desired genes. Once the virus has inserted its genome into the host cell as part of its natural infection process, the host cell will begin to produce the proteins encoded by the gene therapy payload.

Some adenoviruses, due to the relative ease in which they can be genetically modified and their typically strong safety profile, are ideal platforms for developing vaccines, anti-cancer, and gene therapy treatments [24–26]. Adenoviruses are viruses that replicate by inserting their genome into the nucleus of the infected cell. Then, the host cell will produce the payload encoded in the adenovirus genome. Additionally, rather than using genetic manipulation to deliver genes, adenoviruses can be modified to express new proteins for uses in vaccines or be engineered to preferentially infect and kill cancer cells [26–28]. While adenoviruses are ideal vectors for a range of therapies, adenovirus infections ubiquitous in humans result in large proportions of the population possessing anti-adenovirus antibodies [25, 29, 30]. The presence of pre-existing antibodies targeting antigenically similar adenoviruses can seriously impact the efficacy of adenovirus-based therapies [25, 29, 30]. Even in the absence of pre-existing immunity, the first treatment would likely induce a sufficiently-strong level of immunity against the adenovirus vector that would potentially complicate treatment regimens and prevent the development of common platforms. For example, a vaccine clinical trial that started during the 2013–2016 Western African Ebola virus disease outbreak required administration of two doses of a vaccine, one based on an adenovirus and the second on an unrelated virus [31]. The use of two different viral-vectored vaccines came about from non-human primate experiments that demonstrated two doses of vaccine were required. Using the adenovirus a second time was ineffective, possibly due to the immunity against adenoviruses induced during the first vaccine dose [30, 32–34]. Similarly, an experimental adenovirus-based HIV-1 vaccine was found to have significantly reduced efficacy in patients with pre-existing anti-adenovirus antibodies [30]. The high prevalence of pre-existing antibodies against antigenically similar adenoviruses, either through exposure to naturally occurring adenoviruses or through treatment with adenovirus-based therapeutics, is a major roadblock towards effective use of adenoviruses in creating vaccines and other therapies.

The problem of immunity against adenoviruses has led to extensive research on approaches to protect adenoviruses from antibody neutralization by preventing their binding or otherwise inhibiting their effects [35]. Research to prevent neutralization of adenoviruses has focused on two different approaches: protective coatings and modification of target proteins. Protective coatings are designed to prevent binding of neutralizing antibodies without negatively impacting function of the virus. These coatings generally work by physically shielding targets or altering the electrical charge at the surface of the virus particle [36, 37]. An example of a coating is using a process, called PEGylation, in which poly-ethylene glycol (PEG) is chemically attached to the surface of the virus [36, 37]. These PEGylated viruses are not just shielded from antibodies but are also less likely to be identified by certain types of immune cells [38]. Importantly, coatings do not require any genetic modification of the virus. The other approach, modifying proteins targeted by antibodies, relies on the use of genetic engineering to create an altered viral protein that can no longer be bound by antibodies against the original viral protein [39]. More advanced methods shield the virus through forming a coating using a protein present in the host's blood that prevent antibodies from binding or neutralizing the virus [40].

10.2.2 Dual-Use Risks

The dual-use risks of antibody evasion technologies become obvious when one considers the general capability provided: the ability of the virus to evade a critical arm of the immune response. Intended to be used on benign viruses to develop new treatments, BW agents could also utilize antibody evasion technologies. Non-genetic antibody evasion coatings of viruses are particularly vulnerable to abuse as BWs. Modifying or complete deletion of proteins targeted by diagnostic, therapeutic, or host antibodies without significantly altering the pathogen's suitability as a BW may be extremely difficult. Although genome editing technologies have greatly improved, genetically modifying a pathogen in an effective and stable way remains a challenging task. Alteration of pathogen proteins will only confer resistance against antibodies targeting the modified protein or even the specific region modified. Antibodies targeting different proteins of a pathogen are unlikely to be impacted by modifications in other regions of proteins. The required genetic manipulations could tip off an adversary that may develop new antibodies to overcome engineered resistance. However, coatings that require no genetic modification may be less likely to negatively impact BW-critical characteristics, would be free of any potential genetic instability, and could provide broad protection against all or many types of antibodies. The absence of any requirement for genetic modification means coatings could be used in a pathogen-agnostic manner, making them particularly attractive to BW scientists.

The same technologies that may one day produce viable virus-based therapies could be redirected to develop BW resistant to vaccines and antibody therapies. Antibodies are in some ways the backbones of biodefense, they are critical for prevention, diagnosis and even efficacious treatment of many infectious diseases. Many diagnostic assays work by detecting the presence of antibodies to a pathogen elicited during infection or by directly detecting the pathogen by using an antibody coupled to a visual readout such as fluorescence. Many pathogens, such as the causative agents of anthrax, plague and tularemia must be treated rapidly to obtain improved outcomes [41–43]. Biochemically, antibiotics and antivirals would not be impacted by antibody-evading coatings but their usefulness as a countermeasure to a bioweapon attack will be reduced by slow deployment caused by delayed diagnosis. Escaping detection can significantly enhance the lethality of a BW without any need to alter the pathology of the agent used in that BW. In a similar fashion, many vaccines provide protection against infection by eliciting an antibody response against the pathogen [44]. Vaccines are a critical component of a biodefense program as they are preventative rather than curative, reducing the potential stress on any public health response to a BW attack. Additionally, efforts have been made to make BW agents resistant to therapeutics, such as antibiotic-resistant *B. anthracis* (the causative agent of anthrax). Vaccines provide biosecurity in BW attack scenarios when other countermeasures may not work [14, 15].

Antibodies are increasingly being used as curative treatments. Botulism toxin, Ebola virus, *B. anthracis*, and *Yersinia pestis* can be targeted with purified antibodies or even by using a gene therapy vector to cause the patient to produce antibodies [28, 45–47]. Therapeutic antibodies are becoming critical components of biodefense programs, and any technologies that could allow pathogens to escape these treatments is a serious threat [48, 49]. The potential for antibody-evasion coatings to be applied to BW agents is an enormous risk to all aspects of biodefense.

Indeed, similar antibody-evasion approaches were explored in the Soviet BW program to develop enhanced BWs. *Y. pestis*, the causative agent of plague, has long been exploited as a BW across history from the fourteenth century to the twentieth century [13, 14, 50]. The *Y. pestis* F1 antigen is a protein expressed on the outside of the cell wall of the bacterium and plays a role in *Y. pestis* pathogenesis [51]. The F1 antigen has also been used to develop vaccines that generate antibodies against the F1 antigen allowing the body to better defend against *Y. pestis* infection [42, 44]. Additionally, given its potential use as a BW, rapid diagnostic tests have been designed to detect the presence of *Y. pestis* via detecting the F1 antigen [52]. Both the F1-based vaccine and diagnostics were a critical part of American military biodefense [14, 44, 52]. Given the importance of the F1 antigen to American anti-plague biodefense measures, the Soviet BWs program embarked on an effort to create a strain of *Y. pestis* that lacked the F1 antigen to an unknown level of success [14]. The F1-based vaccine would no longer be effective as the antibodies it generates would not neutralize F1-deleted *Y. pestis*. Similarly, the diagnostic would be rendered useless as it relied on the presence of F1 to detect *Y. pestis*. Now vulnerable to infection, victims of an attack with a F1-deleted *Y. pestis* would be unable to receive the critically time-sensitive treatment needed as the remaining diagnostic methods would be too slow to be effective [52]. Together, this loss of protection and effective diagnosis could prove to be extremely devastating.

Francisella tularensis, the causative agent of tularemia, is an extremely infectious bacterium that can cause a severe and highly lethal respiratory disease if inhaled [43, 53]. The extremely low infectious dose and high lethality associated with inhalational infections make *F. tularensis* an ideal BW pathogen leading to extensive weaponization work by both the American and Soviet BW programs [13, 14, 43]. Present in the bacterial cell wall of *Staphylococcus aureus*, protein A, protects against immune system detection and attack by strongly binding antibodies that prevent killing or neutralization of the organism [54]. Exploiting the immune evasion mechanisms of *S. aureus*, a Soviet BW subprogram was initiated to utilize protein A coatings to protect *F. tularensis* bacteria from antibodies, potentially degrading the performance of vaccines or therapeutic vaccines [14]. A BW that could similarly evade destruction mediated by antibody binding would likely have enhanced pathogenic effects. Similar to current research on adenoviruses, both non-genetic and genetic approaches were explored: one approach relied on covalently coating bacteria with protein A, whereas the other approach relied on genetic modification to express protein A [14]. Both coating approaches proved successful in the goal of increasing lethality of *F. tularensis* and likely would have impacted the efficacy of available vaccines and diagnostics against the organism [14].

B. anthracis was extensively explored by the United States, Soviet and United Kingdom BW programs [13, 14]. Soviet BW scientists undertook a project to produce a more severe disease course of *Bacillus anthracis* infection by incorporating additional genes from a related bacterium, *B. cereus* [14]. *B. cereus* is capable of causing blindness and other serious effects in humans [55]. Soviet BW scientists identified the genes in *B. cereus* believed to be responsible for these severe clinical signs, called cereolysins, and inserted them into *B. anthracis* with the hopes of imparting the pathogenic effects of the cereolysins [14, 56]. The modified variant was unsuccessful in the original goal of creating additional clinical signs of *B. anthracis* infection. However, the added genes did have another alarming effect: imparting resistance to available vaccines [57]. Although the vaccine-resistant form of *B. anthracis* was not generated through abusing insights or information on the immune system, vaccine resistance can be generated by seemingly unrelated changes in the proteins expressed by a pathogen. An adversary could potentially obfuscate the presence of antibody evasion modifications in an agent by utilizing indirect approaches that rely on seemingly unrelated proteins. Alternatively, such effects could be encountered through civilian research that relies on genetic manipulation of viruses or bacteria. Obfuscation of any enhanced characteristics of a BW may further hinder responses to an attack by delaying the detection of antibody evasion properties.

10.2.3 Conclusion

Civilian research on developing antibody evasion technologies very closely parallels approaches explored in the Soviet BW program. Importantly, both civilian and military efforts heavily explored the use of coatings that do not require genetic modification of microorganisms. Modern approaches being developed for virus-based therapies, such as polymer coatings, could prove to be a cheaper and more efficient means of enhancing agents, and may circumvent problems experienced during Soviet efforts to develop non-genetic protein A coatings from *S. aureus* [14]. However, protein A has since been found to be a virulence factor [58] and is also widely produced for use in purifying antibodies [59, 60]. The industrial scale production of protein A and the increased significance of protein A in pathogenesis suggest protein A remains a viable option for antibody evasion coatings. While the Soviet efforts with the causative agents of anthrax, tularemia, and plague demonstrate the risk antibody-shielding technologies pose to biodefense vaccines, consideration should also be given to anti-pathogen therapeutic antibodies. Antibodies that can be directly administered to a patient, rather than elicited through a vaccine response, are becoming important fixtures in treating infectious diseases and biodefense [45–49]. The reliability of therapeutic anti-pathogen antibodies in biodefense is similarly jeopardized by developments in antibody evasion approaches.

Although different in their goals and motivations, both the former Soviet BW program and ongoing civilian research efforts have been exploring a range of antibody evasion methods. From deleting known regions targeted by antibodies to utilizing proteins or chemicals that promote bacterial or viral evasion by the immune system, a number of paths towards imparting antibody-resistance to microorganisms now exist. The viability of coatings for antibody evasion for both bacterial and viral agents along with the potential advantages of non-genetic coatings may make them especially attractive to BW scientists. Such a project could, in theory, develop a single coating technology that could be utilized across a range of both bacterial and viral BWs. The flexibility and diversity of antibody evasion technology represent serious biodefense risks, as predicting, detecting, and countering the exact approach used by an adversary may be difficult. Antibody evasion research and development are undeniably at the extreme of dual-use risk as such research has been demonstrated to be an effective weaponization technology. In addition, antibody evasion research has the potential to destroy two cornerstones of biodefense: vaccines and therapeutic antibodies.

10.3 Cancer Virotherapy

10.3.1 Therapeutic Background and Motivation

Cancer, a class of over 100 different diseases, is one of the largest causes of death in the world [61, 62]. Until recently, cancers were almost exclusively treated with chemical agents that imprecisely target tumor cells due to dysregulated metabolic rates, but unfortunately, non-specifically kill healthy cells that damage healthy tissues [63]. Additionally, tumors may only be vulnerable to a small number of drugs and can develop or have pre-existing resistance to available drugs [64]. Among patients with the same type of tumor, intra-tumor phenotypic heterogeneity may be associated with genetic mutations that can cause significant variability in treatment [65]. The relatively poor efficacy of traditional cancer treatment has driven research to identify more efficient means of eradicating tumors. Virotherapy is an emerging avenue for treating cancers. Virotherapy is based on two approaches, used either independently or simultaneously, to achieve tumor killing. The first approach, oncolysis, relies on the ability of a virus to kill tumor cells it infects, thereby shrinking or killing the tumor. The second approach, immunotherapy, relies on an immunomodulatory payload in the virus to stimulate the body's anti-tumor immune response [66]. This stimulatory effect is performed by host immunomodulatory proteins that have been encoded in the virus.

Virotherapy aims to kill tumor cells directly through effects of viral replication, indirectly by stimulating the immune system's attack on the tumor or both. Direct killing of infected cells relies on the virus by triggering apoptosis or causing the cell to die through some other means such as cytopathic effects [66]. All oncolytic viruses are thought to be capable of increasing tumor inflammation through the very act of infection and replication and of causing the release of new tumor-associated antigens by lysing cells [67]. However, a more common approach is to incorporate host protein-encoding genes with immunomodulatory effects into the virus genome. One example of this approach, an FDA-approved treatment talimogene laherparepvec for melanoma, is a herpesvirus that has been modified to encode the cytokine granulocyte macrophage colony stimulating factor (GM-CSF). Cells infected by talimogene laherparepvec will secrete GM-CSF that attracts immune cells to the site of the tumor to aid in the anti-tumor response [68]. GM-CSF has also been used in JX-594, a recombinant vaccinia virus, to enhance anti-tumor immunity [69].

A more complex approach under exploration is a combinatorial experimental vaccine consisting of rilimogene galvacirepvec, a recombinant vaccinia virus and rilimogene glafolivec, a recombinant fowlpox virus to treat prostate cancer. This treatment regimen relies on these two distinct poxviruses that have been modified to express a host protein associated with prostate cancer called prostate-specific antigen (PSA) and three host pro-inflammatory molecules known collectively as TRICOM (LFA-3, ICAM-1, and CD80). A prime vaccination with rilimogene galvacirepvec followed by a boost with rilimogene glafolivec induces the immune system to attack cells expressing PSA. This vaccine is further enhanced by the inclusion of molecules that act to greatly stimulate the immune system cells that are specific to PSA [12]. Unlike other virotherapy approaches, this vaccine does not infect or directly kill tumor cells. Instead it is designed to be captured by the immune system and incite the immune system to more strongly attack tumor cells [12]. The TRICOM platform can be altered to include other antigens and has been explored for other types of cancers, indicating the approach is flexible and can be readily retargeted [70, 71].

10.3.2 Dual-Use Issues

Viruses modified to have immunomodulatory effects are profoundly powerful tools. From non-specifically stimulating infection to steering immune responses against specific healthy tissues, these approaches could be used to enhance or develop new BWs. The approach used to develop rilimogene galvacirepvec and rilimogene glafolivec vaccine could be misused to cause damage to healthy tissues by utilizing a different antigen. Including immunomodulatory proteins similar to this vaccine would likely enhance the virulence of a BW. Many autoimmune diseases can take a relatively long time to develop into clinical disease. Using an approach similar to rilimogene galvacirepvec and rilimogene glafolivec vaccine, a BW may be hard to detect as the clinical signs may be detected long after exposure to the auto-immune trigger. The flexibility in modifying the rilimogene galvacirepvec and rilimogene glafolivec vaccine platform to attack other targets for treatment suggests the immunomodulatory capacity of modified viruses could provide a route to flexible BW platforms. Simpler approaches relying on the inclusion of a single pro-inflammatory protein have the potential for enhancing pathology. At high doses, JX-594 was shown to cause increased inflammation and pathology over its unmodified counterpart demonstrating the potential for single-item payloads to significantly enhance pathology [69]. Including immunomodulatory proteins in pathogens traditionally used for BW could enhance or alter the disease they cause that could impact prevention, treatment or diagnosis of disease.

The ability to create novel pathology through the introduction of immunomodulatory proteins creates a dangerous scenario: an adversary with this capability is no longer limited to pathogens that naturally cause disease. With this technology in hand, an adversary would have far more flexibility in the pathogens they choose to deliver an immunomodulatory payload. Pathogens that are endemic or are not normally considered BW agents could be chosen to help mask an attack. An adversary could choose an organism based on BW criteria such as ease of production, environmental hardiness, or the ability to survive aerosolization. A BW designer need not pick a pathogenic microorganism if sufficient immunomodulatory capabilities can be engineered into the organism they choose. The number of possible platforms, tissue targets, self-antigens, and immunomodulatory payloads creates a wide array of potential BW agents in a world that still struggles with handling natural disease outbreaks during peacetime.

Unfortunately, these capabilities were not lost on the scientists of the Soviet BW program. This program launched intense efforts to identify immunomodulatory proteins and incorporate them into a range of pathogens. Soviet BW scientists explored using tumor necrosis factor, both type 1 and type 2 interferons, endorphins and interleukins [14]. Many pathogens considered for BW are poorly controlled by the immune system in their natural forms, and often many of the clinical signs are the result of the body's ineffectual immune response. Altering pathogens to express pro-inflammatory markers may not hinder the pathogen's ability to replicate but could seriously enhance the detrimental effects of the resulting immune response. One such effort involved modifying Venezuelan equine encephalitis virus (VEEV) to have enhanced immunomodulatory properties. VEEV can cause a severe and sometimes lethal neurological disease in humans and has been explored, in its natural form, for use as a BW [7, 14, 72]. The Soviet BW effort involved modifying VEEV to express interleukin-2, a potent immunomodulatory factor. Infection of the central nervous system by VEEV leads to profound inflammation known as encephalitis [73]. Interleukin-2 is a cytokine naturally formed during the immune response capable of promoting inflammation [74]. VEEV modified to express interleukin-2 would likely increase the severity of encephalitis seen during infection and may increase lethality or cause permanent neurological damage. The effects of these modified pathogens and the progress made weaponizing them are not known, but it is unlikely they moved far from the laboratory given the program ended soon thereafter.

One of the most alarming projects of the Soviet BW program was weaponizing pathogens expressing myelin basic protein. Myelin basic protein is a vital component to all nerve cells. In a fashion similar to the rilimogene galvacirepvec and rilimogene glafolivec dual vaccine incorporating PSA and TRICOM, myelin basic protein or fragments thereof were inserted into a pathogen [14]. The immune response against infection by this modified pathogen would trigger an immune response against the proteins contained in the pathogen, including myelin basic protein. A BW of this type would cause an autoimmune reaction against nerve cells present throughout the body. The suspected goal of the research was to create a multiple sclerosis-like disease that would run its course in days or weeks rather than years [14]. Eventually, efforts using *Legionella pneumophila* (causative agent of Legionnaire's disease) and *Y. pseudotuberculosis* (etiologic agents of pseudotuberculosis) proved successful [14]. Not only was this effort able to radically alter the pathogenesis of both *L. pneumophila* and *Y. pseudotuberculosis* but neither agent has been typically considered for BW. In other words, the Soviet program created a BW causing a previously unknown pathology using agents that may go undetected by most biodefense efforts.

10.3.3 Conclusion

Virotherapy, although a potential source of significant advances in cancer treatment, is fraught with theoretical and realized dual-use risks. As research and development of these treatments expands, risky discoveries or information directly applicable to BW development may be accidentally discovered. This risk is enshrined in the case of addition of interleukin-4 to ectromelia virus that unexpectedly allowed this poxviruses to escape the immunity induced by the smallpox vaccine [75]. This information would be incredibly useful to an adversary with a desire to utilize poxvirus BW against a vaccinated population. The biodefense consequences of the ectromelia interleukin-4 work did not go unnoticed: the United States Army responded by engineering a related virus, cowpox virus, to express interleukin-4 to further characterize the role of interleukin-4 and poxvirus immunity [76]. As development of therapies and basic research delve deeper into immunomodulation, one can reasonably expect similar incidents where controversial biodefense efforts are triggered by intentionally or unintentionally dangerous experiments. The increasing complexity in the approaches used to initiate or enhance anti-tumor immune responses provides additional routes to enhancing, or even creating, pathologies caused by BW agents. From the enhanced inflammation of JX-594 to the specific targeting of a delivered antigen by PROSTVAC, a range of potentially weaponizable products and information are now available. Even without further accidental discoveries like the ectromelia virus-interleukin-4 work, ongoing civilian therapeutic research will continue to provide information directly abusable by a BW program by discovering and characterizing modes of enhancing and targeting inflammation.

10.4 Immunomodulatory Antibodies

10.4.1 Therapeutic Background and Motivation

Diseases of the immune system are the result of overreaction or inaction. The immune system can worsen, or even cause, disease by if it responds too strongly or attacks the wrong targets. Similarly, diseases can be worsened or permitted if the immune system is unable to respond or fails to respond strongly enough. Allergic reactions are the result of a massive response to a harmless antigen. In arthritis, joints are damaged by an inappropriate response against the body's own tissues. Immune system cells present in transplanted tissues or the recipient's native immune system cells can cause disease and even cause the transplant to fail [77, 78]. Some diseases caused by infections may be the result of the immune system's inaction against invading pathogens. Still other diseases may be the result of overzealous immune responses during an infection. Tumors can grow and spread by evading and suppressing the immune system in a myriad of ways [8, 10, 79–81].

As with other types of diseases, treatment of diseases of the immune system is centered on restoring balance from inappropriate immune activation or suppression. Unlike diseases involving other parts of the body, our ability to treat diseases centered in the immune system has been relatively crude. Controlling inappropriate immune responses has typically relied on chemical therapies, such as steroids, that non-specifically suppress the immune system [6]. Such nonspecific suppression is effective at reducing the inappropriate attack on healthy tissues but can leave patients vulnerable to infection since the immune system is also unable to attack pathogens. In the case of allergies, the best therapy is usually prevention by avoiding specific allergens. Emergency medicines are used to handle breakthrough events despite allergen avoidance. Detrimental inflammation caused by infections is often only dealt with by supportive care. Failures to clear pathogens rapidly (or at all) are, when possible, treated with agents that prevent further replication [82]. Efforts to stimulate the immune system have usually been limited to vaccines or by removing the source of the suppression (e.g., tumors, immunosuppressive pathogens) [34, 83, 84].

An ability to manipulate the specific mechanisms or components of the immune system would offer greatly improved treatments for immunologically driven diseases. However, the immune system relies on a dizzying array of soluble signaling molecules, small compounds, and cell surface receptors driven by a complex network of cell types, tissues, and organs. To control the diseases caused by and within the immune system, one must be able to modulate specific parts of the immune system. Immunomodulatory antibodies have been able to deliver highly specific modulation not previously obtainable with chemical therapies [85]. These antibodies work by binding a specific protein to prevent signaling along pathways involving that protein. Additionally, antibodies may directly activate immune system cells by signaling through the target receptor [5] or by marking target cells for destruction [86]. The combination of a diverse set of approaches, from targeting proteins to cells, along with the vast repertoire of immune-related signaling molecules and receptors promotes both diversity and specificity in immunomodulation.

Preventing or dampening an overactive immune system typically involves antibodies that bind pro-inflammatory proteins, preventing them from driving inflammation. For example, a potent pro-inflammatory cytokine, interleukin 6, can be neutralized by an antibody that prevents the cytokine from sustaining or increasing inflammation [87]. Another approach relies on targeting a cell type responsible for inflammation for destruction. CD20, a surface marker of B-cells, can be targeted by an antibody (i.e., rituximab) that is in turn recognized and destroyed by natural killer cells [86, 88]. Antibody-mediated destruction of B-cells has been used to treat cancer and rheumatoid arthritis and other autoimmune diseases [89–92].

Immunomodulatory antibodies are also able to stimulate the immune system by blocking inhibitory signals that would prevent or subdue a given immune response. Cancer immunotherapy is one area that has benefitted from this pro-inflammatory capability to detect and kill tumor cells. Tumors can suppress anti-tumor immune responses by directly subduing responding cells or encouraging populations of anti-inflammatory cells [93, 94]. Cancer immunotherapies are aimed at restoring and promoting inflammation in the tumor site by blocking the direct (tumor surface) and indirect (anti-inflammatory cell surface) tumor-driven inhibitory signals [94]. For example, PD-1, a receptor present on T-cells capable of killing tumors, can be engaged by both tumors and anti-inflammatory cells. T-cells that have been signaled through PD-1 begin have a reduced capability to kill tumor cells. Antibodies that bind PD-1 prevent inhibitory signaling, and allow T-cells to continue killing tumor cells [94].

Genome editing has also been explored to develop recombinant viruses that express antibodies by including the genetic sequences that encode the antibodies into the viruses' genomes. One such example is a vaccinia virus modified to express an antibody aimed at inhibiting the spread of blood flow to tumors [95]. When the antibody is expressed as part of the viral genome, production of the antibody is dependent on infection of cells by the virus expressing the antibody [1, 95]. Antibody production will halt when infected cells are no longer available. The necessity for active infection or viral production is a significant limitation when long-lasting antibody production is desired [96]. To circumvent the need for active infection, approaches using adenovirus gene therapy vectors to deliver antibody encoding genes into the host cells' nuclei have been developed. Once inserted into the host cells' nuclei, antibody production will occur so long as the targeted cells remain alive. Gene-therapy delivery of antibodies has been explored for delivering anti-cancer, anti-viral, and immunomodulatory antibodies [27, 28, 97, 98].

10.4.2 Dual-Use

CD28 is a protein on the surface of T-cells that plays a large role in regulating immunological functions of the cell. Potently activating CD28 is known to induce T-cells to take on a regulatory role that could lead to raising or lowering inflammation to beneficial levels [99]. TGN1412, an immunomodulatory antibody, is a CD28 super-agonist due to its ability to potently activate CD28. The goal is to utilize TGN1412s properties to enhance inflammation in cancer and reduce it in rheumatoid arthritis [5, 99]. In 2006, a first-in-humans trial of TGN1412 was started. Within minutes of starting the infusion, all six participants began to experience severe non-specific immune activation and were hospitalized [5]. TGN1412 had unexpectedly activated an incredibly potent form of T-cell immunity rather than the expected modulating response [5]. Use of TGN1412 demonstrates that immunomodulatory antibodies have the capability to cause profound harm. Under similar circumstances, some patients undergoing antibody-based cancer immunotherapies must be treated with anti-inflammatory agents to combat therapy-driven immunological toxicities [94].

No evidence is available that immunomodulatory antibodies have ever been used or considered for offensive purposes. Compared to chemical compounds, antibodies are larger and relatively fragile molecules, and delivery is limited to injection or infusion [100], features that are typically incompatible with development of BW. However, two areas of research could alter the dual-use risk associated with immunomodulatory antibodies: aerosol delivery and vector-based expression.

Researchers considering aerosol delivery of antibodies have been motivated by a desire to improve the specificity of drugs delivered for lung cancer treatment [101]. Results of antibody aerosol research have demonstrated that delivery of immunomodulatory antibodies to the lung compartment is possible, but that little to no systemic delivery occurs [100]. Even without systemic delivery, aerosolized antibodies could enhance a BW by "priming" the lungs to be more vulnerable to infection. Priming the pulmonary space could be achieved by destroying immune cells present in the lung with antibodies conjugated to toxins [102] or by disrupting the signaling pathways required to coordinate an immune response. "Doping" a BW with antibodies or antibody conjugates may enhance disease or lethality without generating a tell-tale genetic signature. It may not be obvious that an antibody additive is the cause of any enhanced disease, and identifying the target of an antibody would be time consuming and limited by sample availability. A delay or failure to identify an additive antibody would further enhance the efficacy of a BW by delaying or preventing an effective response against it. Overall, aerosol delivery of immunomodulatory antibodies or antibody-toxin conjugates could significantly enhance the lethality of unmodified BW agents and would be extremely difficult to identify or develop countermeasures against the BW.

Vector-based delivery of antibodies escapes the limitations of directly delivering antibodies since the viruses used to deliver antibodies can be delivered via aerosol even without causing disease [103, 104]. Incorporation of immunomodulatory antibody-encoding sequences in BW agents could significantly modify, or even create disease caused by an infectious agent. Already pathogenic agents could be modified to worsen their effects, and both pathogenic and benign agents could be imparted with new disease effects. The disease effects of a BW could possibly be caused by expressed antibodies, and the pathogen would simply acts as a vehicle for delivery. Through gene-therapy delivery of antibodies using viruses that do not cause disease, an enemy could induce a target population to express antibodies without detection. Antibodies that would promote infection by subduing immune signaling, destroying critical cell populations, or neutralizing vaccines could be deployed prior to or in parallel with a pathogenic biological weapon or agent to enhance infectivity or disease.

10.5 Discussion

Near the end of the Soviet BW program, its researchers had begun to explore and harness the human immune system to enhance the suitability of pathogens by altering their immunological profiles [14]. Pathogens were modified to be invisible to the immune system, express proteins that would exacerbate the deadly inflammation they caused, and to create new diseases through modulating the immune system [14]. While it is unlikely that these modified pathogens ever progressed beyond developmental stages, they represented a new way of thinking about BW. Even with relatively crude technology, information and ideas about the immune system are easily diverted from their intended uses to aid and understand human health. This chapter demonstrates that civilian research efforts have the potential to unwittingly produce information or technology directly applicable to BW development. Attempts to modulate the immune response using antibodies and recombinant viruses provide clear examples in which scientific advances that could potentially be directly weaponized were discovered during research designed to benefit humanity. As the size and complexity of immunomodulatory research continues to grow, further dangerous advances and developments will be likely.

A significant development in the Soviet BW project was the development of *L. pneumophila* expressing myelin basic protein. Unlike the other research aimed at evading vaccines or increasing pathology, this research was the first designed with the goal of causing a specific immunity-driven disease. Soviet BW scientists wanted to create an infectious multiple sclerosis-like disease. The BW was reverse-engineered, the desired pathology was identified, a potential mechanism was found through immunology, and the researchers explored a variety of approaches to achieve the desired outcome. Unlike previous BW based on modifications designed to enhance disease caused by the natural form of a pathogen, this BW utilized the bacterium as an immunomodulatory agent to cause a new artificial disease. These scientists had begun to work in reverse: rather than letting the pathogen define the disease, they aimed to shape the pathogen to cause a specific pre-defined disease. Through reverse engineering, proteins, molecules and pathogens can be different tools for modulating the immune system to cause a specific disease. The only thing that separates a BW designer from a civilian researcher developing therapies is intent.

Large scale research, development, and commercialization of immunomodulatory technology also contribute to proliferation and international security risks. One of the main challenges in developing a BW program is scale. Growing small amounts of pathogen in a laboratory is relatively simple but moving to the liters and kilograms needed to sustain a BW program is a massive endeavor. Scaling issues may worsen rapidly when a BW program is aimed at incorporating advanced immunology research and development. So far only the Soviet BW program, an enormous undertaking [14], is known to have seriously pursued applying immunology to the development of BW. Although serious efforts to incorporate immunomodulatory aspects into BW agents were attempted, development of such BW past concepts or highly experimental phases is highly unlikely [14]. Improvements in biotechnology since the end of the Soviet BW program, such as new genetic engineering methods, may allow for greater ease and complexity in engineering biological weapons agents but will likely do little to aid in production, formulation or in developing delivery systems.

Efforts to research and commercialize immunology-based therapies may provide several paths towards a BW program. An example of the dual-use risk of industrialization is found in the production of biopesticide *Bacillus thuringiensis*, a bacterium that produces toxins that are lethal to insects that damage a number of crops [105]. Production of *B. thuringiensis*, a very close relative of *B. anthracis* [106], requires many of the same technologies, facilities, and skills required to grow bacterial agents in amounts needed for a BW program [107]. In a similar fashion, the ability to create, test, and produce large amounts of immunomodulatory viruses or stealth-coated vaccines for healthcare uses would provide a similar breakout hedge. Given that the majority of immunomodulatory research is not currently considered to have high dual-use risks, domestic research programs may attract little attention from the international community. Nations currently unable to support their own research programs would likely be able to send out scientists internationally to receive advanced training and experience. Unlike jobs working with dangerous pathogens that require extensive background checks and security clearances, such as the United States' Centers for Disease Control Select Agent Program (Centers for Disease Control and Prevention and United States Department of Agriculture), training or employment centered on most of the civilian technologies described in this chapter would not likely require any extensive background checks or security clearances.

For nations that are already able to support vast and complex biomedical research and industrialization, a different proliferation risk is associated with pursuing immunomodulatory research: an enhanced BW breakout capability. Through supporting biodefense research, domestic vaccine, virotherapy or biopesticide production, and advanced research on immunology and immunomodulation, a nation may be able to maintain a pool of science, technology, scientists, and facilities/equipment needed to rapidly establish an offensive BW program. This breakout time may be further reduced in nations with advanced biodefense programs that are supported by extensive skills, scientific understanding, technologies, and facilities that could aid in BW development.

Biodefense efforts against pathogens with modified immunological profiles can further drive risks by creating or proliferating technologies and information, and further blurring the line between defensive and offensive research. *B. anthracis* was an important agent in the Soviet BW program and was subjected to extensive production and weaponization efforts [14]. One such effort was to develop a strain of *B. anthracis* that would be resistant to the anthrax vaccine used by the United States [14]. In 1997, a paper was published in the journal *Vaccine* demonstrating that scientists previously associated with the Soviet BW program had created a vaccine-resistant *B. anthracis* strain [57]. The United States were unable to obtain samples of the vaccine-resistant strain, and in the late 1990s the United States Defense Intelligence Agency began Project Jefferson to reconstruct the Soviet vaccine-resistant strain to determine if the United States medical countermeasures would be effective in treating infection [108, 109]. A similar effort was conducted in response to the ectromelia virus interleukin-4 experiments. Cowpox virus, a relative of ectromelia virus, was engineered to conduct experiments to determine the threat against the smallpox vaccine caused by interleukin-4 expression by poxviruses [76]. Both actions were direct responses to threats created by both purposeful and accidental discoveries with direct implications to biodefense. Together these cases demonstrate that biodefense mechanisms of nations may perceive immunological developments in both the military and civilian arenas as serious threats. The response to these perceived threats may involve research that blurs the line between defensive and offensive research. Nations may perceive another nation's well-meaning civilian research or responses to dangerous discoveries as a threat, triggering complementary research efforts. These cycles of discovery and response could erode norms of what is acceptable biodefense research and drive proliferation of BW-related technology and information gained through work mounted in response to high-risk discoveries.

Using antibodies, proteins, and viruses, humanity is rapidly establishing a capability to begin to exert control over the immune system in increasingly precise and complex ways. While these modulatory approaches are crude compared to the staggering complexity of the human immune system, they provide improvements in treating human disease. As with almost all areas of biomedical research, opportunity for the information and technologies derived from immunology to be diverted to cause harm is ample. Like most endeavors in technology and science, immunology and the developments it drives has a history of being abused for developing enhanced BW both with, and without, genetic engineering. In providing so many different paths towards a common goal, immunology offers any would-be BW designers a substantial toolbox for creating and enhancing BW. The expanding size of this toolbox worsens an already overwhelming portfolio of potential BW technologies. As with most dual-use issues in the life sciences, transparency, regulation and education appear to be the obvious approaches to address the issues raised by immunomodulatory research and technology. Through transparency, ensuring that research and development is performed openly and honestly, reduction in inappropriately risky research may be possible by ensuring all stakeholders (e.g., scientific community and the public) are informed and voice concerns on research. Regulation can act to place mechanisms to prevent risky research, ensure transparency, and improve physical and personnel security. Regulation and transparency can only be effective if all stakeholders (e.g., public, academia, industry and government) are properly educated.

The core challenge facing any transparency, policy, or education efforts is in defining the time point when risk of a given instance of actual or potential misuse of immunomodulatory research or technology becomes great enough to warrant action. Developing approaches to detect the less obviously risky experiments or technologies will be critical to any efforts to approach the risks in immunomodulatory research and development. Take, for example, an experiment that modified an already highly virulent pathogen to express some pro-inflammatory factor leading to even greater virulence. It is easy to see that enhancing a pathogen's virulence is incredibly risky work, but the core dual-use product is the information that the pro-inflammatory factor in question may enhance virulence. Suppose the experiment in question had been inspired by two previous works. Mice were infected with the wild-type pathogen that demonstrated high levels of the factor was associated with lethality, and an unrelated clinical trial attempting to use the same factor to stimulate the immune system to attack tumors failed because the factor caused damaging inflammation. If the goal is to enhance a BW, all three of these experiments provide strong evidence that modifying a BW agent to express the pro-inflammatory factor could enhance the weapon's effects. It is critical not to underestimate the intelligence or ingenuity of those who wish to engage in a BW program or to rely on security through obscurity. The true level of risk posed by an experiment is limited only by the person who intends to abuse the experiment's information or products. At the same time, whatever scope is defined must be both practical (what can realistically be done) and grounded (complex threats versus fantasy).

Regulation is probably the most powerful tool of policy but is also the most challenging to effectively implement. The main regulatory mechanism available is to restrict which immunomodulatory research and development can be performed. Restrictions on research can be explicit, such as outlawing or legally limiting certain types of experiments, or it can be implicit, such as requiring work with specific agents take place laboratories with high levels of physical security. An example of explicit regulation exist in the United States Select Agent Program, which stipulates that modifying any Select Agent pathogen to be resistant to available countermeasures is illegal unless prior approval is obtained [110, 111]. Implicit restrictions limit the research and development that can be performed by implementing physical (e.g., secure buildings) and non-physical (e.g., background checks) requirements that must be met before experiments or even possession of regulated items can be obtained. Implicit requirements work by creating an "entry fee." To be able to perform research with Select Agents, an organization must be able to afford the initial (constructing a secure building) and ongoing (training and background investigations) costs. By increasing the number of requirements, the associated entry fee increases thereby reducing the number of entities capable of meeting and sustaining the required funding.

In the case of immunomodulatory research and development, some degree of restrictions exists. Experiments that would alter the immunomodulatory capacity of dangerous pathogens are already subject to restrictions and regulation [112, 113]. Some experiments, such as those designing immune-evasion coatings for gene therapy vectors, should be considered under these rules even though they do not involve dangerous pathogens. Even when provided to relatively safe infectious organisms like gene therapy vectors, imparting impunity on any infectious agent is unquestionably dangerous from a biosafety perspective. However, nearly all immunomodulatory research and development do not involve infectious organisms. More often the information, and not the products themselves, are most readily abusable. A viable BW program will already need the ability to obtain, produce and manipulate pathogens with or without immunomodulatory modifications. A BW program would presumably test any potential BW extensively. An incredibly small amount of genetically modified influenza A virus or immunomodulatory antibody in the freezer of a laboratory in a foreign nation is of little use given the extensive production and testing that would still be needed to make a BW. What is useful, however, is the information those experiments or technologies generate.

For example, TGN1412, an immunomodulatory antibody, causes severe and damaging inflammation in humans. One can assume that the inflammation caused by TGN1412 could seriously exacerbate the already harmful inflammation caused by a pathogen. To implement a TGN1412-like effect in a BW, an adversary would need to produce the antibody or alter a pre-existing agent to express the antibody. Production of TGN1412 in quantities relevant for weapons and developing delivery methods are challenging tasks that would not be adequately supported by theft or obtaining small research or therapeutic amounts of the substance. Experiments need not have such dramatic results to be useful in designing BW. Experiments that identify a mechanism that protects against infection, a factor associated with poor prognosis during infection, or even the function of basic components of the immune system all provide a BW designer with options and information.

Regulating immunological experiments based on the risk they create should be based primarily on the information and not the products generated. Unfortunately, the true degree of risk may only be known to the person designing the BW. It is impossible to understand the risk of an experiment in all but the most obvious cases. Therefore, excluding civilian biosafety concerns or obvious dual-use experiments, regulation of immunomodulatory or basic immunology research is unlikely to have a meaningful impact. An exception is in the case of experiments that may broadly impact the efficacy of vaccines or other immunology-based countermeasures to dangerous pathogens. Given the existential threat to vaccines and therapeutic anti-pathogen antibodies, antibody-evasion approaches should be halted, and alternatives such as non-viral vectors should be explored. However, as these technologies have already been publicly established, and if independently discovered by full-scale BW program, taking an active stance and exploring therapies and vaccines not reliant on pathogen-neutralizing antibodies may be prudent to consider. Unfortunately, a similar case exists for (non-coated) virotherapies and immunomodulatory antibodies. The information is already publicly available, and there is no going back, making a proactive defense a viable option.

Similarly, implicit regulations, those that restrict research by implementing financial or technical barriers to entry, are unlikely to work. Requiring enhanced biosafety, physical security or personnel security to perform immunomodulatory research and development are unlikely to have a meaningful impact. The majority of immunomodulatory research and development do not require infectious organisms, making requirements for enhanced biosafety unnecessary. Although not able to replace laboratory experimentation, computational approaches require no biological material. Unless coupled with publishing restrictions, requiring high levels of biosafety in immunomodulatory research would not stop the proliferation of dual-use information. In terms of physical or personnel security, background checks are not needed. Why would the effort be made to surreptitiously obtain data when one can wait for the publications or simply ask for the data? Many of the techniques used in immunomodulatory research are shared across other subjects of the life sciences, making training readily available through unrelated disciplines. Obtaining small amounts of antibody or even a modified pathogen may save an adversary time but is not required if the adversary is able to pursue advanced BW. Disqualifying someone from access to these technologies may be tantamount to accusing the person, and by extension his/her nationality or other affiliations, of taking part in an active BW program, which could have significant ramifications. An exception does exist for experiments during which infectious agents, even benign, have their immunological profiles modified. Experiments performed with "stealth" viruses, or those that are designed to trigger potent immune responses against tumors, may pose a risk to the laboratory workers and even the public in the event of exposure or release.

Science has long been conducted in a transparent fashion. Scientists publish experimental results in publicly (though not always freely) available journals, present at conferences, and typically openly share data and or reagents with others. Even in the pharmaceutical industry, in which companies face extremely intense competition to develop novel compounds and approaches to combatting disease, data on the function, mechanism, and composition of a drug are available to the public. As information is the core risk of immunomodulatory research and development, the wisdom of defaulting to maximum transparency comes into question. Freely publishing and sharing information that is likely dual-use information is the primary mechanism by which proliferation may occur.

In some cases, information is risky enough that putting it into the public space may allow an adversary to make a particularly dangerous BW. The flaw in this thinking is deriving security through obscurity, meaning that an adversary can only learn of the data if it is divulged. Censorship faces the same challenge of regulation: the seriousness of an experiment's risk is masked by how obvious is the risk. Censorship is likely to catch most of the too risky experiments, but the bulk of the risks may be not obvious enough to be detected or to justify censorship. There is also the implementation of censorship: How are the results, data and physical objects retained safely, and who guides these decisions by which criteria? A classified biodefense program is likely hardened against physical or electronic attack, but the typical academic laboratory does not have protections against electronic eavesdropping or hacking. If the information was not discovered inside a secure environment, one can assume that the data may already be compromised when proposing censorship.

Even after the information is transferred to a classified setting, the scientists who first created the work may have shared the data without malice at conferences, seminars and casual discussions with other scientists before censorship. A significant insider threat may come from the people who performed work that becomes censored or classified. As implementation of the strict background investigations typically required for access to classified material is not feasible or reasonable, any retroactively censored or classified material would be at risk. Financial motivation from being underpaid and overworked, threats to family members abroad, and ideological objectives may drive scientists to divulge or sell any censored information.

Transfer of work into classified settings may be viewed as nefarious by other nations especially if that nation was unaware of the civilian work in the first place. In this scenario, officials from one nation noticed that another nation moved an academic's work to a classified biodefense facility. How are these officials to judge the intent of that decision? On one hand, it may be that the information is being withheld till a countermeasure can be developed, but a countermeasure against a BW cannot be tested without that BW or a suitable model. How do officials from one nation know if the development of this model is within the confines of defensive work? If a discovery is dangerous enough to warrant such extreme efforts, why would that information not be shared to protect global health?

A possible solution to this dilemma may be through improved transparency of ongoing civilian research and more importantly, sharing the plans and criteria for censoring findings or moving research into non-public settings. One such mechanism may be to include civilian immunomodulatory research programs and dual-use policies in the confidence building measures (CBMs) of the Biological Weapons Convention (BWC). Put into force in 1975, the BWC is an international treaty that outlaws the development, production, or stockpiling of BW [114]. In 1986 at a conference on the BWC, the CBMs were introduced as a mechanism to enhance the effectiveness of the treaty and international cooperation in biosecurity [115]. Although the CBMs are intended to reduce ambiguities in defensive research, such measures are not currently suited for modern dual-use research issues [116]. The massive amount of potentially dual-use research may lead to a flood of information and would be difficult to fully capture without extensive efforts on behalf of signatory nations. Furthermore, nations may not want to define any part of their biodefense programs or put valuable pharmaceutical developments at risk for industrial espionage. Together these risks could drive nations away from CBM reporting altogether, which are already in steep decline [116]. Finally, expanded CBMs would still do nothing to solve the issue of intent versus interpretation.

The leading argument for publishing highly risky research is that public health preparedness and countermeasures can be developed. In the case of obvious high-risk experiments, the link between open publication and the development of specific countermeasures is tenuous. The closest success of the "publish and fix" model was in the case of the identification of a new form of botulism toxin not neutralized by standard doses of a critical antibody-based countermeasure [117, 118]. Produced by the bacterium *Clostridium botulinum*, botulism toxins are proteins that are extremely toxic to humans and have been explored as BW agents [14, 119, 120]. Treating botulism relies on the use of antibodies which neutralize the toxins [119], and the knowledge of a treatment-resistant toxin is a significant biosecurity issue. Although the discovery triggered efforts to identify a countermeasure to the novel toxin, the research was not published openly as the protein sequence of the toxin and the genome of the strain that produced it were withheld [121]. In the end, the new toxin was neutralized by high doses of the standard countermeasure [122]. The experience with the novel botulism toxin is mixed: on one hand it was identified, published, and a countermeasure was identified. On the other hand, critical information was with-held likely slowing progress.

The novel botulism toxin experience is not directly applicable to cases involving pathogens with modified immunological properties. The cowpox interleukin 4 experiments performed in response to ectromelia virus expressing interleukin 4, and the experiments involving the replication of vaccine-resistant *B. anthracis* were met with significant criticism. The success of these experiments is not publicly known [108, 109, 123]. The difference between the outcome from the botulinum toxin experiment and the poxviruses and *B. anthracis* experiments appears to be in the perception of the work, rather than intent or scientific merit. In the case of the novel botulism toxin, the perception was one of identifying a countermeasure against something naturally occurring. In contrast, the cowpox virus and *B. anthracis* experiments were perceived as creating dangerous products that would not otherwise exist. Although the modified pathogens were indeed created, the vaccine-resistant *B. anthracis* had already been generated and was only recreated because of Russian refusal to share samples. Furthermore, the research that inspired them was openly available. The vaccine-resistant *B. anthracis* had likely been explored as a BW agent [14, 57, 75]. These experiences highlight the importance in understanding that the bulk of the risk does not arise from the physical products of experiments, but from the information obtained. Perhaps if the public had a better understanding of the reality of the threats, and the existence of at least one of these agents, the responses to these defensive efforts may have been different. Failure by any group of stakeholders to understand the nature of the risks created by immunomodulatory research and development may jeopardize the success of countermeasure development to evolving threats when, and whether development is warranted.

Perhaps the best path forward is to begin with education. All stakeholders must assume dual-use risks are inherent to all aspects of immunomodulatory research and development, and that every research effort or technological development that can positively impact human health also has the capability to harm it. We should try asking: What is the general capability afforded? How can such capability lead to benefit? How can the capability can lead to harm? We must not limit our view of what may be harmed to human health: agriculture, defense, policy, society, and the environment are also at risk. Our ability to improve treatments for a range of diseases is highly coupled with our capacity to control and manipulate the immune system. At each step, progress in aiding human health will come with the risk of creating or proliferating BW. Scientific progress must continue but should not be unrestrained by the risks such progress creates. Do we help or save the lives of few at the cost of putting the health of many at risk, however small that risk likely is? What should be done about the technologies and research already in existence, such as those covered in this chapter? Should threats be actively searched for or identified passively? When a threat is identified, should countermeasures be developed at the risk of generating more dual-use technologies and/or further blurring the line between offensive and defensive research? In the end, the answers to these questions are mostly not technological but moral. As a society, we must decide what costs we are willing to pay, and what risks we are willing to live with. Satisfactory answers can only be found if educated scientists, government and society participate in the discussion.

Acknowledgments This work was funded in part through the NIAID Division of Intramural Research. The content of this publication does not necessarily reflect the views or policies of the US Department of Health and Human Services (DHHS) or of the institutions and companies affiliated with the authors. Mention of trade names, commercial products, or organizations do not imply endorsement by the U.S. Government.

References

1. Pulendran B, Ahmed R. Immunological mechanisms of vaccination. Nat Immunol. 2011;12: 509–17.
2. Becher B, Spath S, Goverman J. Cytokine networks in neuroinflammation. Nat Rev Immunol. 2017;17:49–59. https://doi.org/10.1038/nri.2016.123.
3. Brownlie RJ, Zamoyska R. T cell receptor signalling networks: branched, diversified and bounded. Nat Rev Immunol. 2013;13:257–69. https://doi.org/10.1038/nri3403.

4. Worbs T, Hammerschmidt SI, Forster R. Dendritic cell migration in health and disease. Nat Rev Immunol. 2017;17:30–48. https://doi.org/10.1038/nri.2016.116.
5. Hunig T. The storm has cleared: lessons from the CD28 superagonist TGN1412 trial. Nat Rev Immunol. 2012;12:317–8. https://doi.org/10.1038/nri3192.
6. Smolen JS, Aletaha D, McInnes IB. Rheumatoid arthritis. Lancet. 2016;388:2023–38. https://doi.org/10.1016/s0140-6736(16)30173-8.
7. Weaver SC, Ferro C, Barrera R, Boshell J, Navarro JC. Venezuelan equine encephalitis. Annu Rev Entomol. 2004;49:141–74. https://doi.org/10.1146/annurev.ento.49.061802.123422.
8. Pauken KE, Wherry EJ. Overcoming T cell exhaustion in infection and cancer. Trends Immunol. 2015;36:265–76. https://doi.org/10.1016/j.it.2015.02.008.
9. Touzot F, Hacein-Bey-Abina S, Fischer A, Cavazzana M. Gene therapy for inherited immunodeficiency. Expert Opin Biol Ther. 2014;14:789–98. https://doi.org/10.1517/14712598.2014.895811.
10. Vinay DS, et al. Immune evasion in cancer: mechanistic basis and therapeutic strategies. Semin Cancer Biol. 2015;35(Suppl):S185–98. https://doi.org/10.1016/j.semcancer.2015.03.004.
11. Zaiss AK, Muruve DA. Immunity to adeno-associated virus vectors in animals and humans: a continued challenge. Gene Ther. 2008;15:808–16. https://doi.org/10.1038/gt.2008.54.
12. Singh P, Pal SA, Alex A, Agarwal N. Development of PROSTVAC immunotherapy in prostate cancer. Future Oncol. 2015;11:2137–48.
13. Franz DR, Parrott CD, Takafuj ET. The U.S. biological warfare and biological defense programs. In: Medical aspects of chemical and biological warfare. Washington, DC: TMM Publications; 1997. p. 425–36.
14. Leitenberg M, Zilinskas RA, Kuhn JH. The Soviet biological weapons program. Cambridge: Harvard University Press; 2012.
15. Stepanov AV, Marinin LI, Pomerantsev AP, Staritsin NA. Development of novel vaccines against anthrax in man. J Biotechnol. 1996;44:155–60. https://doi.org/10.1016/0168-1656(95)00092-5.
16. Flyak AI, et al. Mechanism of human antibody-mediated neutralization of Marburg virus. Cell. 2015;160:893–903. https://doi.org/10.1016/j.cell.2015.01.031.
17. Iannello A, Ahmad A. Role of antibody-dependent cell-mediated cytotoxicity in the efficacy of therapeutic anti-cancer monoclonal antibodies. Cancer Metastasis Rev. 2005;24:487–99. https://doi.org/10.1007/s10555-005-6192-2.
18. Meyer S, Leusen JH, Boross P. Regulation of complement and modulation of its activity in monoclonal antibody therapy of cancer. MAbs. 2014;6:1133–44. https://doi.org/10.4161/mabs.29670.
19. Williams SC. The other microbiome. Proc Natl Acad Sci U S A. 2013;110:2682–4. https://doi.org/10.1073/pnas.1300923110.
20. Wylie KM, Weinstock GM, Storch GA. Emerging view of the human virome. Transl Res. 2012;160:283–90. https://doi.org/10.1016/j.trsl.2012.03.006.
21. Cooper NR, Nemerow GR. The role of antibody and complement in the control of viral infections. J Investig Dermatol. 1984;83:S121–7. https://doi.org/10.1038/jid.1984.33.
22. Parenti G, Andria G, Ballabio A. Lysosomal storage diseases: from pathophysiology to therapy. Annu Rev Med. 2015;66:471–86. https://doi.org/10.1146/annurev-med-122313-085916.
23. Kaufmann KB, Buning H, Galy A, Schambach A, Grez M. Gene therapy on the move. EMBO Mol Med. 2013;5:1642–61. https://doi.org/10.1002/emmm.201202287.
24. Athanasopoulos T, Munye MM, Yanez-Munoz RJ. Nonintegrating gene therapy vectors. Hematol Oncol Clin North Am. 2017;31:753–70. https://doi.org/10.1016/j.hoc.2017.06.007.
25. Lasaro MO, Ertl HC. New insights on adenovirus as vaccine vectors. Mol Ther. 2009;17:1333–9. https://doi.org/10.1038/mt.2009.130.

26. Liu TC, Thorne SH, Kirn DH. Oncolytic adenoviruses for cancer gene therapy. Methods Mol Biol. 2008;433:243–58.
27. Dias JD, et al. Targeted cancer immunotherapy with oncolytic adenovirus coding for a fully human monoclonal antibody specific for CTLA-4. Gene Ther. 2012;19:988–98. https://doi.org/10.1038/gt.2011.176.
28. Sofer-Podesta C, Ang J, Hackett NR, Senina S, Perlin D, Crystal RG, Boyer JL. Adenovirus-mediated delivery of an anti-V antigen monoclonal antibody protects mice against a lethal Yersinia pestis challenge. Infect Immun. 2009;77:1561–8. https://doi.org/10.1128/IAI.00856-08.
29. Qiu Q, Xu Z, Tian J, Moitra R, Gunti S, Notkins AL, Byrnes AP. Impact of natural IgM concentration on gene therapy with adenovirus type 5 vectors. J Virol. 2015;89:3412–6. https://doi.org/10.1128/JVI.03217-14.
30. Zak DE, et al. Merck Ad5/HIV induces broad innate immune activation that predicts CD8(+) T-cell responses but is attenuated by preexisting Ad5 immunity. Proc Natl Acad Sci U S A. 2012;109:E3503–12. https://doi.org/10.1073/pnas.1208972109.
31. Milligan ID, et al. Safety and immunogenicity of novel adenovirus type 26- and modified vaccinia Ankara-aectored Ebola vaccines: a randomized clinical trial. JAMA. 2016;315: 1610–23. https://doi.org/10.1001/jama.2016.4218.
32. Croyle MA, et al. PEGylation of a vesicular stomatitis virus G pseudotyped lentivirus vector prevents inactivation in serum. J Virol. 2004;78:912–21.
33. Heider S, Kleinberger S, Kochan F, Dangerfield JA, Metzner C. Immune protection of retroviral vectors upon molecular painting with the complement regulatory protein CD59. Mol Biotechnol. 2016;58:480–8. https://doi.org/10.1007/s12033-016-9944-z.
34. Stanley DA, et al. Chimpanzee adenovirus vaccine generates acute and durable protective immunity against ebolavirus challenge. Nat Med. 2014;20:1126–9. https://doi.org/10.1038/nm.3702.
35. Cattaneo R, Miest T, Shashkova EV, Barry MA. Reprogrammed viruses as cancer therapeutics: targeted, armed and shielded. Nat Rev Microbiol. 2008;6:529–40. https://doi.org/10.1038/nrmicro1927.
36. Choi JW, Lee YS, Yun CO, Kim SW. Polymeric oncolytic adenovirus for cancer gene therapy. J Control Release. 2015;219:181–91. https://doi.org/10.1016/j.jconrel.2015.10.009.
37. Zeng Q, Han J, Zhao D, Gong T, Zhang Z, Sun X. Protection of adenovirus from neutralizing antibody by cationic PEG derivative ionically linked to adenovirus. Int J Nanomedicine. 2012;7:985–97. https://doi.org/10.2147/ijn.S27526.
38. Doronin K, Shashkova EV, May SM, Hofherr SE, Barry MA. Chemical modification with high molecular weight polyethylene glycol reduces transduction of hepatocytes and increases efficacy of intravenously delivered oncolytic adenovirus. Hum Gene Ther. 2009;20:975–88. https://doi.org/10.1089/hum.2009.028.
39. Coughlan L, Alba R, Parker AL, Bradshaw AC, McNeish IA, Nicklin SA, Baker AH. Tropism-modification strategies for targeted gene delivery using adenoviral vectors. Viruses. 2010;2:2290–355. https://doi.org/10.3390/v2102290.
40. Rojas LA, Condezo GN, Moreno R, Fajardo CA, Arias-Badia M, San Martin C, Alemany R. Albumin-binding adenoviruses circumvent pre-existing neutralizing antibodies upon systemic delivery. J Control Release. 2016;237:78–88. https://doi.org/10.1016/j.jconrel.2016.07.004.
41. Friedlander AM, et al. Postexposure prophylaxis against experimental inhalation anthrax. J Infect Dis. 1993;167:1239–43.
42. Sabhnani L, Manocha M, Tomar D, Shashikiran D, Rao DN. Yersinia pestis F1 antigen: a correlation between antibody titres and subclass distribution with differential avidity in different inbred mouse strains. Int Immunopharmacol. 2003;3:1413–8. https://doi.org/10.1016/s1567-5769(03)00138-3.
43. Skyberg JA. Immunotherapy for tularemia. Virulence. 2013;4:859–70. https://doi.org/10.4161/viru.25454.

44. Price JL, Manetz TS, Shearer JD, House RV. Preclinical safety assessment of a recombinant plague vaccine (rF1V). Int J Toxicol. 2013;32:327–35. https://doi.org/10.1177/1091581813497405.
45. Adekar SP, et al. Neutralization of botulinum neurotoxin by a human monoclonal antibody specific for the catalytic light chain. PLoS One. 2008;3:e3023. https://doi.org/10.1371/journal.pone.0003023.
46. Chen Z, Moayeri M, Purcell R. Monoclonal antibody therapies against anthrax. Toxins (Basel). 2011;3:1004–19. https://doi.org/10.3390/toxins3081004.
47. Qiu X, et al. Reversion of advanced Ebola virus disease in nonhuman primates with ZMapp. Nature. 2014;514:47–53. https://doi.org/10.1038/nature13777.
48. Froude JW, Stiles B, Pelat T, Thullier P. Antibodies for biodefense. MAbs. 2011;3:517–27. https://doi.org/10.4161/mabs.3.6.17621.
49. Marston HD, Paules CI, Fauci AS. Monoclonal antibodies for emerging infectious diseases – borrowing from history. N Engl J Med. 2018;378:1469–72. https://doi.org/10.1056/NEJMp1802256.
50. Wheelis M. Biological warfare at the 1346 siege of Caffa. Emerg Infect Dis. 2002;8:971–5.
51. Du Y. Role of Fraction 1 antigen of Yersinia pestis in Inhibition of phagocytosis. Infect Immun. 2002;70:1453–60. https://doi.org/10.1128/iai.70.3.1453-1460.2002.
52. Sha J, et al. Characterization of an F1 deletion mutant of Yersinia pestis CO92, pathogenic role of F1 antigen in bubonic and pneumonic plague, and evaluation of sensitivity and specificity of F1 antigen capture-based dipsticks. J Clin Microbiol. 2011;49:1708–15. https://doi.org/10.1128/JCM.00064-11.
53. Oyston PC. Francisella tularensis: unravelling the secrets of an intracellular pathogen. J Med Microbiol. 2008;57:921–30. https://doi.org/10.1099/jmm.0.2008/000653-0.
54. Falugi F, Kim HK, Missiakas DM, Schneewind O. Role of protein A in the evasion of host adaptive immune responses by Staphylococcus aureus. MBio. 2013;4:e00575–13. https://doi.org/10.1128/mBio.00575-13.
55. Bottone EJ. Bacillus cereus, a volatile human pathogen. Clin Microbiol Rev. 2010;23:382–98. https://doi.org/10.1128/CMR.00073-09.
56. Ramarao N, Sanchis V. The pore-forming haemolysins of bacillus cereus: a review. Toxins (Basel). 2013;5:1119–39. https://doi.org/10.3390/toxins5061119.
57. Marninin LI, Mockov YV, Staritsin NA, Pomerantzev AP. Expression of cereolysine AB genes in Bacillus anthracis vaccine strain ensures protection agains experimental hymolytic anthrax infection. Vaccine. 1997;15:1846–50.
58. Palmqvist N, Foster T, Tarkowski A, Josefsson E. Protein A is a virulence factor in Staphylococcus aureus arthritis and septic death. Microb Pathog. 2002;33:239–49. https://doi.org/10.1006/mpat.2002.0533.
59. Bolton GR, Mehta KK. The role of more than 40 years of improvement in protein A chromatography in the growth of the therapeutic antibody industry. Biotechnol Prog. 2016;32:1193–202. https://doi.org/10.1002/btpr.2324.
60. Goding JW. Use of staphylococcal protein A as an immunological reagent. J Immunol Methods. 1978;20:241–53.
61. National Cancer Institute. What is cancer? 2015. https://www.cancer.gov/about-cancer/understanding/what-is-cancer. Accessed 20 Sept 2017.
62. World Health Organization. The top 10 causes of death. 2017. http://www.who.int/mediacentre/factsheets/fs310/en/. Accessed 20 Sept 2017.
63. National Cancer Institute. Chemotherapy. 2015. https://www.cancer.gov/about-cancer/treatment/types/chemotherapy#1. Accessed 20 Sept 2017.
64. Nicolini A, Ferrari P, Duffy MJ. Prognostic and predictive biomarkers in breast cancer: Past, present and future. Semin Cancer Biol. 2017;52(Pt 1):56–73. https://doi.org/10.1016/j.semcancer.2017.08.010.
65. Marusyk A, Polyak K. Tumor heterogeneity: causes and consequences. Biochim Biophys Acta. 2010;1805:105–17. https://doi.org/10.1016/j.bbcan.2009.11.002.

66. Russell SJ, Peng KW, Bell JC. Oncolytic virotherapy. Nat Biotechnol. 2012;30:658–70. https://doi.org/10.1038/nbt.2287.
67. Yarchoan M, Johnson BA, Lutz ER, Laheru DA, Jaffee EM. Targeting neoantigens to augment antitumour immunity. Nat Rev Cancer. 2017;17:209–22. https://doi.org/10.1038/nrc.2016.154.
68. Rehman H, Silk AW, Kane MP, Kaufman HL. Into the clinic: Talimogene laherparepvec (T-VEC), a first-in-class intratumoral oncolytic viral therapy. J Immunother Cancer. 2016;4:53. https://doi.org/10.1186/s40425-016-0158-5.
69. Lun X, et al. Efficacy and safety/toxicity study of recombinant vaccinia virus JX-594 in two immunocompetent animal models of glioma. Mol Ther. 2010;18:1927–36. https://doi.org/10.1038/mt.2010.183.
70. Aarts WM, Schlom J, Hodge JW. Vector-based vaccine/cytokine combination therapy to enhance induction of immune responses to a self-antigen and antitumor activity. Cancer Res. 2002;62:5770–7.
71. Madan RA, Bilusic M, Heery C, Schlom J, Gulley JL. Clinical evaluation of TRICOM vector therapeutic cancer vaccines. Semin Oncol. 2012;39:296–304. https://doi.org/10.1053/j.seminoncol.2012.02.010.
72. Hawley RJ, Edward EM. Biological weapons – a primer for microbiologists. Annu Rev Microbiol. 2001;55:235–53.
73. Zacks MA, Paessler S. Encephalitic alphaviruses. Vet Microbiol. 2010;140:281–6. https://doi.org/10.1016/j.vetmic.2009.08.023.
74. Hoyer KK, Dooms H, Barron L, Abbas AK. Interleukin-2 in the development and control of inflammatory disease. Immunol Rev. 2008;226:19–28.
75. Jackson RJ, Ramsay AJ, Christensen CD, Beaton S, Hall DF, Ramshaw IA. Expression of mouse interleukin-4 by a recombinant ectromelia virus suppresses cytolytic lymphocyte responses and overcomes genetic resistance to mousepox. J Virol. 2001;75:1205–10. https://doi.org/10.1128/JVI.75.3.1205-1210.2001.
76. MacKenzie D. US develops lethal new viruses. 2003. https://www.newscientist.com/article/dn4318-us-develops-lethal-new-viruses/. Accessed 20 Sept 2017.
77. Quaranta S, et al. Autoantibodies in human chronic graft-versus-host disease after hematopoietic cell transplantation. Clin Immunol. 1999;91:106–16.
78. Sayegh MH, Turka LA. The role of T-cell costimulatory activation pathways in transplant rejection. N Engl J Med. 1998;338:1813–21.
79. Dutsch-Wicherek M, Kazmierczak W. Creation of a suppressive microenvironment by macrophages and cancer-associated fibroblasts. Front Biosci (Landmark Ed). 2013;18:1003–16.
80. Jiang Y, Li Y, Zhu B. T-cell exhaustion in the tumor microenvironment. Cell Death Dis. 2015;6:e1792. https://doi.org/10.1038/cddis.2015.162.
81. Oleinika K, Nibbs RJ, Graham G, Fraser AR. Suppression, subversion and escape: the role of regulatory T cells in cancer progression. Clin Exp Immunol. 2013;171:36–45. https://doi.org/10.1111/j.1365-2249.2012.04657.x.
82. Draenert R, Seybold U, Grutzner E, Bogner JR. Novel antibiotics: are we still in the pre-post-antibiotic era? Infection. 2015;43:145–51. https://doi.org/10.1007/s15010-015-0749-y.
83. Alter G, Sekaly RP. Beyond adjuvants: antagonizing inflammation to enhance vaccine immunity. Vaccine. 2015;33(Suppl 2):B55–9. https://doi.org/10.1016/j.vaccine.2015.03.058.
84. Calik Basaran N, Ascioglu S. Epidemiology and management of healthcare-associated bloodstream infections in non-neutropenic immunosuppressed patients: a review of the literature. Ther Adv Infect Dis. 2017;4:171–91. https://doi.org/10.1177/2049936117733394.
85. Weiner LM, Surana R, Wang S. Monoclonal antibodies: versatile platforms for cancer immunotherapy. Nat Rev Immunol. 2010;10:317–27. https://doi.org/10.1038/nri2744.
86. Maury S, et al. Rituximab in B-lineage adult acute lymphoblastic leukemia. N Engl J Med. 2016;375:1044–53. https://doi.org/10.1056/NEJMoa1605085.

87. Shetty A, et al. Tocilizumab in the treatment of rheumatoid arthritis and beyond. Drug Des Devel Ther. 2014;8:349–64. https://doi.org/10.2147/DDDT.S41437.
88. Rudnicka D, et al. Rituximab causes a polarization of B cells that augments its therapeutic function in NK-cell-mediated antibody-dependent cellular cytotoxicity. Blood. 2013;121: 4694–702. https://doi.org/10.1182/blood-2013-02-482570.
89. de Lemos LLP, et al. Rituximab for rheumatoid arthrits treatment: a systematic review. Rev Bras Reumatol Engl Ed. 2014;54:220–30. https://doi.org/10.1016/j.rbre.2013.08.003.
90. Dierickx D, Kentos A, Delannoy A. The role of rituximab in adults with warm antibody autoimmune hemolytic anemia. Blood. 2015;125:3223–9. https://doi.org/10.1182/blood-2015-01-588392.
91. Maloney DG, et al. IDEC-C2B8 (Rituximab) anti-CD20 monoclonal antibody therapy in patients with relapsed low-grade non-Hodgkin's lymphoma. Blood. 1997;90:2188–95.
92. Mok CC. Current role of rituximab in systemic lupus erythematosus. Int J Rheum Dis. 2015;18:154–63. https://doi.org/10.1111/1756-185X.12463.
93. Noy R, Pollard JW. Tumor-associated macrophages: from mechanisms to therapy. Immunity. 2014;41:49–61. https://doi.org/10.1016/j.immuni.2014.06.010.
94. Pardoll DM. The blockade of immune checkpoints in cancer immunotherapy. Nat Rev Cancer. 2012;12:252–64. https://doi.org/10.1038/nrc3239.
95. Adelfinger M, Bessler S, Frentzen A, Cecil A, Langbein-Laugwitz J, Gentschev I, Szalay AA. Preclinical testing oncolytic vaccinia virus strain GLV-5b451 expressing an anti-VEGF single-chain antibody for canine cancer therapy. Viruses. 2015;7:4075–92. https://doi.org/10.3390/v7072811.
96. Suscovich TJ, Alter G. In situ production of therapeutic monoclonal antibodies. Expert Rev Vaccines. 2015;14:205–19. https://doi.org/10.1586/14760584.2015.1001375.
97. Adachi Y, Yoshio-Hoshino N, Aoki C, Nishimoto N. VEGF targeting in mesotheliomas using an interleukin-6 signal inhibitor based on adenovirus gene delivery. Anticancer Res. 2010;30: 1947–52.
98. Tutykhina IL, et al. Passive immunization with a recombinant adenovirus expressing an HA (H5)-specific single-domain antibody protects mice from lethal influenza infection. Antivir Res. 2013;97:318–28. https://doi.org/10.1016/j.antiviral.2012.12.021.
99. Tyrsin D, Chuvpilo S, Matskevich A, Nemenov D, Römer PS, Tabares P, Hünig T. From TGN1412 to TAB08: the return of CD28 superagonist therapy to clinical development for the treatment of rheumatoid arthritis. Clin Exp Rheumatol. 2016;34:S45–8.
100. Respaud R, et al. Effect of formulation on the stability and aerosol performance of a nebulized antibody. MAbs. 2014;6:1347–55. https://doi.org/10.4161/mabs.29938.
101. Storti C, Le Noci V, Sommariva M, Tagliabue E, Balsari A, Sfondrini L. Aerosol delivery in the treatment of lung cancer. Curr Cancer Drug Targets. 2015;15:604–12.
102. Mathew M, Verma RS. Humanized immunotoxins: a new generation of immunotoxins for targeted cancer therapy. Cancer Sci. 2009;100:1359–65. https://doi.org/10.1111/j.1349-7006.2009.01192.x.
103. Roy CJ, et al. Aerosolized adenovirus-vectored vaccine as an alternative vaccine delivery method. Respir Res. 2011;12:153. https://doi.org/10.1186/1465-9921-12-153.
104. White AD, et al. Evaluation of the safety and immunogenicity of a candidate tuberculosis vaccine, MVA85A, delivered by aerosol to the lungs of macaques. Clin Vaccine Immunol. 2013;20:663–72. https://doi.org/10.1128/cvi.00690-12.
105. Melo AL, Soccol VT, Soccol CR. Bacillus thuringiensis: mechanism of action, resistance, and new applications: a review. Crit Rev Biotechnol. 2016;36:317–26. https://doi.org/10.3109/07388551.2014.960793.
106. Erlendur H, et al. Bacillus anthracis, Bacillus cereus, and Bacillus thuringiensis—one species on the basis of genetic evidence. Appl Environ Microbiol. 2000;66:2627–30.
107. Hanham M. Kim Jong Un tours pesticide facility capable of producing biological weapons: a 38 North special report. 2015. http://www.38north.org/2015/07/mhanham070915/. Accessed 7 Feb 2018.

108. Federation of American Scientists. US department of defense news briefing. 2001. https://fas.org/sgp/news/2001/09/dod090401.html. Accessed 25 Sept 2017.
109. Miller, J, Engelberg, S, Broad, WJ. U.S. germ warfare research pushes treaty limits. 2001.
110. CDC/USDA. Federal Select Agent Program.
111. CDC/USDA. Restricted experiments guidance. 2017.
112. Centers for Disease Control and Prevention, United States Department of Agriculture. Federal select agent program. 2017. https://www.selectagents.gov/. Accessed 3 Oct 2017.
113. Government US. United States government policy for institutional oversight of life sciences dual use research of concern. 2014. Online.
114. UNODA. Biological weapons. 2018. https://www.un.org/disarmament/wmd/bio/. Accessed 18 May 2018.
115. UNODA. Guide to participating in the confidence-building measures of the biological weapons convention. 2015.
116. Koblentz GD, Chevrier MI. Modernizing confidence-building measures for the biological weapons convention. Biosecur Bioterror. 2011;9:232–8. https://doi.org/10.1089/bsp.2011.0023.
117. Barash JR, Arnon SS. A novel strain of Clostridium botulinum that produces type B and type H botulinum toxins. J Infect Dis. 2014;209:183–91. https://doi.org/10.1093/infdis/jit449.
118. Dover N, Barash JR, Hill KK, Xie G, Arnon SS. Molecular characterization of a novel botulinum neurotoxin type H gene. J Infect Dis. 2014;209:192–202. https://doi.org/10.1093/infdis/jit450.
119. Arnon SS, et al. Botulinum toxin as a biological weapon: medical and public health management. JAMA. 2001;285:1059–70.
120. Hooper DC, Hirsch MS. Novel clostridium botulinum toxin and dual use research of concern issues. J Infect Dis. 2014;209:167. https://doi.org/10.1093/infdis/jit528.
121. Keim P. A novel botulinum neurotoxin and how it tested our scientific institutions. J Infect Dis. 2016;213:332–4. https://doi.org/10.1093/infdis/jiv328.
122. Maslanka SE, et al. A novel botulinum neurotoxin, previously reported as serotype H, has a hybrid-like structure with regions of similarity to the structures of serotypes A and F and is neutralized with serotype A antitoxin. J Infect Dis. 2016;213:379–85. https://doi.org/10.1093/infdis/jiv327.
123. Broad WJ. Gene-engineered anthrax: is it a weapon? The New York Times. 1998.

Synthetic Biology: Biosecurity and Biosafety Implications

11

Gigi Kwik Gronvall

11.1 Introduction

Synthetic biology is a scientific discipline that aims to make biology more useful, and easier to engineer [1]. Synthetic biology techniques make it easier to genetically manipulate living organisms, such as using the relatively new Clustered Regularly Interspaced Short Palindromic Repeat (CRISPR) to "find and replace" sections of genetic material. A variety of synthetic biology applications are being developed for use in medicine, manufacturing, environmental monitoring, and manufacturing, and synthetic biology products are available, including detergents, adhesives, perfumes, and flavorings [2]. Despite the beneficial advances that may be made with synthetic biology, however, there are security and safety issues to consider: synthetic biology may be misused to increase the range and types of biological weapons that can be developed, and there are new safety considerations for these powerful technologies.

11.2 Biosecurity Considerations of Synthetic Biology

Synthetic biology is a dual-use technology, as many other areas of biology are; as such, entirely preventing the misuse of this field is not possible. There are also reasons for hope as a result of using synthetic biology techniques: if biological weapons are used, synthetic biology holds the potential to speed up the medical response including developing of drugs, vaccines, and other medical

G. K. Gronvall (✉)
Department of Environmental Health and Engineering, Johns Hopkins Bloomberg School of Public Health, Baltimore, MD, USA

Johns Hopkins Center for Health Security, Baltimore, MD, USA
e-mail: ggronvall@jhu.edu

S. K. Singh, J. H. Kuhn (eds.), *Defense Against Biological Attacks*,
https://doi.org/10.1007/978-3-030-03053-7_11

countermeasures more quickly. It is also important to remember that developing a biological weapon does not require synthetic biology or any advanced biotechnologies. Bioweapons were made by state weapons programs prior to synthetic biology technologies being developed, and they may still be developed using classical microbiological techniques. Yet, even though a great deal of progress has been made in biosecurity, a biological weapon could present a major challenge for public health and for national security.

Synthetic biology poses problems for biodefense, as these techniques may help to produce a pathogen "from scratch" in a laboratory, versus requiring that a nefarious actor acquire a sample of a pathogen, obtained from a laboratory, the environment, or from a clinical sample. Access controls to pathogens are a central tenet of biosecurity regulations, so this is a concerning development for biosecurity. In addition, the tools of synthetic biology are making the tools for biological weapons development more accessible to a wide range of people (i.e. the technology is being democratized), and costs for access to these technologies continue to drop.

In the US, as well as in many other countries, the ability to acquire pathogen samples has restrictions, particularly if the pathogen is regulated as a Select Agent, or is on the Australia Group List [3–5]. Multiple levels of specific clearances are required to access a Select Agent in the US, including a security risk assessment (SRA), an electronic records check performed by the Department of Justice (DOJ), and a laboratory inspection performed by the Center for Disease Control and Prevention (CDC) or the US department of Agriculture/Animal and Plant Health Inspection Service (APHIS). Transferring samples of pathogens between laboratories also requires documentation.

Although other methods to acquire pathogens are certainly available, including isolation of a pathogen during an ongoing disease outbreak, or finding another environmental source, the Select Agent rules do make accessing pathogens more difficult [6]. The requirement for an existing sample changes with synthetic biology, however. In 2002, a *Science* magazine article published by researchers at State University of New York (SUNY) at Stony Brook demonstrated the synthetic creation of an infectious poliovirus using commercially supplied pieces of DNA, a process that took 3 years [7, 8]. The synthesis of this poliovirus alarmed policymakers, especially as the genomic sequence of that poliovirus, like those of other viruses, is available on the Internet through the National Center for Biotechnology Information (NCBI) databases.

Only 2 years after the poliovirus synthesis, J. Craig Venter Institute (JCVI) researchers synthesized and rescued (also referred to us "booting up") a bacteriophage from its genetic sequence [9]. This process took 2 weeks, demonstrating that this technology was getting much more powerful in a short period of time. Additional synthesis projects followed. In 2005, a 1918 H1N1 influenza A virus was reconstructed [10]. More recently, a group of Canadian researchers synthesized and booted up horsepox virus, a cousin of variola virus (cause of the feared and now eradicated smallpox), causing additional controversy [11–14].

Although many viruses may be synthesized and booted up, other organisms are not as easily made from scratch. In 2010, J. Craig Venter and colleagues published the synthesis of an entire bacterial genome, but that was not a trivial undertaking [15–17]. It took the JCVI researchers 15 years and $40 million to do this [18].

In 2010, the US Department of Health and Human Services (HHS) issued guidance for US suppliers of double-stranded DNA to screen their customers' orders for prohibited genetic matches, including to regulated pathogens like variola virus; this guidance was adhered to by most global suppliers of gene synthesis products [19]. The HHS guidance also called for enhanced customer screening. Some synthesis companies formed the International Gene Synthesis Consortium (IGSC) to screen their gene synthesis orders; this group claims their 11 DNA providers consists of 80% of the gene synthesis market [20]. The screening protocol consists at least of the US Select Agent list, as well as the Australia Group list agents, the US Commerce Control List and regulated European sequences [21]. Unfortunately, companies that perform screening are currently at a commercial disadvantage as costs for gene synthesis are dropping, but costs for screening remain constant and may increase, particularly as this process requires human judgement for particularly problematic sequences [22]. This has led to concerns that some companies will stop screening [23].

11.2.1 Determining Whether New Advances Raise Biosecurity Concerns: The Imperiale Report

Synthetic biology and other powerful technological advances will continue to raise biosecurity concerns for the foreseeable future, even as they have clear beneficial applications. Given the tremendous volume of scientific information that is published yearly, and other biotechnology developments, a process has been needed to evaluate whether a scientific advancement or publication raised biosecurity concerns.

To address this need, the US Department of Defense Chemical and Biological Defense Program funded the National Academies of Sciences, Engineering, and Medicine to "address the changing nature of the biodefense threat in the age of synthetic biology" [24]. NASEM appointed a committee to develop a framework to identify potential biodefense vulnerabilities posed by synthetic biology. An interim framework has been published and may be used as a tool to aid analysts who are, for example, analyzing biotechnologies to evaluate security concerns about those biotechnologies; to understand how different biotechnologies relate to each other in raising biosecurity concerns; to identify key bottlenecks or barriers that if addressed or removed, would change the level of biosecurity concern for the entire technology; and for horizon scanning or future threat analyses to project how biosecurity concerns may change over time. The framework does not address the *intent* of an actor to use biological weapons. Use of the framework does not require intelligence information, though that may be an important part of some biosecurity analyses.

To assess the capability of malicious use of a biotechnology advance, an analyst should consider the **use of the technology**. This includes its *ease of use* (if a technology is not challenging to use, it is thought to be more likely to be used), the *rate of development of the technology* (a signal that a technology may eventually become a biosecurity concern in the future), the *barriers to use* (technological barriers, such as the inability to predict function from sequence), *synergy with other technologies*, and *cost*. The analyst would also need to consider the potential for a technology to be **used as a weapon**. Such considerations include the capacity for *production and delivery* of the technology, the *scope of casualty* if used as a weapon, the *predictability of result* (describing the degree that a malicious actor could be confident that if a technology would be used to develop a weapon that it would have the intended result), the *testing* required, as well as the *fidelity* of the technology, i.e. whether it is predictable and reliable. The framework also addresses the **attributes of actors**, such as their *access to expertise*, *access to resources*, and whether the technology has *organizational footprint requirements*.

Any biological attack would depend upon both the actor being able to use a technology in the development of a weapon as well as **mitigation factors**, the target's capability to deter, detect, or respond to an attack. The interim report lists some of the *deterrence and prevention* capabilities, asking, for example, "Can the use of the technology be controlled or prevented through regulation or other means?" and "Is the technology geographically centralized, or widely distributed?" Other factors include the *capability to recognize an attack* as well as the capability for *attribution or consequence management*. By analyzing the myriads of aspects that contribute to biosecurity risks of a new biotechnology, analysts may judge whether the biotechnology is concerning for biosecurity.

11.3 Safety Concerns of Synthetic Biology

Synthetic biology has not been associated with any reported biosafety lapses, but the possibility of an accident involving synthetic biology needs to be examined, planned for, and mitigated. This section describes the safety risks that synthetic biology may pose to people, animals, and the environment. Addressing biosafety concerns within synthetic biology will require a range of new training programs and policies, as well as International norms for how nations should train, monitor, and fund biosafety.

There are several potential safety concerns within synthetic biology. The first is that there are synthetic biology products and applications that are necessarily "outside the laboratory" applications. For example, potential applications of synthetic biology that include mosquito control, agriculture, pollution remediation, mining, medications, biofuels, or the re-creation of extinct or endangered animals would not be contained within traditional biocontainment [25]. One approach to biocontainment in these circumstances is called "intrinsic" biosafety—biosafety controls are built into a synthetic organism, so that the synthetic organism cannot "escape" the boundaries set for them. Early attempts to develop intrinsic biosafety often had a single point of failure, and so they often failed, but newer approaches rely on combinatorial complexity to increase the number of failures required before an organism could escape containment. This reduces the rates of failure [26].

Intrinsic biosafety has benefits beyond safety, such as the ability to design a signature or "watermark" into the genome of a synthetic organism, which is helpful for attribution. A DNA "watermark" was put into *Mycoplasma mycoides* JCVI-syn1.0, the synthetic bacterial cell designed by researchers from the J. Craig Venter Institute [27]. There were coded messages placed within the bacterial genome, such as the names of all 46 authors on the publication, an email address, and quotations, including from physicist Richard Feynman, "What I cannot create, I cannot understand" [28, 29]. It is possible that decoy circuits could also be put into organisms which would confuse people intent on stealing synthetic organisms. Possibly, genomes may include sequences that would help the organism self-destruct if certain genetic manipulations are attempted [26].

Another biosafety concern for synthetic biology is that people who have little experience with biosafety and biocontainment may inadvertently cause a problem in their use of synthetic biology tools out of inexperience. One reason why synthetic biologists may not have received biosafety training is because they may be amateurs; such as participants of DIY ("do-it-yourself") Bio. The DIY Bio movement is dwarfed in size by the traditional scientific community, but there are over 4400 subscribers to a DIY Bio online forum, where practitioners may communicate about their projects and arrange for meet ups with other "bio-hackers." Many DIY Bio activities are educational, fun, and safe. Indeed, the DIYbio.org organization was created in 2008 to establish a "vibrant, productive and safe community of DIY biologists." [30] Yet, while DIY Bio projects are generally not advanced, tools for advanced work are accessible. For about $150, an amateur can purchase a gene-editing kit featuring CRISPR—the technology that is revolutionizing the biosciences right now, and which is only a few years old [31]. DIY Bio practitioners may work in community laboratories. Some well-established ones are found in Brooklyn, San Francisco, Baltimore, and Manchester, England but there are DIY Bio community groups all over the world, including New Delhi, Tel Aviv, and São Paulo [30]. Rules for community laboratories tend to be strict, and most operate only at Biosafety Level 1 (the lowest level) of containment [32]. Another amateur synthetic biology environment is iGEM, the International Genetically Engineered Machine Competition [33]. iGEM began as an undergraduate class at the Massachusetts Institute of Technology in 2003 to build synthetic biological systems from standard, interchangeable parts called BioBricks, and to operate them in living cells. Biosafety is integrated into the iGEM competition, and there is a safety committee that evaluates iGEM projects.

There is also a general concern about synthetic biology and biotechnologies more broadly: due to the democratization of powerful technologies, biosafety errors or "bio-errors" may be expected to occur more frequently. If a biosafety accident occurs with a transmissible pathogen, the consequences could spread and potentially cause an epidemic, or other harm to people, animals, or the environment.

The possibility that a novel pathogen could escape containment was a major reason for the controversy surrounding so-called gain-of-function (GOF) influenza A virus research. In this type of research, pathogens are deliberately altered to introduce traits and functions not found in the wild; the research aims, as described by the US government, "to increase the ability of infectious agents to cause disease by enhancing its pathogenicity or by increasing its transmissibility" [34]. The research initially caused controversy in 2011, when two laboratories in the US and Netherlands sought to make H5N1 avian influenza A virus more transmissible between mammals to determine how and whether this transition was likely to occur in nature, and which genetic mutations would indicate that the virus was evolving to become more dangerous to humans. The GOF influenza A virus work that sparked the initial controversy is not technically considered to be synthetic biology, because the experiments were set up so that the virus could *evolve* into a more transmissible pathogen. The virus was not specifically *designed* to be more transmissible. However, the research became emblematic of potential synthetic biology and advanced biotechnology concerns, including the risks of democratization of the technology [35]. Although the original experiments were performed by world-class experts in world-class facilities, synthetic biology techniques could allow such influenza A virus strains to be created in laboratories with less-robust safety systems, health monitoring, and extensive experience with dangerous pathogens. As trends towards democratization and deskilling continue, this may result in more people with the capacity to cause an epidemic, even if only by accident.

Biosafety could be improved if it were specifically studied [36]. Right now, there are not only gaps in knowledge about which incidents and accidents occur in laboratories, but also in how to make scientific research safer [36]. Few studies about how to work more safely in the laboratory are undertaken, to determine which laboratory procedures are more safe, or to investigate which laboratory management techniques may increase safety performance. These gaps in knowledge became evident in the risk-benefit analysis of GOF research: many data gaps were uncovered that could theoretically be addressed by scientific experimentation [37].

More may be done internationally to promote biosafety as well. Biosafety is often not perceived as important enough to afford distinct resources that may be dedicated towards training, oversight, equipment, and standards intended to increase biosafety oversight. Biosafety is also not usually categorized as an important political priority, though when there is a lapse it is politically damaging [38]. Having international expectations for biosafety could be reassuring the next time there is concern about research as there was about gain-of-function research. It would be helpful to know whether the research took place in an environment where there are national standards for the work, such as for equipment maintenance, worker safety training, health monitoring, surveillance, and other activities to help keep the researchers and the larger public safe. It would also be helpful to know whether a particular nation has an adequate surveillance system in place to identify and limit any outbreaks that could result from such accidents. Currently, there is great variability from one research institution to another, even within a nation [39]. Implementing nationwide biosafety norms could begin to address this deficit.

In summary, synthetic biology is a fast-moving field that holds exciting potential, but there are biosafety and biosecurity implications that should be identified and addressed. Doing so will minimize the risks inherent in synthetic biology and to safeguard its enormous promise.

References

1. Synthetic Biology Project. What is synthetic biology? Woodrow Wilson International Center for Scholars. Accessed 31 Jan 2018.
2. Doudna JA, Charpentier E. Genome editing. The new frontier of genome engineering with CRISPR-Cas9. Science. 2014;346(6213):1258096.
3. Uniting and Strengthening America by Providing Appropriate Tools Required to Intercept and Obstruct Terrorism (USA PATRIOT ACT) Act of 2001, §Sec. 817. Expansion of the biological weapons statute. 2001.
4. Federal Select Agent Program. Select agents and toxins list. 2017. http://www.selectagents.gov/SelectAgentsandToxinsList.html. Accessed 31 Jan 2018.
5. The Australia Group. Common control lists. 2017. http://www.australiagroup.net/en/controllists.html. Accessed 16 July 2017.
6. Rambhia KJ, Ribner AS, Gronvall GK. Everywhere you look: select agent pathogens. Biosecur Bioterror. 2011;9(1):69–71.
7. Cello J, Paul AV, Wimmer E. Chemical synthesis of poliovirus cDNA: generation of infectious virus in the absence of natural template. Science. 2002;297(5583):1016–8.
8. Wimmer E. The test-tube synthesis of a chemical called poliovirus. The simple synthesis of a virus has far-reaching societal implications. EMBO Rep. 2006;7(Spec No S3):3–S9.
9. Smith HO, Hutchison CA 3rd, Pfannkoch C, Venter JC. Generating a synthetic genome by whole genome assembly: phiX174 bacteriophage from synthetic oligonucleotides. Proc Natl Acad Sci USA. 2003;100(26):15440–5.
10. Tumpey TM, Basler CF, Aguilar PV, et al. Characterization of the reconstructed 1918 Spanish influenza pandemic virus. Science. 2005;310(5745):77–80.
11. Noyce RS, Lederman S, Evans DH. Construction of an infectious horsepox virus vaccine from chemically synthesized DNA fragments. PLoS One. 2018;13(1):e0188453.
12. Koblentz GD. The de novo synthesis of horsepox virus: implications for biosecurity and recommendations for preventing the reemergence of smallpox. Health Secur. 2017;15(6):620–8.
13. Inglesby T. Important questions global health and science leaders should be asking in the wake of horsepox synthesis. The Bifurcated Needle; 2017.
14. DiEuliis D, Berger K, Gronvall G. Biosecurity implications for the synthesis of horsepox, an Orthopoxvirus. Health Secur. 2017;15(6):629–37.
15. Hutchison CA, Chuang R-Y, Noskov VN, et al. Design and synthesis of a minimal bacterial genome. Science. 2016;(6280):351.
16. Venter JC, Gibson D. How we created the first synthetic cell. The Wall Street Journal; 2010, May 26 (Opinion).
17. J Craig Venter Institute. First self-replicating synthetic bacterial cell frequently asked questions. 2010. http://www.jcvi.org/cms/research/projects/first-self-replicating-synthetic-bacterial-cell/faq. Accessed 4 Oct 2013.
18. Pollack A. His corporate strategy: the scientific method. NY Times. 2010;5:BU1.
19. US Department of Health and Human Services. Screening framework guidance for providers of synthetic double-stranded DNA. Fed Regist. 2010;75(197):62820–32.
20. International Gene Synthesis Consortium (ICSC). Harmonized screening protocol: gene sequence & customer screening to promote biosecurity. 2009.
21. International Gene Synthesis Consortium. International gene synthesis consortium updates screening protocols for synthetic DNA products and service. 2018, January 3.
22. Carter SR, Friedman RM. DNA synthesis and biosecurity: lessons learned and options for the future. San Diego: J. Craig Venter Institute; 2015.
23. Dieuliis D, Carter SR, Gronvall GK. Options for synthetic DNA order screening, revisited. mSphere. 2017;2(4)
24. National Academies of Sciences E, Medicine. A proposed framework for identifying potential biodefense vulnerabilities posed by synthetic biology: interim report. 2017.

25. Church G. De-extinction is a good idea. Scientific American; 2013.
26. Cai Y, Agmon N, Choi WJ, et al. Intrinsic biocontainment: multiplex genome safeguards combine transcriptional and recombinational control of essential yeast genes. Proc Natl Acad Sci USA. 2015;112(6):1803–8.
27. Wade N. Synthetic bacterial genome takes over a cell, researchers report. NY Times; 2010, May 20, A17(L).
28. Grant B. News in a nutshell. Scientist; 2011.
29. J. Craig Venter Institute. Press release: first self-replicating synthetic bacterial cell. 2010, May 20.
30. DIYBio.org web page. Accessed 22 Feb 2016.
31. Zayner J. Learn science by doing. 2016. https://www.indiegogo.com/projects/diy-crispr-kits-learn-modern-science-by-doing#/. Accessed 25 Mar 2016.
32. Baltimore Under Ground Science Space (BUGSS). BUGSS Safety Manual. 2012, July 1.
33. iGEM: Synthetic Biology Based on Standard Parts. 2015. http://igem.org/About. Accessed 1 Apr 2016.
34. White House Office of Science and Technology Policy. Doing diligence to assess the risks and benefits of life sciences gain-of-function research. 2014, October 17.
35. Gronvall GK. H5N1: a case study for dual-use research. Council on Foreign Relations Working Paper, 2013, July.
36. Ritterson R, Casagrande R. Basic scholarship in biosafety is critically needed to reduce risk of laboratory accidents. mSphere. 2017;2(2):e00010–7.
37. Gryphon Scientific. Risk and benefit analysis of gain of function research. 2015.
38. Gronvall GK, Rozo M. Addressing the gap in international norms for biosafety. Trends Microbiol. 2015;23(12):743–4.
39. Gronvall GK, Shearer M, Collins H. National biosafety systems. UPMC Center for Health Security; 2016, July.

Is CRISPR a Security Threat? 12

Sonia Ben Ouagrham-Gormley and Shannon R. Fye-Marnien

12.1 Introduction

When a team of American and European researchers discovered in 2012 that they could reprogram a natural bacterial process called CRISPR and use it to edit potentially any organism, the world became mesmerized by the new technology. Many hailed it for all of the promises that it seems to hold for finding cures for diseases like cancer and HIV/AIDS to eradicating infectious diseases like malaria. Others feared it for the potentially disastrous consequences that deliberate or accidental misuse of the technology can bring about. The rapid speed at which new discoveries are being made, and the fierce competition existing between various leading scientific teams in the U.S., Europe, and China, fuel fears that scientists and governments will not take the time required to evaluate ethical and moral dilemmas associated with altering the human genome, or introducing edited animals that may have a negative and lasting impact on our environment and ecosystem. Security experts and the intelligence community worry that CRISPR might fall into the wrong hands (terrorist groups or hostile states) who could use this technology to design novel harmful agents or bioweapons. In the past 20 years, several other new technologies have been both lauded and feared, but CRISPR seems to be different because of its versatility and apparent simplicity. In this chapter, we review the ethical, safety, and security concerns raised by this new technology and evaluate them in light of the current regulatory framework. We also review the challenges and limitations of the technology. We start by providing an overview of CRISPR's

S. Ben Ouagrham-Gormley (✉)
Schar School of Government and Policy, George Mason University, Arlington, VA, USA
e-mail: sbenouag@gmu.edu

S. R. Fye-Marnien
The Tauri Group, Alexandria, VA, USA
e-mail: shannon.marnien@taurigroup.com

S. K. Singh, J. H. Kuhn (eds.), *Defense Against Biological Attacks*,
https://doi.org/10.1007/978-3-030-03053-7_12

history and its current use. We then address the technology's promises and perils, and conclude with an evaluation of the safety and security threats.

12.2 CRISPR's History

Although much public attention has focused on the 2012 discovery that the CRISPR-Cas9 system could be used as a tool to edit the genome of virtually any organism, CRISPR's discovery dates back to the late 1980s, and has involved a dozen researchers worldwide who have each elucidated a portion of the CRISPR puzzle.

The discovery of CRISPR can be dated to 1987 when a team of Japanese scientists working on *Escherichia coli* first observed the genomic DNA repeats that are the hallmark of the prokaryotic CRISPR system [1]. In subsequent years, similar repeats were found in other bacteria: *Shigella dysenteriae* and *Salmonella Typhimurium* [2], *Mycobacterium tuberculosis* [3], halobacteria [4], cyanobacteria [5], and *Streptococcus* [6]. The origin and function of these repeats was still a mystery at the time of discovery, but the development of genome sequencing technology and the resulting increased number of bacterial sequencing material allowed researchers to observe their presence in a large number of prokaryotes [7], and identify their unique characteristics. In 2000, Spanish researcher Francisco Mojica, who played a major role in characterizing CRISPR, described them as "repeated short elements ... occurring in clusters... regularly spaced by unique intervening sequences of constant length." [8]. Multiple names had been used to describe these repeats until the term CRISPR was introduced by Ruud Jansen in a 2002 publication in which he also identified four CRISPR-associated (Cas) genes, which he observed were always located adjacent to a CRISPR locus [9].

A better understanding of CRISPR's function began to emerge 3 years later. Then, three teams working independently—two in France and one in Spain, including Francisco Mojica—observed that the spacer sequences were derived from foreign genetic material such as that found in prokaryotic-invading viruses or plasmids. When added to host genomes after invasion, the spacer sequences allow their hosts to become immune to future attack [10–12]. One of the French teams also identified three additional Cas genes [13]. These results suggested that the CRISPR system was part of the bacterial immune system that protected against viral and plasmid invaders. A year later, another study made the same hypothesis and offered some details about CRISPR's mechanism. The authors hypothesized that CRISPR operated like a prokaryotic small interfering RNA (siRNA) and protected against viral or plasmid invasion "by base-pairing with the target messenger RNA (mRNAs) and promoting their degradation or translation shutdown" [14].

The theory that CRISPR constitutes a defensive mechanism, however, would not be confirmed until 2007, when a team led by Randolph Barrangou, while working on bacteria used in the dairy industry, tested the theory experimentally. The authors showed that indeed when bacteria were attacked by viruses, they "integrated a new spacer derived from the phage genomic sequence...and removal or addition of particular spacers modified the phage-resistance phenotype of the cell" [15]. Much credit is given to this publication for launching a new era in CRISPR research by demonstrating CRISPR's mechanism [7]. Although credit is well deserved, some believe that it diminishes the impact of the 2005 publications by the French and Spanish teams who established the hypothesis in the first place [16]. Following Barrangou's 2007 publication, other researchers further clarified the CRISPR mechanism of action. In 2008, an international team of European and American scientists highlighted the role of mature CRISPR RNAs, which "serve as small guide RNAs" allowing the Cas protein to interfere with virus proliferation [17]. Another study published the same year demonstrated that CRISPR targets DNA. Subsequent studies further characterized CRISPR's mechanism by showing that the various Cas genes operate differently and achieve different cleavage outcomes. To date, six types of CRISPR-Cas systems have been identified (Types I–VI) together with 19 sub-types, organized into two classes, depending on whether they use a multiprotein complex (Class 1—types I, III, and IV) or a single protein (Class 2—types II, V, VI) for targeting [7]. For example, type I and type III systems use a multiprotein complex for CRISPR RNA (crRNA) targeting, whereas type II system such as CRISPR-Cas 9 uses only one protein for RNA-guided targeting and cleavage of the invader's DNA [18]. On the other hand, type IV systems target RNA instead of DNA [19].

A new milestone in CRISPR's history was achieved in 2012, when an international team led by Jennifer Doudna and Emmanuelle Charpentier demonstrated that Cas9 is directed by not just one but two short RNAs (crRNA and tracrRNA), which their team reprogrammed into a single RNA guide to cut specific DNA sites. This action induced cellular endogenous DNA repair pathways [7, 20, 21]. In other words, the team had created a simpler, more precise tool that could potentially be used to disrupt, insert, or repair specific genes in any organism.

This discovery launched a race to demonstrate the efficiency of CRISPR-Cas9 in human cells, and a year later, three teams—Doudna's at the University of California (UC), Berkeley, George Church's at Harvard University, and Feng Zhang's at the Massachusetts Institute of Technology (MIT) and Harvard—published their findings the same month, showing that CRISPR-Cas9 could indeed be used to edit human and mouse cells [22]. This race also resulted in a very public patent dispute between Doudna's and Zhang's institutions. Although the Broad Institute at MIT, where Zhang's laboratory is located, filed a patent application several months after Doudna's laboratory at Berkeley deposited their request, Zhang's was granted a patent first. UC Berkeley claimed that Zhang's patent interfered with Doudna's application and that his work was an extension of the work done by Doudna, not a new invention. As a result, UC Berkeley argued that Zhang's patent should be redacted. In early 2017, the U.S. Patents Trial and Appeals Board decided in favor of Zhang on the grounds that while Doudna had demonstrated that CRISPR-Cas9 can cut DNA in vitro, Zhang's laboratory showed that it works in human cells, thus making his patent distinct from hers [23]. Two months later, UC Berkeley appealed the decision; However, in September 2018, a U.S. Federal court sided again with the Broad institute, judging that Zhang's patent did not interfere with Doudna's. UC Berkely could in theory pursue the matter further and possibly bring the case before the Supreme Court. But legal experts consider that the September ruling effectively puts an end to the patent dispute [24].

While the case seems to have reached its conclusion in the U.S., Doudna and Emmanuelle Charpentier, her associate now at the Max Planck Institute for Infection Biology, Germany, are winning in the international arena: the European Patent Office, the UK Intellectual Property Office, and China's State Intellectual Property Office have all granted Doudna and Charpentier the patent for CRISPR-Cas9 in cellular and non-cellular settings [24–27].

Meanwhile, the scientific and business communities did not wait for the patent dispute to be resolved. The authors of the three 2013 articles—Doudna and Charpentier on the one hand, Zhang and Church on the other—created companies and start-ups both for licensing purposes and to exploit their discoveries for medical, scientific, and agricultural purposes. Worldwide, scientists started using CRISPR-Cas9 in a variety of studies, ranging from investigating genetic diseases, finding cures for cancer or HIV/AIDS, developing more efficient ways to generate biofuels, non-browning mushrooms, and preventing the spread of insect-borne diseases like malaria.

12.3 Current CRISPR Research

Currently, researchers worldwide are using CRISPR for a variety of applications that could potentially benefit human health, agriculture, and the environment. CRISPR research to date has primarily focused on applications like correcting genetic disorders; preventing and treating cancer; treating chronic viral infections; conferring disease resistance and other desirable traits in crops or animals; creating new or improving established disease models; and optimizing the functionality of cells to produce pharmaceuticals, biofuels, and other beneficial substances more efficiently. Although the majority of research is limited to the laboratory, some projects have already advanced to clinical trials. The US National Library of Medicine's Clinical Trials database shows several CRISPR-related human trials are either under way or pending. So far, the majority of the trials test cancer treatments, and are taking place in China [28]. The New Scientist reported in mid-2017 that as many as 20 human trials would soon be under way, primarily in China [29, 30].

12.3.1 Correcting Genetic Disorders

Researchers are using CRISPR-Cas9 to correct causative mutations for genetic disorders. In 2017, researchers at the University of Texas Southwestern Medical Center used CRISPR to correct mutations causing Duchenne muscular dystrophy (DMD) in patient-derived stem cells, as well as in a mouse model of DMD [31].[1] In 2016, U.S. researchers used CRISPR to correct a mutation responsible for sickle cell

[1]DMD is an inherited disease caused by mutations in the gene that encodes for dystrophin, a protein that is required for muscle fiber integrity.

disease in patient-derived stem and progenitor cells [32]. The treated cells were transplanted into mouse bone marrow, where they survived for at least 16 weeks, demonstrating that it is possible to reverse mutations in cells.

12.3.2 Preventing and Treating Cancer

There is much interest within the research community to use CRISPR for cancer prevention and treatment. Researchers are using CRISPR to target genes in cancer cells, and to modify immune cells so they can attack tumors more effectively. The first CRISPR clinical trial began in 2016, when scientists at Sìchuān University of China in Chéngdū injected CRISPR-edited immune cells into a patient with metastatic non-small-cell lung cancer [33]. The researchers had used CRISPR-Cas9 to disable a gene that encodes the protein PD-1, which arrests a cell's immune response. Cancers exploit the arrested immune response afforded by PD-1 to proliferate. Without PD-1, the researchers hypothesized that the CRISPR-edited immune cells would attack the cancer cells. The trial was expected to be completed in April 2018 [34]. The Clinical Trials database shows that at least six other studies that use CRISPR to target PD-1 are planned or underway in China [35]. In 2018, human trials were expected to commence for a Chinese study that aims to target and destroy human papillomavirus (HPV) genes that cause cervical cancer [36]. CRISPR's utility in targeting HPV genes has already been demonstrated. For example, in 2014, researchers used it to target HPV-16 oncogenes, resulting in significantly reduced proliferation of cervical cancer cells in vitro [37]. However, the 2018 clinical trial is unique because, instead of editing cells outside the body, researchers will instead apply a gel containing CRISPR-derived genetic material to participants' cervixes. The researchers hypothesize that the CRISPR-derived genetic material will destroy viral genes in HPV-infected cervical cells [36].

12.3.3 Treating Chronic Viral Infections

Researchers have been using CRISPR to target genes that are responsible for chronic viral infections. For example, Chinese researchers in 2015 used CRISPR-Cas9 to target portions of the hepatitis B virus (HBV) genome that are responsible for viral replication [38]. In 2016, Chinese researchers used CRISPR-Cas9 to effectively target EBNA1, a gene that initiates Epstein-Barr Virus (EBV) replication in cancer cells [39]. And in 2015, another group of researchers at Temple University removed the entire integrated human immunodeficiency virus (HIV)-1 genome from previously infected human immune cells [40].

12.3.4 Conferring Disease Resistance and Other Desirable Traits in Crops and Livestock

Researchers are using CRISPR to improve crops, livestock, and organisms serving as disease models by removing undesirable genes or adding desirable genes. In crops, researchers in Israel used CRISPR-Cas9 to make cucumber plants resistant to cucumber vein yellowing virus infection in 2016 [41]. In 2014, Chinese researchers used CRISPR-Cas9 to establish herbicide resistance in rice plants [42]. In addition to transgenic crops, CRISPR has also been used to create transgenic livestock. Researchers at the College of Veterinary Medicine, Northwest A&F University in Shanghai successfully produced live cows with increased resistance to bovine tuberculosis [43].

12.3.5 Creating New or Improving Established Disease Models

CRISPR-Cas9 is also being used to generate organisms that are more useful for disease modeling. In 2013, researchers demonstrated the utility of CRISPR-Cas9 in generating laboratory mice that carry genetic mutations in multiple genes, a process that could expedite in vivo studies of human diseases [44]. The approach greatly simplified traditional methods for generating such mutations, which typically involve recombination of embryonic stem cells or crossing of mice with a single mutation. In addition to improving traditional models for disease, CRISPR (coupled with advances in stem cell research) has also enabled scientists to grow mini-organs, or "organoids," in vitro using a patient's own cells. These structures enable researchers to model tissues, disease processes, and modes of treatment in three dimensions, potentially providing a more accurate representation of pathophysiological processes, pharmacokinetics, and pharmacodynamics. To date, multiple organ structures have been recapitulated using organoids, including gut and kidney [45, 46].

12.3.6 Optimizing Cell Functionality

CRISPR has been shown to be effective in a broad array of cells and organisms, including yeast, tobacco, rice, wheat, laboratory and wild-derived mice, rats, frogs, and silkworms [47]. Researchers are using it to create more efficient metabolic processes within these and other organisms, with the goal of improving their ability to produce products that are medically, environmentally, and industrially useful. For example, in 2017, researchers at US-based Synthetic Genomics, Inc. used CRISPR-Cas9 and other gene editing tools to double the lipid production of an industrial microalga. The lipid could potentially be processed into a biofuel [48]. Korean researchers in 2016 used CRISPR-Cas9 to knock out unwanted genes in a different microalga strain that could also be used to produce biofuels [49]. And in 2016, researchers at UC Riverside used CRISPR-Cas9 to edit the genome of *Yarrowia lipolytica*, a yeast with industrial potential due to its ability to synthesize and store intracellular lipids, but one that has historically been difficult to manipulate genetically [50]. This research demonstrated CRISPR's utility in knocking out and introducing new genes even into organisms that are difficult to manipulate.

Another line of research permitted by CRISPR is the development of "gene drives," a process by which scientists can override the rules of natural reproduction by forcing an altered gene into an animal or plant population, eventually altering the entire population. Under a gene drive, scientists insert an altered gene in an animal along with the CRISPR-Cas9 system (the drive cassette). When the altered animal mates with a wild animal, the offspring inherits 50% of its genome from the wild parent and 50% from the altered parent. The CRISPR-Cas9 system inherited from the altered parent then cuts the wild gene inherited from the wild parent and inserts in its place the altered gene and the CRISPR-Cas9 system, thus allowing the gene alteration to replicate itself in subsequent generations. Although much of the work involving gene drives has been limited to laboratory studies, gene drives could in theory spread a gene alteration in an entire population and possibly over the members of an entire species.

Whereas CRISPR research is broad in its application, gene drive research thus far is primarily focused on vector and pest control. Much of gene drive research is dedicated to controlling the spread of malaria-transmitting mosquitoes. In 2015, researchers at the University of California (UC), Irvine used a CRISPR-induced gene drive to introduce a malaria resistance gene into *Anopheles stephensi* populations, preventing these mosquitoes from transmitting the plasmodia that cause the disease [51]. In 2010, UC Irvine and Oxford University researchers used gene drives to produce female *Aedes aegypti* mosquitoes that could not fly in the hopes of suppressing the spread of dengue viruses to local populations [52]. In addition, the Bill and Melinda Gates Foundation, known for its focus on malaria prevention, is using gene drives to reduce the population of female plasmodium-transmitting mosquitoes. In 2016, these researchers demonstrated that a gene drive aiming to edit the mosquito genome to affect the fertility of female mosquitoes, resulted in a 99% transmission rate [53].

12.4 Promises and Perils of CRISPR-Cas9 and Gene Drives

12.4.1 Potential for Curing or Eradicating Diseases

Although much of the work involving CRISPR is still at an early stage of development, current research shows that it has the potential to positively impact health, agriculture, and the environment. It could facilitate the direct, in vivo correction of genetic defects that lead to genetic disorders like DMD, sickle cell disease, and cystic fibrosis. Because CRISPR enables researchers to correct these defects in somatic cell lines, there would be no need to manipulate embryonic tissue to correct defects [54].

Current research into cancer and infectious disease could lead to improved diagnostics and therapeutic strategies, including treatments that are tailored for specific patients. Current advances in agriculture could result in crops and livestock that are increasingly resistant to infection, herbicides, and environmental austerity [54].

Improved disease models could provide a better understanding of human diseases, as well as large-scale screening for drug targets [18]. In addition to improving disease modeling, researchers could use mini-organs to develop patient-specific treatments or replace defective tissue in patients. Efforts to optimize cell functionality using CRISPR could result in biofuels, synthetic materials, drug precursors, and other useful materials [54].

CRISPR might also address the shortage of organs available for human transplantation by preventing cross-species viral transmission. In 2017, a team of researchers used it to successfully inactivate porcine endogenous retroviruses (PERVs), viruses that could potentially infect human cells following pig-to-human xenotransplantation [55].

By controlling or eradicating the animal population that transmits diseases, gene drives also have the potential to reduce the occurrence of infectious diseases that affect millions of people worldwide. According to the World Health Organization, malaria alone kills an average of 500,000 people per year.[2] The long-term effects of a gene drive could also lead to environmental benefits by reducing the use of insecticide and rodenticide, which have been generally ineffective and have caused some animal populations to develop resistance to them.

12.4.2 Ethical, Safety, and Security Concerns

For all of its potential benefits, CRISPR has also been the subject of a lot of anxieties. Due to its versatility, relatively low cost, and apparent simplicity compared to other gene-editing techniques, such as those that make use of zinc finger nucleases (ZFNs) and transcription activator-like effector nucleases (TALENs), CRISPR's use has spread all over the world and is being used not only by practicing scientists working at university or government laboratories, but also by amateur scientists and the do-it-yourself (DIY) community. The incredible possibilities offered by CRISPR and its rapid democratization have raised ethical, safety, and security concerns among scientists, security experts, and the intelligence community.

12.4.2.1 Ethical and Moral Concerns

On the ethical front, scientists have been the first to raise the alarm about the potential unintended consequences of using CRISPR to introduce permanent changes in the human germline during disease research [56]. A key question debated

[2]World Health Organization, "10 Facts on Malaria," http://www.who.int/features/factfiles/malaria/en/

within the scientific community is: under which conditions would it be ethically acceptable to repair a disease-causing mutation? Even though the benefits of such manipulations seem obvious, there is a lot we do not know about how genes operate, how they interact with each other and the environment, and about the pathways of diseases. As a result, editing one disease-causing gene could have numerous unintended consequences in the human body, in addition to the longer-term risks of modifying the human germline. In 2015, American and European pioneers of CRISPR met in Napa Valley, California, at a meeting convened by Jennifer Doudna to discuss the ethical and safety implications of CRISPR use for genome editing. One of their recommendations was to tread cautiously and investigate the safety and efficacy of the technology before attempting human engineering and clinical trials [56]. Their advice fell on deaf ears: research involving human embryos commenced in 2015 in China and soon moved from the use of genetically abnormal non-viable embryos to normal embryos and, eventually, to human testing.

A scientific and ethical boundary was crossed for the first time in April 2015 when Chinese researchers at the Sun Yat-Sen University in Guangzhou announced that they had used CRISPR to edit human embryos in an attempt to correct mutations in the gene that causes the disease beta-thalassemia, a blood disorder that results in abnormal formation of hemoglobin [57]. A year later, a separate group of Chinese researchers attempted to render human embryos resistant to HIV infection using CRISPR [58]. The results of both projects were somewhat disappointing. Of the 86 embryos that were edited in the 2015 study, only 4 contained the genetic material intended to correct the genes causing beta-thalassemia—an efficiency rate that was too low to make the method practical. Furthermore, the edited embryos were mosaic, comprised of both edited and unedited cells [57]. These results called into question CRISPR's utility for genetically engineering human beings, owing in part to the high number of off-target effects observed in the experiment, i.e., CRISPR cut DNA at the wrong spots. The 2016 experiment was also inefficient: only 4 of the 26 human embryos that were targeted had been modified successfully, and not all of the embryos' chromosomes contained the gene that had been mutated with CRISPR [58].

Nevertheless, in spite of the low efficiency, the 2016 experiment served as a proof of principle for what would be required to generate HIV-1-resistant humans [59]. In addition, the 2015 and 2016 experiments paved the way for new research with increasingly serious ethical and moral implications. Indeed, the Chinese teams who conducted the 2015 and 2016 embryo experiments used genetically abnormal, non-viable embryos, which might have affected the outcome of the experiments [59]. To determine whether CRISPR-Cas9 could correct disease-causing mutations in genetically normal embryos, a third team of Chinese researchers fertilized eggs donated by people undergoing in vitro fertilization with sperm from men with genetic diseases [60]. This time, while two of the six embryos were mosaics, mutations were corrected in three of the six embryos, a dramatic 50% success rate, suggesting that CRISPR repair may have been inefficient in previous experiments because of the use of non-viable embryos [61].

In 2016, the British government granted UK scientists permission to edit the genomes of healthy human embryos donated by patients who had undergone in vitro fertilization. The researchers used CRISPR to disable a gene believed to play a key role in early development. Their results confirmed that the gene did indeed play a key role, a finding that could help researchers understand the causes of miscarriages [29, 30]. In 2017, an international research team led by Shoukhrat Mitalipov of Oregon Health and Science University, claimed to have used CRISPR to efficiently correct a mutation leading to hypertrophic cardiomyopathy in dozens of viable, pre-implanted human embryos [62]. The study was hailed as significantly improving the efficiency of CRISPR-related human embryo research, with a low rate of mosaics [63]. However, later that year, six researchers—including George Church—called into question the way Mitalipov and colleagues interpreted their results, i.e., how the embryos' DNA changed after editing [64].

According to Mitalipov and colleagues' paper, the embryos lacked the genetic sequences the researchers had inserted into cells as templates for repair. The researchers hypothesized that the embryos had used a healthy gene from the egg donor as a template for repairing the Cas9-induced DNA breaks [62]. However, this mechanism has not been observed in previous CRISPR experiments, and the six researchers questioning the study's interpretation believe it runs counter to the conventional understanding that DNA from sperm and egg are not located closely enough to interact or exchange genes immediately after fertilization [64].

Going even further, Chinese scientists were the first to test CRISPR-edited organisms in humans, with the 2016 clinical trial during which a patient with metastatic non-small-cell lung cancer was injected CRISPR-edited immune cells [34]. And as noted earlier, more clinical trials are in the pipeline, notably in the US and Europe, with pending human trials of edited T-cells to fight cancer [65], repair a rare form of blindness [66], and treat beta-thalassemia [67]. Ironically, the last two trials will be conducted by companies founded by some of the pioneers of CRISPR who in 2015 recommended a cautious approach to human trials until the technology is proven to be safe. In February 2017, in a move that would have been unimaginable a year ago, an advisory panel of the U.S. National Academy of Sciences and the National Academy of Medicine issued a report endorsing non-heritable human clinical trials and heritable genome editing in human embryos. The report cautions that such work should be done for certain applications, including to prevent inheritance of serious diseases and disability when no other alternative is available, and should be carried out with strong oversight [68]. Some, however, view the report as another step on the slippery slope of human enhancement, i.e. that CRISPR could be used to produce so-called 'designer babies' having certain traits—such as eye color, height, intellectual abilities—or to create individuals with specialized skills, such as those that could be harnessed for war fighting [69]. And in October 2018, a Chinese scientist announced that he had edited viable embryos to prevent the transmission of the HIV virus from one of the carrier-parents to the offspring, and that the experiment led to the birth of twin girls. As of December 2018, the experimental results have not been published, so it is not clear whether the experiment was performed as stated by the author, whether it succeeded, and what effect it will have on the twin girls [70].

12.4.2.2 Safety Concerns

The ethical concerns about CRISPR are compounded by safety anxieties. On the one hand, scientists have warned that the CRISPR tool is still not safe; there are still too many off-target effects observed, and there may be other off-target effects that scientists are not aware of. Currently, scientists can help anticipate which areas of the genome CRISPR may accidentally cut by using algorithms that identify other regions in the genome that have the same string of nucleotides as the target area. However, these algorithms are based on reference genomes that do not necessarily reflect unique mutations present in each individual [71]. Therefore, it is possible that several off-target effects remain ignored. As human trials are moving along, there is a risk that any off-target effects associated with a CRISPR-based treatment could cause or exacerbate diseases instead of curing them. Further, if the CRISPR system is inserted in a patient's cells via a virus, the virus may continue producing the Cas9 protein for an extended period, thus creating the potential for more unintended DNA cleavage [72].

Safety issues are particularly concerning considering that regulatory frameworks worldwide have not caught up with the fast pace of CRISPR research [73, 74]. For example, in the US CRISPR research falls under the rules that regulate gene-therapy treatments, and human trials require several levels of review and permission, including by the federal biosafety and bioethics committee and the Food and Drug Administration (FDA). However, there is no law in the U.S. that specifically bans germline modifications, and research done with private funding does not fall under the same level of scrutiny [74]. Similarly, in China, current legislation does not ban clinical use of germline genetic engineering, but it is prohibited by guidelines issued by the Ministry of Health [75]. Human trials in China are also subject to review but the process tends to be much faster than in the U.S., which might question the level of scrutiny applied. Europe, too, is struggling with the challenges posed by gene editing, but it seems to have moved faster to update its regulations: the European Medicine Agency launched in the summer of 2017 a review of its 2012 guidelines on products containing genetic modifications to adapt them to the increasing use of CRISPR-Cas9. A revised document was expected by the end of 2018 [76].

In addition to CRISPR, safety is also a major concern with gene drives, which are particularly challenging from the regulatory standpoint. In the U.S., for example, NIH biosafety guidelines do not provide specific guidance for containment of gene-drive research in the laboratory [77]. Therefore, when a gene drive product is ready for marketing or field-testing, identifying the relevant regulatory agency is more complicated than it seems. Three agencies are involved in regulating field trials of commercial or medical goods: the FDA if the good is a drug; the Environmental Protection Agency (EPA) for insecticides; and the US Department of Agriculture (USDA) for plant pests. Yet, this apparently neat distribution of responsibilities does not fit CRISPR-edited goods. For example, in 2016, the USDA, which regulates genetically modified organisms (GMOs), ruled that it had no authority to regulate a

CRISPR-edited mushroom that resists browning, because the genetic modification did not introduce DNA from plant pests (as is the case for traditional GMOs) [78]. The same year, the FDA authorized field-testing of a gene-edited mosquito to fight viral infections in Florida, implying that the mosquito itself is a drug [79]. Some believe that the EPA would be more competent in this instance to evaluate the environmental impact of the gene-edited insect [77]. This regulatory confusion fuels fears that CRISPR-edited goods will continue to bypass existing regulations or allow scientists to work under unsafe conditions, accidentally releasing an edited organism in nature that might eradicate the entire membership of a species or jump from one population to another, with potentially disastrous effects.

Gene drive scientists have attempted to address these concerns by working on molecular safety mechanisms that would prevent an edited organism from wreaking havoc in the wild. For example, 'reverse drives'—an alteration made in an organism that would reverse a previous drive—showed a 99% success rate in laboratory tests [80]. Another safety mechanism called 'Daisy drive' would involve placing the different elements of a drive at different places in the genome, so the various elements are shed by natural selection in successive generations. As a result, the CRISPR-induced alterations would be temporary and limited to a certain number of generations [81, 82]. We will not know, however, whether these mechanisms actually work until they are tested in nature. But testing cannot occur without first updating the regulatory framework.

12.4.2.3 Security Concerns

Whether CRISPR can be used to cause deliberate harm is another major cause for concern. The rapid adoption of the technology by amateur and practicing scientists around the world has fostered fears that it might be used for malevolent purposes, such as producing bioweapons. Already, CRISPR's use has spread to the do-it-yourself (DIY) bio-community, where it is used in projects that range from editing yeast to de-engineering blue carnations to return them to their original white color [80]. CRISPR kits are also available on the Internet to help individuals with no scientific knowledge to use the technique.[3] In light of these developments, some argue that the low cost, availability, and apparent ease of use of CRISPR could allow terrorist groups such as ISIS or Al Qaeda to engineer pathogens to target the food supply chain [83]. Others claim that CRISPR-aided advances in finding treatment for diseases such as cancer could be hijacked to create instead weapons that can target only individuals with specific genetic mutations [84].

These are not the fantasies of people who have no expertise working with CRISPR. CRISPR scientists have themselves advanced several scenarios for misuse, including the use of gene drives by terrorists to spread modified pathogens altered to increase lethality or infectiousness [85]. Other scenarios involve adding a toxin-

[3]For example, see: Josiah Francisco, "DIY CRISPR Kits, Learn Modern Science by Doing," https://www.indiegogo.com/projects/diycrispr-kits-learn-modern-science-by-doing#/; see also "DYI Bacterial Gene Engineering CRIPSR Kit," ODIN.com, http://www.the-odin.com

making gene to the salivary glands of malaria-transmitting mosquitoes to transfer both the disease and the deadly toxin, or altering mosquitoes so they spread outside of their natural habitat and transmit pathogens such as plasmodia or dengue viruses in new areas [86]. Some scientists have also argued that the expertise required to develop gene drives can be acquired relatively easily [87], and that an amateur scientist could set up a laboratory in his or her garage and develop a gene drive undetected [88]. It seemed that the fears of potential misuse were justified when, in 2016, college students from the University of Minnesota attempted to create a gene drive by downloading the drive's genetic sequence, sending pieces of the drive to DNA synthesis companies, and stitching the pieces together during the international biotechnology fair iGEM [89]. Although contamination and storage issues ultimately stifled the students' success, the experiment cemented existing concerns that gene drives could be created easily.

US government agencies have also been warning against the potential CRISPR threat. In February 2016, Director of National Intelligence (DNI) James Clapper explicitly designated gene editing as a security threat in his annual worldwide threat assessment report to Congress. Clapper warned, "[g]iven the broad distribution, low cost, and accelerated pace of development of this dual-use technology, its deliberate or unintentional misuse might lead to far-reaching economic and national security implications" [90]. Then, late in 2016, the President's Council of Advisors on Science and Technology (PCAST) called for a new biodefense strategy that would address emerging biotechnologies like CRISPR that open up new possibilities for misuse [91]. On the prevention front, the US Defense Advanced Research Projects Agency (DARPA) has developed a new initiative called the Safe Genes program to develop biosafety measures for gene editing with the goal of protecting against intentional or accidental misuse.[4] Similarly, the Intelligence Advance Research Projects Activity (IARPA) has launched a program to develop new technical means that will help detect organisms modified using gene editing techniques and prevent misuse [92].

At the international level, members of the Biological Weapons Convention (BWC), which bans the development of biological weapons, have been debating how to assess the impact of emerging biotechnologies, including gene editing, on biosecurity and bioweapons developments [92]. Other UN agencies are struggling with the question of how to regulate CRISPR and gene-drives [93, 94].

12.5 Evaluation of the Threat

Considering the current fragmented regulatory framework and the general absence of CRISPR-specific regulations worldwide, fears that an altered organism might be released accidentally in nature with harmful consequences for humans, the ecosystem, or the environment do not seem to be far-fetched. In addition, the race that

[4]See: http://www.darpa.mil/program/safe-genes

CRISPR scientists have engaged in to make scientific headways or bring a CRISPR-related product to the market might also lead some to cut corners and ignore safety for the sake of obtaining faster results as demonstrated by the 2018 experiment that led to the birth of twin girls in China. This could be detrimental to humans but also to the scientific community as a whole. Therefore, it is important that governments, in cooperation with the scientific community and the public, tackle the biosafety challenge by not only issuing clearer safety regulations at the national level, but also agreeing internationally on a set of best practices to guide national regulatory changes. Achieving an international regulation on germline alternations is probably unlikely considering that each country has different ethical sensitivities, and while one may view a certain type of modification acceptable, another may ban it outright, as was the case with stem cell research [74]. Therefore, an agreement on a limited set of safety requirements and best practices that focus on process for different types of CRISPR/gene drive work could be more easily achieved and desirable [74].

Discussions about developing internationally-accepted best practices should also take into account the varying technical capabilities of states. A set of requirements that necessitates states acquire expensive technologies would place an undue burden on developing states, and might not be implemented. Moreover, such a move might be perceived as a discriminatory measure meant to push potential competitors out of the market. The fact that US and European pioneers of CRISPR advocate caution in CRISPR research on the one hand, but on the other hand have created companies that seem to be on a race to get approval for clinical trials and bring their products to the market, has raised some suspicion about their true motivations [95].

On the security front, the likelihood that a terrorist without expertise might be able to use CRISPR to design a new virus or release a harmful gene drive in nature seems far-fetched, particularly considering the current state of the technology. CRISPR is not the simple and precise tool that the press makes out to be: it still presents a high rate of off-target cleavage. CRISPR scientists have focused much of their energy trying to solve this problem, or find other enzymes that are more easily controlled than Cas9. Yet, there is still a lot that we do not know about how CRISPR-Cas9 truly operates. CRISPR-Cas9 has proved to be a good tool for disabling genes but it has been less successful with editing DNA. This is because Cas9 only performs the cutting part of the process; the editing itself is carried out by the cells' natural DNA repair mechanism, which scientists do not control [7]. In addition, Cas9 cuts blunt ends, i.e., both strands of the DNA are cut at the same position, creating strands that are the same length [96], which complicates the editing process; the cell's repair process is more likely to put the two ends back together than to insert a new DNA sequence, thus causing many errors [97]. Designing the RNA guide is another challenging aspect of the technology. Research has shown that some RNA guides are less efficient than others, or even inactive, and in the absence of a good RNA guide, multiple off-target effects can occur [98]. While numerous in silico tools are available for designing the RNA guide and predicting off-target effects, research has shown that they have limited efficiency and miss a lot of off-target effects [99]. Delivery of the CRISPR system into a cell is another important challenge of the technology, and each delivery medium, such as plasmid or viruses, brings about a variety of additional technical challenges and potential for errors [98].

In gene drives, a key challenge is the insertion of the drive cassette into the animal. For example, in one malaria-related study, only 2 males out of over 25,000 mosquitoes carried the gene drive, and females transmitted the gene drive at a rate only slightly higher than under natural reproduction [100]. Cage studies have also shown that wild females prefer to mate with wild males rather than altered males [88]. It is not clear in these conditions whether the gene drive could propagate and survive in nature.

With so many variables to juggle, it is unlikely that an inexperienced terrorist might succeed in using CRISPR for malevolent purposes. The University of Minnesota students who attempted to create a gene drive at iGEM in 2016 failed in their efforts because of such mundane issues as contaminated equipment and a refrigerator that was not cold enough [89]. Therefore, even minor problems can become insurmountable for individuals lacking the proper skills and experience.

States might be in a better position to use CRISPR in a bioweapons effort. However, they too would face the same technical challenges as current researchers. In addition, the study of past state and terrorist bioweapons program has shown that bioweapons development involves a whole set of technical, social, and organizational challenges [101–103], and that technology does not eliminate one of the key challenges of bioweapons development: the mercurial nature of living organisms, which makes them difficult to control [101].

References

1. Ishino Y, et al. Nucleotide sequence of the *iap* gene, responsible for alkaline phospatase isozyme conversion in *Escherichia coli*, and identification of the gene product. J Bacteriol. 1987;169:5429–33.
2. Nakata A, et al. Unusual nucleotide arrangements with repeated sequences in the *Escherichia coli* K-12 chromosome. J Bacteriol. 1989;171:3553–6.
3. Groenen PM, et al. Nature of DNA polymorphism in the direct repeat cluster of *Mycobacterium tuberculosis*; application for strain differentiation by a novel typing method. Mol Microbiol. 1993;10:1057–65.
4. Mojica FJ, et al. Long stretches of short tandem repeats are present in the largest replicons of the Archaea *Haloferaz mediterranei* and *Haloferax volcanii* and could be involved in replicon partitioning. Mol Microbiol. 1995;17:85–93.
5. Masepohl B, Gorlitz K, Bohme H. Long tandemly repeated (LTRR) sequences in the filamentous cyanbacterium Anabaena sp. PCC 7120. Biochim Biophys Acta. 1996;1307:26–30.
6. Hoe N, et al. Rapid molecular genetic subtyping of serotype M1 group a *Streptococcus* strain. Emerg Inf Dis. 1999;5:254–63.
7. Barrangou R, Horvath P. A decade of discovery: CRISPR functions and applications. Nat Microbiol. 2017;2(1709):1–9.
8. Mojica FJ, et al. Biological significance of a family of regularly spaced repeats in the genomes of Archaea, Bacteria and mitochondria. Mol Microbiol. 2000;36:244–6.
9. Jansen R, et al. Identification of genes that are associated with DNW repeats in prokaryotes. Mol Microbiol. 2002;43(6):1565–75.
10. Mojica F, et al. Intervening sequences of regularly spaced prokaryotic repeats derive from foreign genetic elements. J Mol Evol. 2005;60(2):174–82.
11. Marx J. New bacterial defense against phage invaders identified. Science. 2007;315:1650–1.

12. Poucel C, et al. CRISPR elements in *Yersinia pestis* acquire new repeats by preferential uptake of bacteriophage DNA, and provide additional tools for evolutionary studies. Microbiology. 2005;151:653–63.
13. Bolotin A, et al. Clustered regularly interspaced short palindromic repeats (CRISPR) have spacers of extrachromosomal origin. Microbiology. 2005;151:2551–61.
14. Makarova K, et al. A putative RNA-interference-based immune system in prokaryotes: computational analysis of the predicted enzymatic machinery, functional analogies with eukaryotic RNAi, and hypothetical mechanisms of action. Biol Direct. 2006;1:7.
15. Barangou R, et al. CRISPR provides acquired resistance against viruses in prokaryotes. Science. 2007;315:1709–12.
16. Morange M. What history tells us XXXVII. CRISPR-Cas: the discovery of an immune system. J Biosci. 2015;40(2):221–3.
17. Brouns S, et al. Small CRISPR RNAs guide antiviral defense in prokaryotes. Science. 2008;321:960–3.
18. Doudna JA, Charpentier E. Genome editing. The new frontier of genome engineering with CRISPR-Cas9. Science. 2014;346(6213):1–9.
19. Ledford H. Five big mysteries about CRISPR's origins. Nature. 2017;541:280–2.
20. Jinek M, et al. A programmable dual-RNA-guided DNA endonuclease in adaptative bacterial immunity. Science. 2012;337(6096):816–21.
21. Jinek M, et al. RNA-programmed genome editing in human cells. Elife. 2013. https://doi.org/10.7554/eLife.00471.001
22. Cong L, et al. Multiplex genome engineering using CRISPR/Cas systems. Science. 2013;339 (6121):819–22.
23. Begley S. 6 takeaway from the CRISPR patent decision. STAT. 2017, February 16. https://www.scientificamerican.com/article/6-takeaways-from-the-crispr-patent-decision/
24. Cross R. Broad prevails over Berkely in CRISPR patent dispute. C&En, September 10, 2018. Available at: https://cen.acs.org/policy/litigation/Broad-prevails-over-Berkeley-CRISPR/96/web/2018/09
25. Babcock B. The continuing CRISPR patent battle: The Broad Institute loses a key European patent. MedCityNews. 2018, January 23.
26. Buhr S. China sides with Emanuelle Charpentier and Jennifer Doudna in CRISPR patent war. TechCrunch. 2017. https://techcrunch.com/2017/06/19/china-sides-with-emmanulle-charpentier-and-jennifer-doudna-in-crispr-patent-war/
27. Cohen J. Europe says University of California deserves patent for CRISPR. Science. 2017. https://doi.org/10.1126/science.aal0969
28. Clinical Trials Database. CRISPR. 2017. https://clinicaltrials.gov/ct2/results?term=CRISPR&type=&rslt=&age_v=&gndr=&cond=CRISPR&intr=&titles=&outc=&spons=&lead=&id=&cntry1=&state1=&cntry2=&state2=&cntry3=&state3=&locn=&sfpd_s=&sfpd_e=&lupd_s=&lupd_e=. Accessed 29 Sept 17.
29. LePage M. Boom in human gene editing as 20 CRISPR trials gear up. New Scientist. 2017. https://www.newscientist.com/article/2133095-boom-in-human-gene-editing-as-20-crispr-trials-gear-up/. Accessed 29 Sept 2017.
30. LePage M. Why has a UK team genetically edited human embryos? New Scientist. 2017. https://www.newscientist.com/article/2148057-why-has-a-uk-team-genetically-edited-human-embryos/. Accessed 29 Sept 2017.
31. Zhang Y, Long C, Li H, et al. CRISPR-Cpf1 correction of muscular dystrophy mutations in human cardiomyocytes and mice. Sci Adv. 2017;3(4):1–10.
32. Dever DP, Bak RO, Reinisch A, et al. CRISPR/Cas9 β-globin gene targeting in human haematopoietic stem cells. Nature. 2016;17(539):384–9.
33. Krisch JA. First CRISPR clinical trial commences. Scientist. 2016, November 15. http://www.the-scientist.com/?articles.view/articleNo/47528/title/First-CRISPR-Clinical-Trial-Commences/. Accessed 28 Sep 2017.

34. Clinical Trials Database. PD-1 knockout engineered t cells for metastatic non-small cell lung cancer, Bethesda. 2017. https://clinicaltrials.gov/ct2/show/results/NCT02793856. Accessed 29 Sept 2017.
35. Clinical Trials Database. PD-1 and CRISPR. 2017. https://clinicaltrials.gov/ct2/results?cond=PD-1&term=CRISPR&cntry1=&state1=&recrs=. Accessed 29 Sept 2017.
36. Clinical Trials Database. A safety and efficacy study of TALEN and CRISPR/Cas9 in the treatment of HPV-related Cervical Intraepithelial Neoplasia. 2017. https://clinicaltrials.gov/ct2/show/NCT03057912. Accessed 29 Sept 2017.
37. Zhen S, Hua L, Takahashi Y, et al. In vitro and in vivo growth suppression of human papillomavirus 16-positive cervical cancer cells by CRISPR/Cas9. Biochem Biophys Res Commun. 2014;450(4):1422–6.
38. Dong C, Qu L, Wang H, et al. Targeting hepatitis B virus ccDNA by CRISPR/Cas9 nuclease efficiently inhibits viral replication. Antivir Res. 2015;118:110–7.
39. Yuen KS, Wong NHM, Jin D. Suppression of Epstein-Barr virus infection in nasopharyngeal carcinoma cells through CRISPR/Cas9 targeting of EBNA1, OriP and W repeats. Paper presented at the 17th international symposium on Epstein Barr virus and associated diseases (EBV 2016), University of Zurich, Zurich, Switzerland, 8–12 August 2016.
40. Kaminski R, Chen Y, Fischer T. Elimination of HIV-1 genomes from human T-lymphoid cells by CRISPR/Cas9 gene editing. Sci Rep. 2016;6(22555):1–15.
41. Chandrasekaran J, Brumin M, Wolf D. Development of broad virus resistance in non-transgenic cucumber using CRISPR/Cas9 technology. Mol Plant Pathol. 2016;17(7):1140–53.
42. Xu R, Li H, Qin R, et al. Gene targeting using the agrobacterium tumefaciens-mediated CRISPR-Cas system in rice. Rice. 2014;7(1):5.
43. Gao Y, Wu H, Wang Y, et al. Single Cas9 nickase induced generation of NRAMP1 knockin cattle with reduced off-target effects. Genome Biol. 2017;18(1):13.
44. Wang H, Yang H, Shivalla CS, et al. One-step generation of mice carrying mutations in multiple genes by CRISPR/Cas-mediated genome engineering. Cell. 2013;153(4):910–8.
45. Freedman BS, Brooks CR, Lam AQ, et al. Modelling kidney disease with CRISPR-mutant kidney organoids derived from human pluripotent epiblast spheroids. Nat Commun. 2015;6 (8715). https://doi.org/10.1038/ncomms9715
46. Schwank G, Koo BK, Sasselli V, et al. Functional repair of CFTR by CRISPR/Cas9 in intestinal stem cell organoids of cystic fibrosis patients. Cell Stem Cell. 2013;13(6):653–8.
47. Sander JD, Joung JK. CRISPR-Cas systems for editing, regulating, and targeting genomes. Nat Biotechnol. 2014;32(4):347–55.
48. Ajjawi I, et al. Lipid production in Nannochloropsis gaditana is doubled by decreasing expression of a single transcriptional regulator. Nat Biotechnol. 2017;35:647–52.
49. Shin SE, Lim JM, Koh HG. CRISPR/Cas9-induced knockout and knock-in mutations in Chlamydomonas reinhardtii. Sci Rep. 2016;6:27810. https://doi.org/10.1038/srep27810.
50. Schwartz CM, Hussain MS, Blenner M, Wheeldon I. Synthetic RNA polymerase III promoters facilitate high-efficiency CRISPR-Cas9-mediated genome editing in Yarrowia lipolytica. ACS Synth Biol. 2016;5(4):356–9.
51. Gants VM, Jasinskiene N, Tatarenkova O, et al. Highly efficient Cas9-mediated gene drive for population modification of the malaria vector mosquito Anopheles stephensi. Proc Natl Acad Sci. 2015;112(49):E6736–43.
52. Fu G, Lees RS, Nimmo D, et al. Female-specific flightless phenotype for mosquito control. Proc Natl Acad Sci. 2010;107(10):4550–4.
53. Hammond A, Galizi R, Kyrou K, et al. A CRISPR-Cas9 gene drive system targeting female reproduction in the malaria mosquito vector Anopheles gambiae. Nat Biotechnol. 2015;34 (1):78–83.
54. Hsu PD, Lander ES, Zhang F. Development and applications of CRISPR-Cas9 for genome engineering. Cell. 2014;157(6):1262–78.
55. Niu D, Wei HJ, Lin L, et al. Inactivation of porcine endogenous retrovirus in pigs using CRISPR-Cas9. Science. 2017;357(6357):1303–7.

56. Baltimore D, et al. A prudent path forward for genomic engineering and germline gene modification. Science. 2015;348(6230):36–8.
57. Liang P, Xu Y, Zhang X, Ding C, Huang R, et al. CRISPR/Cas9-mediated gene editing in human trioronuclear zygotes. Protein Cell. 2015;6(5):363–72.
58. Kang X, He W, Huang Y, Yu Q, Chen Y, et al. Introducing precise genetic modifications into human 3PN embryos by CRISPR/Cas-mediated genome editing. J Assist Reprod Genet. 2016;33(5):581–8.
59. Callaway E. Second Chinese team reports gene editing in human embryos. Nature News. 2016. http://www.nature.com/news/second-chinese-team-reports-gene-editing-in-human-embryos-1.19718. Accessed 29 Sept 17.
60. Tang L, Zeng Y, Du H, et al. CRISPR/Cas9-mediated gene editing in human zygotes using Cas9 protein. Mol Gen Genomics. 2017;292(3):525–33.
61. LePage M. Mosaic problem stands in the way of gene editing embryos. New Scientist. 2015. https://www.newscientist.com/article/mg23331174-400-mosaic-problem-stands-in-the-way-of-gene-editing-embryos/. Accessed 29 Sept 2017.
62. Ma H, Marti-Gutierrez N, Park SW, et al. Correction of a pathogenic gene mutation in human embryos. Nature. 2017;548(7668):413–9.
63. Ledford H. CRISPR fixes disease gene in viable human embryos. Nat News. 2017;548:13–4.
64. Servick K. Skepticism surfaces over CRISPR human embryo editing claims. Science Magazine. 2017. https://doi.org/10.1126/science.aap8382
65. Begley S. Federal panel approves first use of CRISPR in humans. STAT. 2016, June 21.
66. LaMotta L. Editas delays CRISPR move to human trials. BioPharmaDive. 2017, May 16.
67. CRISPR Therapeutics. News Release. 2017, March 31. http://ir.crisprtx.com/phoenix.zhtml?c=254376&p=irol-newsArticle&ID=2272286
68. National Academies of Sciences, Engineering, and Medicine. Human genome editing: science, ethics and governance. 2017.
69. Harmon A. Human gene editing receives science panel's support. NY Times. 2017, February 14.
70. He Jiankui defends 'World's first gene-edited babies'. BBC News, November 28, 2018. https://www.bbc.com/news/world-asia-china-46368731
71. Begley S. Do CRISPR enthusiasts have their head in the sand about the safety of gene editing? STAT. 2016, July 18.
72. Begley S. They're going to CRISPR people: what could possibly go wrong? STAT. 2016, June 23.
73. Araki M, Ishii T. International regulatory landscape and integration of corrective genome editing into in vitro fertilization. Reprod Biol Endocrinol. 2014;12:108. https://doi.org/10.1186/1477-7827-12-108.
74. Konig H. The illusion of control in germ-line engineering policy. Nat Biotechnol. 2017;35:502–6.
75. Vogel G. Embryo engineering alarm. Science. 2015;347(6228):1301.
76. Allen & Overy LLP. Regulating CRISPR genome editing in humans: where do we go from there? JDSupra. 2017, August 14. http://www.jdsupra.com/legalnews/august-2017-regulating-crispr-genome-64475/
77. Carter S, Friedman R. Policy and regulatory issues for gene drives in insects. Workshop Report, August 2016, J. Craig Venter Institute, La Jolla, California. 2016. http://www.jcvi.org/cms/fileadmin/site/research/projects/gene-drive-workshop/report-complete.pdf
78. Cyranoski D. CRISPR tweak may help gene-edited crops to bypass biosafety regulations. Nature. 2015. https://doi.org/10.1038/nature.2015.18590
79. FDA (Food and Drug Administration). FDA releases final environmental assessment for genetically engineered mosquito. 2016, August 5. http://www.fda.gov/AnimalVeterinary/NewsEvents/CVMUpdates/ucm490246.htm
80. Ledford H. Biohackers gear up for genome editing. Nature. 2015;524:398–9.

81. Esvelt KM. Daisy drive systems. Sculpting Evolution. http://www.sculptingevolution.org/daisydrives
82. Ledford H. Safety upgrade found for gene-editing technique. Nature. 2015. https://doi.org/10.1038/nature.2015.18799
83. Acharya A, Acharya A. Cyberterrorism and biotechnology: when ISIS meets CRISPR. Foreign Affairs. 2017, June 1.
84. Hern A. There are things worse than death: can a cancer cure lead to brutal bioweapons? Guardian. 2017, July 31.
85. Begley S. Gene drive gives scientists power to highjack evolution. STAT. 2015, November 17.
86. Begley S. Why the FBI and pentagon are afraid of this new genetic technology. STAT. 2015, November 12.
87. Ledford H. CRISPR the disruptor. Nature. 2015;522(7554):20–4. https://doi.org/10.1038/522020a
88. Shaw J. Editing an end to malaria. Harvard Magazine. 2016, May–June. http://www.harvardmagazine.com/2016/05/editing-an-end-to-malaria
89. Swetlitz I. College students try to hack a gene drive and set a science fair abuzz. STAT. 2016, December 14.
90. Clapper J. Worldwide threat assessment of the US Intelligence Community. Senate Select Committee on Intelligence. 2016, February 9.
91. Mullin E. Obama advisors urge action against CRISPR bioterror threat. MIT Technol Rev. 2016, November 17.
92. Mackby J. Dispute mire BWC review conference. Arms Control Today. 2017, January 11.
93. Brown K. The UN just gave scientists the green light to mess with natural selection. GIZMODO. 2016, December 22. https://gizmodo.com/the-un-just-gave-scientists-the-green-light-to-mess-wit-1790413806
94. International Bioethics Committee (IBC). Report of the IBC on updating its reflection on the human genome and human rights UNESCO. 2015. http://unesdoc.unesco.org/images/0023/002332/233258E.pdf
95. Xiong S. Is international CRISPR regulation a pipe dream? Transcripts. 2017, July 24.
96. Doudna JA, Sternberg S. A crack in creation. Boston: Houghton Mifflin Harcourt; 2017.
97. Ledford H. Alternative CRISPR system could improve gene editing. Nature. 2015;526:17.
98. Peng R, et al. Potential pitfalls of CRISPR/Cas9-mediated genome editing. FEBS. 2016;283:1218–31.
99. Chuai G-h, Wang Q-L, Liu Q. In silico meets in vivo: towards computational CRISPR-based sgRNA design. Trends Biotechnol. 2017;35(1):12–21.
100. Saey T. Gene drives spread their wings. Sci News. 2015;18:16.
101. Ben Ouagrham-Gormley S. Barriers to bioweapons: the challenge of expertise and organization for weapons. Ithaca: Cornell University Press; 2014.
102. Leitenberg M, Zilinskas R, Khun J. The soviet biological weapons program: a history. Cambridge: Harvard University Press; 2012.
103. Vogel K. Phantom menace or looming danger. Baltimore: Johns Hopkins University Press; 2013.

Emerging and Enabling Technologies in Biodefense

13

Kavita M. Berger

13.1 Introduction

Biotechnologies provide critical capabilities in the defense against natural and man-made threats. In health, new scientific discoveries and platforms contribute to the development of vaccines and medicines against pathogens, such as influenza A virus or *Bacillus anthracis*. Researchers' ability to understand the relationship between pathogens and their hosts provide the information needed to create targeted vaccines and medicines. Similarly, scientists have begun applying synthetic biology tools and concepts to vaccine development, hoping to reduce the time involved in creating a vaccine against an emerging infectious disease and enabling rapid response efforts. More recently, investments in additive bio-manufacturing have provided new opportunities to print human tissues and organs, which in some laboratories, are being used to study the effects of pharmaceutical candidates before testing in animals or people. In addition, advances in knowledge gained from neuroscience and behavioral sciences provides new opportunities to understand and identify clinical interventions to prevent or treat various psychological disorders, such as post-traumatic stress disorder after return from military conflict.

Beyond health efforts, governments are looking to the life sciences and biotechnology to assist in early warning of infectious disease events, forensics, decontamination of materials, and many other uses. Efforts to gain advanced warning of infectious disease events have leveraged new data analysis technologies (e.g., natural language processing, image analysis, and text analysis) to evaluate a variety of information, including environmental, ecological, genomic, physiological, and other relevant data. Efforts to promote longevity of materials and equipment have leveraged advances in synthetic biology. Efforts to enhance microbial forensics and

K. M. Berger (✉)
Gryphon Scientific, LLC, Takoma Park, MD, USA
e-mail: kberger@gryphonscientific.com

S. K. Singh, J. H. Kuhn (eds.), *Defense Against Biological Attacks*,
https://doi.org/10.1007/978-3-030-03053-7_13

attribution have leveraged advances in microbiology and improved data handing and sharing procedures between public health and law enforcement communities. Enabling this research and data sharing was a commitment by the U.S. government and international organizations to seek solutions for capability gaps in preventing, detecting, and responding to natural and/or man-made infectious disease events of international concern.

13.1.1 Biotechnology Market

The global market for biotechnology is in the tens of billions of dollars. Financial forecasts estimate the size of the global biotechnology market by 2021 to be between $2 billion and $6 billion dollars, suggesting an annual growth rate of 14–39% [1–3]. The variability in estimates results from the pharmaceutical business model wherein a single pharmaceutical product can be worth several billions of dollars [4], but about only 1% of all compounds transition from pre-clinical research stages to licensure. This estimate does not include knowledge generated, products developed, and services provided by academia, amateur biologists, entrepreneurs, and governments, indicating that the total scope of biotechnology research and development may be significantly larger than the forecasts suggest. In fact, the Silicon Valley venture capital community has begun creating new models for promoting innovation among entrepreneurial individuals interested in leveraging biology to address a particular concern. In addition, amateur biologists (DIYBio) are raising tens to hundreds of thousands of dollars from crowdsource funding websites, such as Kickstarter, enabling their increasingly sophisticated research activities.

The biotechnology enterprise contributes to the health, agriculture, energy, environmental, scientific, and defense sectors. The multidisciplinary nature of modern biotechnology drives advancement in the basic and applied sciences, enabling new approaches for addressing health care and public health needs, enhancing production of agricultural commodities, pushing scientific research and analysis, preserving biodiversity and ecological health, and protecting military personnel and materiel. Computational technologies have enhanced the study of complex biological processes and systems [5], including collection, integration, and analysis of large amounts of data on genomes [6–8], epigenomes (elements that are not encoded in DNA, but control expression of genes) [9, 10], proteomes (protein interactions) [11], metabolomes (biosynthetic pathways controlling an organism's natural chemicals), microbiomes (all the commensal microbes on or in an individual) [12, 13], exposome (all the changes that occur after an individual is exposed to a foreign particle) [14, 15], pharmacogenome (genetic changes to a pharmacologic molecule) [16, 17], and other -omics data sets. Better understanding of biological systems has enabled the development of *in silico* models of animals to enable computer-based studies of medicines, proteins structures, and DNA structures [18, 19]. Data analysis technologies have enabled the creation of new computational models for infectious disease surveillance and enhanced efforts to leverage published data better [20]. Engineering and design concepts have led to new approaches for modifying biological systems to enable production of industrial chemicals, illicit drugs, ethanol, precursor chemicals for cosmetics, and other molecules [21, 22]. Advances in material science has enabled 3D bioprinting of tissues and organs, which are being used to test drug efficacy before human trials [23–25]. These are some examples of how advances in biotechnology are enabled by convergence with other scientific disciplines to benefit society.

13.1.2 Drivers of Biotechnology

The major drivers of advances in biotechnology vary by technology and application. Strong health, agriculture, energy, and scientific applications drive continued advancement of biotechnologies. Human health and agriculture are two primary drivers for biotechnology. Better diagnostics, medicines, and tailored health care promote the development and use of new biotechnologies. Agricultural production has led to use of genetic engineering to create hardier, larger, and more pest-resistant crops [26, 27]; data analytics of different types of agriculturally-relevant data sets to improve crop management and reduce of intra-crop variability [28, 29]; genomics and genome association studies in crops and livestock to identify the genetic basis of favorable and deleterious traits [30, 31]; and genome editing tools to modify crops and livestock [32, 33]. In addition, clean energy contributed to early development of synthetic biology approaches to produce organisms that can make or consume ethanol [34, 35]. Although non-biologically-based technologies have overtaken these synthetic organisms as better clean energy sources, synthetic biology has demonstrated utility in other sectors, such as the chemical, pharmaceutical, and cosmetic industries. These industries use biological systems for production of chemicals, and active ingredients in pharmaceuticals and other commercial products. Furthermore, scientific play and education (both formal and informal) contribute to the continued use and development of synthetic biology. Advancement in scientific knowledge and capability serves as a primary driver for the creation of new genetic-manipulation tools, data analysis software, and commercial kits, products, and services to make experimental procedures easier. Examples of drivers of genome editing and personalized health are described below.

Advances in genome editing largely are driven by human health and agriculture. The ability to make specific changes in a human cell's genome to correct a disease, an agriculturally-important animal to eliminate a pesky trait, or in crops to enhance pest resistance provide significant drivers for improving the capability and precision of genome editing tools. In 2015, scientists used a non-CRISPR-based genome editing tool to modify healthy donor cells in the laboratory before injecting them into a 2-year old girl who had leukemia [36]. The cells were edited to reduce the chance that they would be rejected by the girl's body. After treatment with the modified cells, the girl was cured of leukemia. More recently, research groups in Sweden, China, and the U.S. reported successful editing of human embryos [37–39]; though controversial, the U.S. National Academy of Medicine, Engineering, and Sciences stated that editing human embryos may be warranted to correct a debilitating genetic disease [40]. Unlike humans, modification of germ cells of animals, insects, and plants are much less controversial and occurring routinely, in some cases in the commercial sector. For example, Sangamo uses another genome editing tool (specifically, zinc finger nucleases) to produce hornless dairy cows, making milking less dangerous for farmers [41, 42]. The Bill and Melinda Gates Foundation is investing $35 million into CRISPR-enabled gene drives in mosquitoes as part of its initiatives to control the spread of malaria-causing plasmodia [43]. (Gene drives are DNA sequences that rapidly push a specific trait through a population.) Agriculture companies are editing crop seeds to increase resistance to pests.

The ability to tailor health care to individual people (i.e., personalized medicine) is a significant driver of advancement and application of -omics, systems biology, and similar types of research. Private health care providers, such as Inova Health Care, are beginning to collect genomic information of patients to improve research and patient care [44]. In 2015, the U.S. government launched its Precision Medicine Initiative, to collect, integrate, and analyze one million individual genomes for enhancing the scientific foundation for patient-centered health care in the U.S. One year later, China announced its precision medicine initiative with an investment of $9.2 billion [45]. These efforts, as with other similar efforts, include other influencing factors such as daily habits and patterns of family history.

Precision agriculture is much farther along than personalized medicine, in large part because the genomes of crops can be studied with high reliability and reproducibility. Unlike human genetic linkage studies, similar studies in plants have produced highly reliable information. This high degree of reliability enables better identification, prediction, and elimination of harmful disease in crops that have been studied. In addition to genetic information, precision agriculture programs integrate information about environmental, meteorological, and climate conditions in their analyses.

Empowerment of people to take control of their own health has and continues to push advancement and application of personal genomics, mobile health technologies, and biomedical research in amateur community labs. Personal genomics is a relatively new commercial market that provides three main focus areas: (1) mobile health devices, such as health apps for smart phones and fitbits; (2) direct-to-consumer genome sequencing services, such as 23andMe; and (3) genealogy services that provide individuals with information about their ancestry based on their genomic sequences. Going further than personal genomics, members of the DIY Bio community are securing funds to conduct research on their own microbiome, genetic diseases that affect their families, and biotechnologies for self-augmentation or enhancement.

13.1.3 Advances in and Biosecurity Considerations of Biotechnology

Small and large (i.e., transformative) changes influence the advancement and creation of new tools, knowledge, and concepts in biology. Small changes generally occur as part of the normal cycle of fundamental and applied research wherein new discoveries lead to new scientific questions or knowledge. Larger changes tend to occur when new fields of science, disciplines, or conceptual frameworks are combined (often referred to as convergence in science). Recent examples include additive biomanufacturing, which combines technologies of 3D printing with development of "ink" made of living cells [24]; synthetic biology (i.e., combining DNA-based "circuits" to create an organism of known function), which bring engineering and computer science concepts to the field of biology; and systems biology, which involves advanced data analytics and visualization (often considered big data analytics) to study connections and networks within and between molecules, pathways, cells, physiological systems, and life forms. In addition, transformative changes could result from trial-and-error during the discovery process or through careful observation of biological systems. Examples of these changes include genetic engineering, which was enabled by the discovery of proteins that cut DNA internally (called endonucleases), and CRISPR-based genome editing, which emerged from research on immune responses in bacteria.

The convergence of scientific disciplines has elicited concerns among security experts about the potential for new technical capabilities to disrupt national security, health, and prosperity. These concerns often stem from surprise about the development of new technologies and knowledge gained that could be used to cause physical, social, economic, and/or political harms to a country. However, technologies rarely arise without warning. Scientific publication, patent applications, conferences, and online media and blogs shed light on near-term advances taking place within scientific fields and disciplines. Through these resources, security experts can learn about cutting-edge research and applications to address health, agriculture, energy, and other societal needs. Identifying larger changes that occur through the integration and/or leveraging of advances from distinct fields involves an evaluation of technical challenges and/or gaps limiting benefits to social needs, technical challenges that stall advancement within fields, and scientific advances in different fields. For example, 3D bioprinting came about as a result of increasing need for tissues and organs for research (e.g., *in vitro* testing of drug candidates) and clinical purposes (e.g., more effective tissue or organ transplantation), advances in stem cell research and tissue engineering, and advances in additive manufacturing [46]. Another example is neuroprosthetics, which has developed from the need to improve the lives of service members and other individuals who have lost limbs, increasing knowledge in neuroscience, and advances in material science involved in the development of prosthetic limbs [47, 48]. Still another example is the field of precision medicine, which is enabled by improved individualized health care, increasing collection of biological and health data, and new computational analysis and data storage technologies [49]. The convergence of science and new technology development often are driven by beneficial purposes and social need. Continued examination of these needs and scientific activities can prevent or limit the technology surprise that often elicits concern from the security community.

Another phenomenon associated with biotechnology that has raised national security concerns is the increasingly unconventional approach to the life sciences and their key components. A few examples have emerged during the past decade highlighting this phenomenon. One example is the emergence of synthetic biology as a new field of study. At its surface, synthetic biology is genetic engineering. It involves joining together specific DNA sequences to produce a certain biological outcome. But what distinguishes synthetic biology from traditional genetic engineering is the conceptual approach to designing, building, and testing of biological systems. The concept of synthetic biology was created by engineers and computer scientists who asked whether biological systems could be assembled using specific parts to generate expected outcomes. The synthetic biology approach involves the identification of individual "parts", which functionally are DNA (specifically, plasmids containing genes with desired functions), that when placed in a particular order, could produce a desired outcome such as creating organisms that produce small molecules (e.g., synthetic artemisinin for malaria treatment [50], chemicals [51], opioids [52]), or creating bacteria in all the colors of the rainbow [53]. This design-based approach enables new applications in health, agriculture, industry, and national security.

More recently, technology-based companies and academic scientists have begun exploring the use of DNA as a storage medium for image, audio, and video data files. This use of DNA is driven by the increasing need for platforms that can store extremely large data and datasets. The concept of DNA-based storage emerged from the recognition that the human genome, which is 3 billion base pairs long, can store a lot of information stably and with relatively high fidelity for thousands of years. DNA-based storage leverages advances in DNA sequencing and synthesis technologies. The concept that image data can be stored in and retrieved from DNA was demonstrated in 2017, when scientists from Harvard University published their work on using CRISPR-based tools to encode image data of a galloping horse in bacteria [54, 55]. This effort leverages genome editing tools to demonstrate the feasibility of DNA-based storage of data. The use of DNA as a data storage medium also may present new, potentially exploitable, vulnerability. In 2017, scientists described their ability to hack computers via sequences they embedded into DNA [56]. The researchers claimed to encode "malicious software" in "DNA they purchased online," which they used "to gain 'full control' over a computer that tried to process" or analyze the sequence data generated from physically sequencing the purchased DNA [56].

Despite concerns about how advances in biotechnology could be applied to harm national security, these advances can enhance efforts to prevent, prepare for, and respond to incidents involving chemical and biological threats. For example, next generation sequencers and data analysis software allow scientists to collect information about host response following exposures to pathogens or hazardous chemicals, the microbiome that supports normal growth and functioning of the human body, the organisms in the environment, and many other molecules associated with the normal state of the human body and potential risk factors of exposure to chemical or biological hazards. These data can inform surveillance efforts to help determine whether and when a natural or man-made incident has occurred, aiding both detection of potentially harmful incidents and forensic analysis. In addition, the data may inform the development of vaccines, drugs, and diagnostic tools against chemical and biological agents [called medical countermeasures (MCMs)] [57–61]. Another examples include the use of synthetic genomics, which is the chemical synthesis of genomic DNA of organisms, as platform methods for rapid development of vaccines against epidemic pathogens. Synthetic biology is being used to create technology platforms for rapid MCM development against pathogens and for synthesis of pharmaceutical products. Similar technologies can be used to prevent, detect, and respond to incidents involving contamination or infection of animals and agriculturally-important plants. In addition, synthetic biology is being used to develop biological systems that "eat" nerve agent [62, 63], clean up contaminated materials [61], and exert biocontrol in the environment [64, 65]. These efforts are being aided by advances in genome editing, laboratory automation, high-throughput experimental procedures, bioinformatics, and other technologies that improve repeatability and reduce human error during experimentation.

13.2 U.S. Biodefense Needs

Before September 11, 2001, the field of biodefense was small, restricted to a select group of researchers at a few agencies around the world, and focused largely on the threat of biological weapons from state actors or terrorist organizations. The level and scope of biodefense changed completely after the terrorist events of 2001 when anthrax spores were sent in letters through the mail (i.e., Amerithrax) 1 month after terrorists flew planes into the World Trade Center towers. The U.S. government passed a series of laws and issued presidential directives in rapid succession to enhance its ability to anticipate, prevent, detect, prepare for, respond to, and recover from terrorist events. Several of these programs were further enhanced to prevent, detect, and respond to naturally-occurring infectious disease and natural disasters. These changes were driven in large part by the emergence of West Nile virus in Northern America (1999), H5N1 influenza A virus in Asia (1997), severe acute respiratory syndrome coronavirus in Asia and North America (2003), Middle East respiratory syndrome coronavirus in the Middle East (2012), H1N1 influenza A virus in North America (2009), Ebola virus in Western Africa (2013–2016), Zika virus in the Americas (2016–2017), and the delayed federal government response to hurricane Katrina in 2005. Today, biodefense efforts cover threat characterization, risk assessment, MCM research and development, protection of laboratory workers from accidental exposures, detection and surveillance of infectious disease, emergency response, forensics, and protection of public health, agriculture, and health care critical infrastructures. This section highlights biodefense activities that were initiated after 2001, subsequent policy changes associated with those activities, and their science and technology needs.

13.2.1 Medical Countermeasures

For the past 17 years, the U.S. government has been investing in research and development of MCMs, which includes the development of new vaccines, therapeutics, and diagnostic tools to identify, prevent, and treat infectious diseases. These investments involved several complementary activities, including basic research, advanced development, regulatory science, and development of product platforms. These efforts required coordination and communication between the U.S. National Institutes of Health (NIH), Food and Drug Administration (FDA), and newly created programs for advanced development such as the Biomedical Advanced Research and Development Authority (BARDA), which was established in 2006 by the Pandemic and All-Hazards Preparedness Act (USC109§417). Together, these efforts support fundamental research that provides the scientific foundation for rational design and pre-clinical testing of vaccines and drugs, research to develop and/or validate the animal models and safety and efficacy data used to inform product approval by the FDA (often referred to as regulatory science research), and research that supports testing and evaluation of product efficacy in animal models. These efforts focus on developing and maintaining MCMs for use by civilians and first responders in an emergency.

Complementary activities are being supported in the Department of Defense (DoD) for protection of military forces [66]. Having vaccines and drugs against biological agents that may affect a sizable number of the military and/or be used on the battlefield is a priority for the DoD and has resulted in research, development, and manufacturing investments to support this defensive need. In 2006, DoD created the Transformational Medical Technologies Initiative (TMTI) to leverage scientific advances and technology innovation to counter genetically-engineered biological weapons [67, 68]. The TMTI program supported the development of several vaccine and drug candidates, including 2 products against hemorrhagic fever-causing viruses that reached the Investigational New Drug (IND) stage of the FDA approval process, preparation of 2 INDs for broad spectrum antibiotics, 3 broad-spectrum antibiotics, and 20 additional candidates in early research and development. In addition, TMT had begun examining FDA-approved antibiotics for activity against non-biothreat diseases [69]. In 2010, TMTI moved under the DoD Joint Program Executive Office for Chemical and Biological Defense (JPEO-CBD) [70]. In addition to this program, the Defense Threat Reduction Agency (DTRA) Joint Science and Technology Office for Chemical and Biological Defense (JSTO-CBD) supports research on MCMs against biological threats at DoD research facilities (i.e., under the Army's Medical Research and Materiel Command [71]) and at the nongovernmental research institutions [72]. The Joint Project Manager for Medical Countermeasure Systems, which is overseen by the DoD JPEO-CBD office, has issued calls for development of countermeasures against chemical, biological, radiological, and nuclear threats [73]. More recently, the DoD launched the Medical Countermeasures Advanced Development and Manufacturing capability, which is a joint effort between DoD and Nanotherapeutics Inc. that provides advanced development and manufacturing of biopharmaceuticals [74].

Although the operational considerations for the MCM needs for the Department of Health and Human Services (HHS) and DoD differ, the early research and development studies do not differ dramatically by product end-use. Therefore, the relevant HHS and DoD offices established the Integrated Portfolio, poviding opportunities to co-manage investments in MCM research and development of common interest [72, 75].

These programs provide the structure within which new scientific research and technologies can be funded and leveraged for medical countermeasure development. Through a combination of funding programs and technology watch efforts, program managers support innovation in technology development and learn about existing biotechnologies that could be applied to MCM development. For example, the Defense Advanced Research Projects Agency (DARPA) has created several programs to generate the scientific knowledge needed to detect human infection from any pathogen, develop new technologies to conduct field diagnostics with little sample, and create platform methodologies for rapid development of vaccines against outbreak pathogens [76, 77]. Although the results of this research are expected to enhance military force protection against biological threats, the scientific discoveries and technologies may benefit development of MCMs for civilians.

Recognizing potential funding opportunities for MCM development, academic and industrial researchers have undertaken efforts to apply new knowledge and scientific approaches to MCM development. For example, in 2013, the J. Craig Venter Institute and Novartis described the rapid creation of influenza vaccines within a cell-free system, capitalizing on recent innovations in synthetic biology [78]. More recently, scientists have used synthetic genomics (i.e., chemically-synthesized genomes) to generate hard-to-obtain pathogens, such as severe acute respiratory syndrome coronavirus [79, 80], or to develop candidate vaccine vectors against biological agents such as the recently-recreated horsepox virus [81]. Still other scientists have conducted studies to develop research animals and conditions that reflect pathogen infection and onset of disease in humans [70, 82] or alternative approaches for studying product efficacy, such as the "human on a chip" [83]. Although promising, application of emerging biotechnologies to MCM development is in early stages of research and development.

A primary difficulty in harnessing these advances is the up-front process and cost of FDA approval, particularly for products for which little or no commercial market exists. In response to this challenge, NIH and FDA have supported research towards enhancing regulatory science, a field of study designed to generate the foundational science needed to evaluate and validate scientific studies on product safety and efficacy [84–86]. This field emerged from the significant difficulties that the FDA encountered in assessing data submitted as part of the FDA approval process. Leveraging of emerging biotechnologies for MCMs depends on the functionality and capabilities of these technologies, the availability of information to evaluate safety and efficacy of products in a timely manner, and the existence of a steady market that helps to off-set product development costs.

13.2.2 Biosurveillance

Early detection and warning of infectious disease events was recognized as an important part of biodefense capabilities since the mid-1990s. The concept emerged from the need for preventing large-scale, man-made outbreaks, a concern that emerged in the 1990s after the discovery of the Former Soviet Union's massive overt offensive biological weapons program [87] and the release of an vaccine strain of *Bacillus anthracis* by the Japanese apocalyptic cult Aum Shinrikyo [88]. The first biosurveillance platform, ProMed Mail, was developed in 1994 and enables open, internet-based sharing of media reports, official reports, online information, local observations, and other similar sources. Users monitor these reports to identify incidents involving emerging or unusual infectious disease outbreaks or acute exposure to toxins affecting human health. Today, this platform includes 60,000 members from every continent and receives philanthropic funding and donations [89].

Shortly thereafter (1997), Health Canada in partnership with the World Health Organization established the Global Public Health Intelligence Network (GPHIN), which is a curated system that involves human analysis of media reports that are automatically collected [90]. The platform uses text analysis to detect potential infectious disease outbreaks, contamination of food and water, bioterrorism and exposure to chemicals, natural disasters, and safety incidents involving products, drugs, medical device, and agents. Through human analysis of these data, the platform provides users with actionable information about public health threats. Today, GPHIN is part of the WHO's epidemic intelligence system, which seeks to enable global alert and response of potential outbreaks before they become public health emergencies of international concern. The WHO's system focuses on a few infectious diseases and detects food and water contamination and chemical events [91].

More recently-developed systems collect, integrate, and inform users about potential infectious disease events that could cause large-scale epidemics or that are of questionable or suspicious origin. Although these systems include data from clinical and public health sources, they also include data from various other sources such as media reports of unusual illnesses or events, availability of medications in pharmacies, and ecological data of sick animals. Some systems build on the concepts of syndromic surveillance, which is the monitoring of infectious disease infections through surrogate markers (e.g., antibiotic sales), and sentinel surveillance (e.g., illness of animals from a zoonotic pathogen in specific locations). Furthermore, several of these systems leverage computational platforms that enable the collection, integration, and analysis of data of various types, from various sources, and with various degrees of uncertainty. Still others, which no longer exist, relied on crowdsourcing to verify reports of possible illness.

Between 2004 and 2006, four systems were developed: Georgetown University's Argus, the European Union Joint Research Centre's MediSys, the Japanese National Institute of Informatics' BioCaster, and Harvard University's HealthMap. MediSys and BioCaster were fully automated and able to analyze information in multiple languages. MediSys seeks to enable early detection of suspected bioterrorism activities [92]. This system uses a computational tool that scans articles for specific keywords and newsworthy information. BioCaster uses text mining technologies to scan RSS newsfeeds from local and national news in Japan to detect keywords related to infectious disease incidents [93]. This capability systematically indexes relevant information from the collected articles, increasing the rapidity with which users can access information. Argus, which no longer exists, and HealthMap were/are moderated by people. Argus was established to detect the emergence, illnesses, and person-to-person transmission of avian H5N1 influenza A virus [94]. Argus used a variety of data analysis technologies and human analysis to examine and monitor media reports on potential health security incidents. HealthMap was created to collect, integrate, and analyze a variety of data that may indicate a disease outbreak and enable ongoing monitoring and surveillance of infectious disease threats. Users can access and visualize data of interest by applying filters of data sources [95]. A retrospective analysis of the efficacy of HealthMap at detecting the 2014–2016 Western African Ebolavirus outbreak found that the system had identified cases of viral hemorrhagic fever a week before the WHO declared the outbreak [96]. Whether this system can be used to provide public health officials with predictive, actionable information is unclear, especially because the viral hemorrhagic fever cases identified by the platform do not appear to have led to outbreak response. An evaluation of these systems concluded that their efficacies are influenced by involvement of human moderation, sources of data, languages of the collected information, the regions in which incidents occur, and the types of cases collected [97].

The U.S. government provided financial support for the development of Argus and HealthMap. In addition, the U.S. government established several systems to obtain early warning of potential infectious disease outbreaks, allocating about $32 billion dollars between 2001 and 2012 [98]. DoD supported the Electronic Surveillance System for the Early Notification of Community-based Epidemics (ESSENCE), which was a system of systems integrating reportable disease surveillance, sentinel surveillance, syndromic surveillance, and clinical identification [99]. This system sought to detect natural and/or man-made outbreaks through the identification of abnormal trends in the data. ESSENCE involved cooperation between the Walter Reed Army Institute of Research, the Centers for Disease Control and Prevention (CDC), the Emergency Medical Associates of New Jersey Research Foundation, the New York City Department of Health and Mental Hygiene, and Harvard Medical School and Harvard Pilgrim Health Care. In 2003, the CDC created BioSense to enable early detection and assessment of potential illnesses that may be related to bioterrorism incidents [100]. Through this system, the CDC sought to build the capabilities of state and local public health departments to conduct syndromic surveillance. Today, the program is called the National Syndromic Surveillance Program and enables the collection, evaluation, sharing, and storage of health information via a cloud-based system [101]. The CDC also created the Early Warning Infectious Disease Surveillance system, which sought to enhance detection and reporting of transboundary infectious diseases outbreaks with

Canada and Mexico [102]. In 2007, the Department of Homeland Security (DHS) established the National Biosurveillance Integration Center (NBIC) after passage of the Implementing Recommendations of the 9/11 Commission Act [103]. The NBIC integrates and analyzes animal, plant, human, and environmental health data to improve early warning of potential biological incidents and gain situational awareness [104]. Although NBIC was created from a need to detect potential bioterrorism incidents, the Center now includes natural events that may cause significant disruption to the U.S. [105]. In 2012, the DoD invested in the Biosurveillance Ecosystem, which is a cloud-based platform that enables users to plug in data-specific applications to detect anomalies in human and animal infectious disease events [106]. In addition, the platform enables collaboration among users and incorporates advanced data analytics, such as machine learning of health and non-health data. The concept for this platform emerged from the 2009 H1N1 influenza outbreak, which caught many by surprise [107].

Building of these efforts and recognizing the benefits of identifying potentially-devastating infectious events, President Obama issued the National Strategy for Biosurveillance in 2012 [108]. This Strategy had three primary objectives: (1) to "integrate and enhance national biosurveillance efforts" to provide decision-makers with timely information in a crisis; (2) to increase the network of national expertise and capabilities for rapid detection and understanding (i.e., situational awareness) of emerging crises; and (3) to implement new approaches for considering and addressing uncertainty in the decision-making process. Included as core functions are environmental scanning; integration of human, animal, and plant health information; and assessment and forecasting of potential consequences. One month prior to the Strategy's release, the U.S. government established an interagency working group on science and technology for biosurveillance, which was chaired by the White House Office and Science and Technology Policy and DoD. This working group was tasked with developing the research and development priorities for achieving the core functions of the National Strategy for Biosurveillance. The working group discussion culminated in the release of the National Biosurveillance Science and Technology Roadmap in 2013, which defined the key priorities for detecting unusual or above-baseline events, anticipating risk including assessment of the potential consequences of incidents, identification and characterization of threats to human health, and integration of, analysis of, and information-sharing with key decision-makers, responders, and healthcare providers. The roadmap highlights current national activities in each of these areas and identifies capability needs to achieve the described priorities. The priorities encompassed the production of baseline data of natural or man-made human, animal, plant, and environmental health events; assessment of emergence and disease dynamics; development of rapid detection and diagnosis technologies and systems; and improvements in sharing, integration, and communication of biosurveillance data. This strategy and roadmap provided a framework for enhancing U.S. capabilities for early warning of natural and/or man-made biological events, which could be leveraged for detection and early warning against chemical, radiological, nuclear, and environmental incidents.

13.2.3 Infectious Disease Characterization and Modeling

Following the investments for biosurveillance, NIH, DHS, U.S. Agency for International Development (USAID), and DoD initiated programs to enhance understanding of pathogens of national security and public health concern, host-pathogen interactions, and disease dynamics through modeling the emergence and spread of infectious disease. Furthermore, these agencies embraced newer biotechnologies, computational and data science, and convergence of chemical, biological, materials, and other physical sciences to provide innovative solutions for preventing, detecting, and responding to biological threats.

In the aftermath of September and October 2001, NIH convened a Blue Ribbon Panel on Bioterrorism and its Implications for Biomedical Research to define biodefense research priorities for pathogens and toxins that could present the highest risk to national security and public health (NIAID Category A Priority Pathogens [109]). To support early research, build scientific expertise, and provide ready capability in a bioterrorism or public health emergency, NIH invested in Regional and National Biocontainment Laboratories and in research and training consortia through their newly established Regional Centers of Excellence for Biodefense and Emerging Infectious Diseases [110, 111]. NIH priorities primarily focused on developing specialized knowledge and skill for studying biological agents on the Category A and B Priority Pathogen lists [112], which initially did not require the use of emerging biotechnologies. However, more recent research efforts have begun exploring synthetic biology, systems biology and data science, microfluidic technologies, tissues-on-a-chip, and other methodologies that offer new capabilities for studying host-pathogen responses and developing, testing, and validating diagnostic tools and MCMs. For example, NIH supported a Modeling Immunity for Biodefense program in which immunologists, computational biologists, modelers, and microbiologists worked together to model the immunological responses to infection and vaccination [113]. Other examples include the applications of next generation sequencing to characterize newly emerging pathogens using microfluidic technologies [114] and lab-on-a-chip systems for monitoring pathogens in various samples, including environmental samples [115]. In addition, research programs supporting metagenomic studies for examining host-pathogen responses are leveraged to understand biological threat agents better [116].

In 2004, NIH created a collaborative network of researchers to develop new computational, statistical, and mathematical models for studying pathogen dynamics, building technical expertise through education and training, and enabling communication with policy-makers and international partners [117]. This program, called Models of Infectious Disease Agent Study (MIDAS), supports research on the emergence and spread of infectious disease, surveillance, effectiveness of intervention strategies, host-pathogen interactions, and environmental and societal factors involved in emergence and spread of infectious diseases. Although not designed to inform public health control measures directly, the models can inform high-level strategy for detecting and controlling infectious disease outbreaks. For example, some of the influenza models [118] informed the development of the U.S. National

Strategy for Pandemic Influenza in 2006. In addition, modeling approaches can provide insight about public health strategies that would help control pandemics [119].

NIH created a second program, the Research and Policy for Infectious Disease Dynamics (RAPIDD) program that lasted from 2008 to 2016, received funds from the Department of Homeland Security to support infectious disease modeling efforts, and was administered through the Fogarty International Center [120]. The RAPIDD program supported international teams of researchers to develop community-based and/or environmental models and to generate and/or collect the necessary data to study infectious disease dynamics. Some models integrated clinical data with data on local population migration, and hospital location and capacity. Other models focused on the existence of specific pathogens (e.g., viruses that cause hemorrhagic fevers) in wild animals throughout the world, the migration patterns of the animals, and the interactions between the animals and people. Some of these models include data on disease vectors, such as mosquitoes, to study the spread and potential risk of vector-borne diseases. In addition to these modeling efforts, the RAPIDD program includes opportunities to translate these models to decision-makers responsible for public health emergencies of international consequence. In 2015, one research group (Drs. David Hayman of Massey University in New Zealand and James Wood of Cambridge University in the United Kingdom) partnered with Ambassador Bonnie Jenkins (former Ambassador for Cooperative Threat Reduction of the U.S. Department of State) and Gryphon Scientific (Dr. Kavita Berger, the author of this chapter) to explore approaches for improving communication between global health security stakeholders and the research community. This discussion occurred within the context of the 2014–2016 Ebolavirus disease outbreak and the Global Health Security Agenda (launched in 2014), and leveraged Dr. Hayman and Wood's expertise and research on the existence of viral hemorrhagic fevers.

In the early-2010s, the U.S. and U.K. entered into a collaboration to support research for combatting zoonoses (i.e., infectious diseases that infect both humans and animals) in low-income countries [121]. The U.S. program was part of the Ecology of Infectious Disease program, which involved joint funding from NIH and the U.S. National Science Foundation [122]. Funding in the U.K. was provided by the Biotechnology and Biological Sciences Research Council and Economic and Social Research Council [121]. This collaborative effort lasted until November 2017 [123]. The knowledge gained from the research can inform infectious disease surveillance efforts.

In 2009, the USAID Emerging Pandemic Threats (EPT) program funded teams of scientists, One Health experts (including veterinarians), and epidemiologists to build international capacity to prevent, detect, and respond infectious disease incidents in animals and people [124]. Leading up to the creation of the EPT program were several outbreaks (e.g., severe acute respiratory syndrome in 2003, H5N1 influenza from 1997 to present day, and H1N1 influenza in 2009) that emerged unexpectedly, resulting in international concern. This program supports multi-sector teams to implement projects to predict, prevent, identify, and respond to infectious disease threats. The Predict project involves capacity building for strengthening detection and monitoring of, characterizing the drivers for, and improving predictive models for disease transmission between wildlife and people. The Prevent project involves assessment of the risks of associated with disease transmission and strategies to mitigate those risks. The Identify project involves the building laboratory capacity, including training personnel, to diagnose infectious diseases in animals and people. The Respond project involves facilitating networks of public health, veterinary medicine, and environmental health to improve transdisciplinary training and response efforts. In addition to these projects, the EPT program funds CDC to promote the detection and monitoring of zoonotic diseases, including providing support for the Field Epidemiology Training Program.

DoD's DTRA leverages emerging biotechnologies to develop products for development of innovative MCMs (see previous section), detection of biological threats, protection from exposure to chemical and biological agents, and decontamination [125]. For example, the DoD supports research involving synthetic biology to develop decontamination products, develop and test vaccine and drug candidates, and improve product development pathways. The agency supports innovative projects for the development of bio-based sensors for chemical and biological agents and for integration and analysis of data via its Global Biosurveillance Portal. In addition, the DoD coordinates internally and with HHS to leverage investments in science and technology.

DHS supported University-based Centers of Excellence to develop the approaches, technologies, and knowledge needed to enhance U.S. homeland security. Texas A&M University and Kansas State University hosted the Center of Excellence for Zoonotic and Animal Disease Defense, which conducted research to prevent foreign animal and zoonotic disease threats [126]. This Center supported research to develop new vaccine candidates and platforms for animal pathogens, promote educational and training programs, and inform decision-makers about prevention and control of animal and zoonotic diseases. The University of Minnesota leads the Food Protection and Defense Institute, which conducts research to address security vulnerabilities in the food supply [127]. This center supports research on risk analysis and modeling of incidents involving the food system. Purdue University and Rutgers University hosted the Center of Visualization and Data Analytics [128]. This center supported research for integrating and analyzing large datasets involved in detecting security threats. All three centers are in emeritus status; none of the current Centers of Excellence focus on biological threats.

13.2.4 Microbial Forensics

The field of microbial forensics emerged in 2001 when the Federal Bureau of Investigation (FBI) had to develop new sampling, storage, testing, and analysis methods for identifying the source and determining attribution of the *Bacillus anthracis*-laced letters. Microbial forensics and attribution was included in Presidential level strategy for biodefense in the early part of the 21st century, specifically in Homeland Security Presidential Directive-10, *Biodefense for the 21st Century*. A significant investment towards microbial forensics was the establishment of the National Bioforensic Analysis Center (NBFAC), which is part of the National Biodefense Analysis and Countermeasure Center [129]. NBFAC has achieved certification of ISO standards enabling it to provide laboratory support for federal law enforcement investigations that involve biological agents [130]. NBFAC works closely with the FBI and relevant U.S. government agencies on specific cases that warrant sample analysis for agent identification, sample signatures, and attribution. In addition, NBFAC conducts forensic research to address capability gaps to enhance its support for biocriminal and bioterrorism investigations. In addition, the National Institute of Justice of the Department of Justice has provided funding since 2011 for research on application of microbiome research to forensic data collection and analysis [131]. This funding program leverages recent research on the microbiome, which receives significant funding from the NIH. In 2014, the National Research Council (NRC) published a report on scientific needs for microbial forensics [132]. Drawing on the recent advances in cloud computing, data collection, and data analytics, the NRC report primarily focused the potential for bioinformatics, whole genome sequencing, and systems biology analyses to enhance microbial forensic capabilities internationally.

More broadly, the National Science Foundation and Department of Justice have sponsored seminars and research on forensic science. For example, the National Institute of Justice provided over 140 grants since 2004 to support research and development of forensic science, some of which involves analysis of biological materials [133]. In 2014, the National Science Foundation established a forensic science program that supports research and training in seven directorates: Biological Sciences, Computer and Information Science and Engineering, Education and Human Resources, Engineering, Geosciences, Mathematical and Physical Sciences, and Social, Behavioral, and Economic Sciences [134]. Through this program, NSF established a Center for Advanced Research in Forensic Science, which is a consortium of nine universities that have received planning grants in various topics [135]. The University of North Texas and Texas A&M University received planning grants in microbial forensics and California State University received a planning grant in chemistry, biology, and biomechanical engineering. The Center for Advanced Research in Forensic Science enables cooperation between universities, industry, and government to support fundamental research in forensic science and to develop scientists with the knowledge and skills to leverage scientific advances for application to forensic science. In addition to these efforts, the Department of Justice and National Institute of Standards and Technology established the National Commission on Forensic Science to promote scientific progress and improve scientific validity in forensic science [136]. However, this Commission lasted only 4 years, from 2013 to 2017.

13.3 Examples of Creative Application of Emerging Biotechnologies

Although much of the biodefense investment remains focused on microbiological and virological research, U.S. agencies and scientists have leveraged emerging biotechnologies to meet other defensive needs. Multidisciplinary scientific efforts have been used to address military medical needs, global health, and forensic analysis of chemical weapons attacks. This section briefly describes efforts that promoted creativity and innovation to meet practical needs. The approaches used in these and other examples could be applied to leveraging emerging biotechnologies and converging sciences to address current and future biodefense and health security needs.

13.3.1 Neuroprosthetics

Advances in neuroscience and robotics enabled the development of brain-controlled prosthetics, a field which the DARPA pioneered [137–139]. In 2006, DARPA established the Revolutionizing Prosthetics program to restore normal upper limb function to individuals who have lost that capability [140]. At the time, the state of the science for artificial limbs was much less advanced than lower-body prosthetics [141]. To enhance upper-limb prosthetics, DARPA funded engineers and neuroscientists to work together to design and develop prosthetic limbs that can be controlled by the user's thoughts. Through this multidisciplinary effort, engineers developed artificial limbs that had the range of motion like a natural limb and neuroscientists developed user control interfaces that enable the transfer the signals generated by thought to the actual movement and feeling of limbs [140]. The initial prototypes were tested in the laboratory before being tested on amputees. Through this multidisciplinary effort, the DARPA program was able to develop neuroprosthetics that restored fine motor function in the amputees [142, 143].

13.3.2 Predicting Infectious Disease Events

Advances in data science and availability of environmental, geographic, and population density and migration data have ushered in a new age for detecting infectious disease events. This approach draws on studies correlating infectious disease infection, transmission, and emergence in the human population with temperature, humidity, urbanization and city expansion, night lights, local migration, and other environmental and human factors. These data, along with satellite imagery, can be used to identify conditions conducive for pathogen outbreaks. In addition, geolocation data embedded in videos, pictures, and text files can be used to tag the location and time of infection events. In 2009, scientists described the use of satellite imagery to identify climate events (e.g., flooding) with outbreaks of waterborne disease (e.g., cholera, typhoid, hepatitis, and leptospirosis) [144]. They leveraged published research correlating outbreaks of these diseases with societal changes caused by flooding, including overcrowding and lack of sanitation and clean water. The scientists argued that coupling knowledge about human factors with factors observable by satellite imagery also could be used to predict movement of vectors with the emergence and spread of vector-borne diseases. Another research group examined the correlation between malaria incidence and climatic data (i.e., rainfall estimates, vegetation levels, and surface temperature of the land) collected through satellites [145]. More recently, researchers described the use of satellite imagery to monitor nighttime lights to examine the pattern of measles cases in Western Africa [146]. These scientists used nighttime lights as a surrogate for population density, which affects pathogen transmission, to predict outbreaks of measles.

13.3.3 Syrian Chemical Weapons Attack

In 2013, hundreds of men, women, and children were killed in Ghouta, Syria with missiles carrying chemical weapons. In the immediate aftermath of the attacks, Human Rights Watch, working with scientists, examined images and videos posted on social media sites to analyze the types and amounts of weapon used in the attack, which led the organization to attribute the attacks to the Syrian government [147]. They leveraged Facebook and other social media sites to examine the content in image and video files that were being posted by attack victims and Human Rights Watch staff in country. The scientists, some of whom had expertise in chemistry, chemical weapons, and arms control, were able to use the posted images to identify: (a) the signs of disease caused by chemical weapons; (b) the location, timing, and duration of the attacks; and (c) the type of weapon used to deliver the agent.

These examples illustrate different ways by which technology can be used to address societal needs. Similar innovations, technologies, and applications can be leveraged to address biodefense needs ranging from threat assessment through recovery after a significant natural or man-made event.

13.4 Underlying Factors of Leveraging Biotechnologies for Biodefense

During the Obama Administration, the White House Office of Science Technology Policy, National Science and Technology Council, and National Security Staff issued several reports and Executive-level policies encouraging the advancement and application of science and technology (S&T) to address national security objectives, including biodefense objectives. In 2013, the White House released the National Biosurveillance Science and Technology Roadmap, which described current capabilities and S&T gaps and needs to support implementation of the National Biosurveillance Strategy [148]. In 2014, the White House released a report on Strengthening Forensic Science, which highlighted capabilities and implementation challenges for accreditation and certification of forensic science and workforce development of medical and legal personnel [149]. In 2016, the White House issued three reports highlighting S&T needs for disaster preparedness, reduction, and resilience, including The National Preparedness Science and Technology Task Force Report: Identifying Science and Technology Opportunities for National Preparedness [150], The Implementation Roadmap for the Critical Infrastructure Security and Resilience Research and Development Plan [151], and The Homeland Biodefense Science and Technology Capability Review [152]. The Task Force Report highlighted preparedness initiatives that could benefit from advances in S&T. The Implementation Roadmap enumerated the research and development activities that enhance security and resilience of critical infrastructure. The biodefense capability review highlighted the S&T needs for countering biological threats affecting human and animal health.

The Homeland Security Biodefense Science and Technology Capability Review described the results of an interagency process to identify science and technology needs to meet critical biodefense objectives in threat awareness, surveillance and detection, and recovery [152]. The report highlighted the critical needs for: (a) enhancing U.S. risk assessment methodologies; (b) conducting threat characterization research (including assessing the consequences of a deliberate attack on pets and wildlife); (c) examining transmission through food and water; (d) studying avian influenza; (e) linking biosurveillance data to improve decision-making in a biological incident; (f) improving sampling, detection, and surveillance of environmental samples after an attack; (g) improving forensic analysis of a biological incident; (h) enhancing research and laboratory capacity during a crisis to inform modeling of biological attacks; (i) improving MCM shelf-life, production, and storage; (j) enhancing workforce development of responders to biological attacks; and (k) developing innovative decontamination methods. This list of science and technology needs for biodefense highlights the general need for workforce development, research support, and information-sharing with policy-makers. Some of these needs were addressed by the 2018 National Biodefense Strategy, which calls for research and development in a variety of areas important for preventing, detecting, and responding to natual and man-made biological threats [153].

Common to these documents are the recognized need for a proficient workforce, data sharing and communication with other scientists and decision-makers, and research and development to increase knowledge and develop new technologies and methodologies. However, a significant challenge seems to be the translation of operational or strategic need to specific research requirements that can be used to generate information and develop new products. Shortly after 2001, the U.S. government, philanthropic organizations, and universities supported research, education, and communication efforts. In the early 2010s, interest in continuing these efforts waned as time has progressed without the occurrence of a major biological attack. However, during the past few years, particularly since the 2014–2016 Western African Ebolavirus and 2016 Zika virus outbreaks, concern about catastrophic natural and man-made threats have met with renewed interest in funding research on risk assessment, prediction of future infectious disease threats, trends in biotechnology (specifically, horizon scanning[1]), and decontamination. This increased interest in research has not necessarily been met with increased investment in education and training. Although a few formal, degree-granting programs in biodefense have existed, the only known programs that exist today are the Biohazardous Threat Agents and Emerging Infectious Diseases of Georgetown University [154] and Masters in Biodefense at George Mason University [155]. Furthermore, the number of universities that provide laboratory biosecurity training to graduate students in the life sciences has decreased during the past decade, but in their place programs through educational service providers, such as CITI [156], have emerged. In addition, the American Biological Safety Association provides laboratory biosafety and biosecurity training to biosafety professionals [157] the International Federation of Biosafety Associations has a certification program for biorisk management (i.e., methods for analyzing and managing accidental and deliberate biological risks) [158], three organizations host leadership in biosecurity or health security programs [159–161], and the U.S. government supports training activities. Despite these efforts, most scientists are not taught about biosecurity or biodefense issues (the exceptions are researchers working with Biological Select Agents and Toxins or in synthetic biology). This lack of training results in scientists who may not be able to evaluate the broader impact of their research, whether that impact is increased risk of research activities or enhanced benefits to meet national security needs. Without such knowledge, scientists and engineers may not be able to identify and transition knowledge or technologies that can address outstanding biodefense needs, or to identify potential risks associated with their research. This difficulty is pronounced with pathogen research compared to research that does not involve pathogens (e.g., creating synthetic organisms to develop platforms for synthetic antimicrobial products). However, the ability to translate biodefense needs to research projects appears to be more challenging than similar capabilities for translating needs to improve human health, agriculture, or environmental health.

Overcoming these challenges requires a system in which decision-makers are able to convey their needs to funding organizations, which in turn need to convey their needs to the scientific and engineering communities. To be successful, this system requires creativity at all stages to identify solutions that address particular needs and leverage the capabilities generated by advances in physics, chemistry, engineering, material science, computer science, social and behavioral sciences, and the life sciences. Promoting creativity is harder than it appears, often requiring researchers and decision-makers to think differently than they were taught. For example, many scientists, especially life scientists, approach research from a discovery science standpoint, i.e., they develop and test hypotheses to generate knowledge, create new products (e.g., high-throughput drug discovery and pre-clinical vaccine development), and/or develop new technologies to improve research efforts or as product platforms. On the other end of the spectrum are engineers, who often

[1]Horizon scanning is the gathering and analyzing data to forecast trends in society, including technology trends.

conduct design-based research, which involves identifying the specific product needs before developing the methodological approach to achieve the desired output (e.g., synthetic organisms designed to produce industrial chemicals). Both approaches enable innovation within fields and sometimes across fields, but they also may contribute to conceptual barriers to identifying solutions from high-level need. This challenge may be exacerbated for decision-makers whose area of expertise is not research and consequently, may have trouble defining product specifications from the strategic or operational needs.

One potential solution to addressing this issue is to provide more opportunities for students, junior and senior scientists and engineers with opportunities to learn about different approaches to knowledge and product development. This level of awareness-raising can be achieved through training or fellowship opportunities, conferences and symposia, collaborative grants, and/or challenge competitions. In addition, these workforce development opportunities could be enhanced further by including biodefense topics and needs in training curricula and other engagement venues. The goal of these efforts would be to impart the necessary knowledge to the scientific and engineering communities about the critical needs in biodefense that emerging biotechnologies may be able to address. Through these training and career development opportunities, scientists and engineers may gain the necessary knowledge about the problems that need addressing, previously-attempted approaches, and willingness for transitioning new technologies. Furthermore, they may be better positioned to identify which information and technologies may be most effective for addressing a given biodefense need.

In addition to developing a science and engineering workforce that understands the biodefense objectives and needs, educating decision-makers about advances in emerging biotechnology may enhance current or add new capabilities not previously imagined. Although some decision-makers are examining cutting-edge biotechnologies to harness their capabilities for biodefense purposes, many largely are unaware of the benefits afforded by these technologies. In fact, several security experts and decision-makers view emerging biotechnology more as a risk than benefit and consider the realization of risky outcomes to be more plausible that the realization of benefit. This attitude prevents decision-makers and security experts, alike, from seeing the potential of biology and biotechnology for addressing unaddressed capability gaps. Educational efforts directed towards raising awareness of current and near future technology trends and their application to other fields may help decision-makers understand which technologies can be leveraged, which require further investment, and which have associated legal, ethical, and social barriers limiting their application.

These efforts could be enhanced through directed engagement and dialogue between scientists, engineers, and decision-makers, especially if one result is the act of jointly brainstorming potential technology solutions for biodefense science and technology needs. This ideation process could help drive creativity in identifying potential technologies and information that could be harnessed, even if outside the expertise of the scientists involved in the process. Furthermore, this process could have broader benefit by helping to develop research agendas that may generate wider interest among the scientific community. This process could assist with translation of strategic or operational biodefense needs to more defined research and development programs, particularly ones that harness the capabilities of emerging biotechnologies.

13.5 Conclusions

The emergence of new biotechnologies provides great promise for biodefense, especially for key objectives of biosurveillance and early warning, microbial forensics, risk and threat assessment, horizon scanning in biotechnology, and MCM development, scale-up, and delivery. Understanding and leveraging newly-developed capabilities afforded by emerging biotechnologies require knowledge about cutting-edge research and its real or proposed application(s) to analogous needs, the process through which biotechnologies advance, and the educational and research infrastructure that promotes multi-disciplinary science. When combined with knowledge about biodefense gaps that could be addressed through science and technology, creative scientists, engineers, and decision-makers can design research agendas or concepts based on the identified gaps and S&T needs, and that leverage emerging biotechnologies. Innovation in research and technology development are driven by sector-specific needs and the convergence of the physical, chemical, material, computer, engineering, and life sciences. Biotechnologies developed for other sectors could be applied to biodefense, especially if the individuals involved are able to innovate in concept design and development. Of all biodefense objectives, biosurveillance seems to have reaped the most benefit from emerging biotechnologies, specifically the integration and analysis of diverse clinical, biological, demographic, and other relevant data. More recently, scientists have begun applying synthetic biology, genomics, and microfluidics to development of new products and platforms for MCMs. Unlike these initiatives, investments in microbial forensics have been few, limiting its ability to leverage biotechnology advances for collecting and analyzing data. Looking to the future, emerging biotechnologies can provide new opportunities for enhancing biodefense by addressing capability gaps.

References

1. Global Market Insights. Genome editing market. 2016.
2. Market Research Service. Genome editing: Genome Engineering Markets. Research and Markets. 2017.
3. Research and Markets. Genome editing: technologies and global markets. 2016.
4. PM Live Top Pharma List. Top 50 pharmaceutical products by global sales.
5. Bowick GC, Barrett ADT. Comparative pathogenesis and systems biology for biodefense virus vaccine development. J Biomed Biotechnol. 2010; Article ID 236528.
6. Burke W, Psaty BM. Personalized medicine in the era of genomics. JAMA. 2007;298(14): 1682–4.

7. Ginsburg GS, Willard HF. Genomic and personalized medicine, vol. 1. San Diego: Academic; 2008.
8. Hesse H, Höfgen R. Application of genomics in agriculture. In: Hawkesford MJ, Buchner P, editors. Molecular analysis of plant adaptation to the environment. New York: Springer; 2001. p. 61–79.
9. Menachery VD, Schäfer A, Burnum-Johnson KE, Mitchell HD, Eisfeld AJ, Walters KB, et al. MERS-CoV and H5N1 influenza virus antagonize antigen presentation by altering the epigenetic landscape. Proc Natl Acad Sci. 2018;201706928.
10. Schäfer A, Baric RS. Epigenetic landscape during coronavirus infection. Pathogens. 2017; 6(1):8.
11. Kruse U, Bantscheff M, Drewes G, Hopf C. Chemical and pathway proteomics powerful tools for oncology drug discovery and personalized health care. Mol Cell Proteomics. 2008;7(10): 1887–901.
12. Bundy JG, Davey MP, Viant MR. Environmental metabolomics: a critical review and future perspectives. Metabolomics. 2009;5(1):3.
13. Spratlin JL, Serkova NJ, Eckhardt SG. Clinical applications of metabolomics in oncology: a review. Clin Cancer Res. 2009;15(2):431–40.
14. Dennis KK, Auerbach SS, Balshaw DM, Cui Y, Fallin MD, Smith MT, et al. The importance of the biological impact of exposure to the concept of the exposome. Environ Health Perspect. 2016;124(10):1504.
15. Rappaport SM. Implications of the exposome for exposure science. J Expo Sci Environ Epidemiol. 2011;21(1):5.
16. Evans WE, Relling MV. Pharmacogenomics: translating functional genomics into rational therapeutics. Science. 1999;286(5439):487–91.
17. Phillips KA, Veenstra DL, Oren E, Lee JK, Sadee W. Potential role of pharmacogenomics in reducing adverse drug reactions: a systematic review. JAMA. 2001;286(18):2270–9.
18. Raunio H. In silico toxicology–non-testing methods. Front Pharmacol. 2011;2:33.
19. Viceconti M, Clapworthy G, Jan SVS. The virtual physiological human—a European initiative for in silico human modelling. J Physiol Sci. 2008;58(7):441–6.
20. Hay SI, George DB, Moyes CL, Brownstein JS. Big data opportunities for global infectious disease surveillance. PLoS Med. 2013;10(4):e1001413.
21. Khalil AS, Collins JJ. Synthetic biology: applications come of age. Nat Rev Genet. 2010; 11(5):367.
22. National Research Council. Positioning synthetic biology to meet the challenges of the 21st century: summary report of a six academies symposium series. Washington, DC: National Academies Press; 2013.
23. Chang R, Nam J, Sun W. Direct cell writing of 3D microorgan for in vitro pharmacokinetic model. Tissue Eng Part C Methods. 2008;14(2):157–66.
24. Murphy SV, Atala A. 3D bioprinting of tissues and organs. Nat Biotechnol. 2014;32(8):773.
25. Nguyen DG, Funk J, Robbins JB, Crogan-Grundy C, Presnell SC, Singer T, Roth AB. Bioprinted 3D primary liver tissues allow assessment of organ-level response to clinical drug induced toxicity in vitro. PLoS One. 2016;11(7):e0158674.
26. Moeller L, Wang K. Engineering with precision: tools for the new generation of transgenic crops. AIBS Bull. 2008;58(5):391–401.
27. Voytas DF, Gao C. Precision genome engineering and agriculture: opportunities and regulatory challenges. PLoS Biol. 2014;12(6):e1001877.
28. Davis G, Casady WW, Massey RE. Precision agriculture: an introduction. Extension Publications (MU); 1998.
29. Schmaltz R. What is precision agriculture? AgFunder News. 2017. https://agfundernews.com/what-is-precision-agriculture.html
30. Huang X, Han B. Natural variations and genome-wide association studies in crop plants. Annu Rev Plant Biol. 2014;65:531–51.

31. Sharma A, Lee JS, Dang CG, Sudrajad P, Kim HC, Yeon SH, et al. Stories and challenges of genome wide association studies in livestock—a review. Asian Australas J Anim Sci. 2015; 28(10):1371.
32. Georges F, Ray H. Genome editing of crops: a renewed opportunity for food security. GM Crops Food. 2017;8(1):1–12.
33. Laible G, Wei J, Wagner S. Improving livestock for agriculture–technological progress from random transgenesis to precision genome editing heralds a new era. Biotechnol J. 2015;10(1): 109–20.
34. Adesina O, Anzai IA, Avalos JL, Barstow B. Embracing biological solutions to the sustainable energy challenge. Chem. 2017;2(1):20–51.
35. Ferry MS, Hasty J, Cookson NA. Synthetic biology approaches to biofuel production. Biofuels. 2012;3:9–12.
36. Reardon S. Leukaemia success heralds wave of gene-editing therapies. Nature. 2015; 527(7577):146–7. https://doi.org/10.1038/nature.2015.18737
37. Bowler J. A Swedish scientist is using CRISPR to genetically modify healthy human emryos. Science Alert. 2016.
38. Ma H, Marti-Gutierrez N, Park SW, Wu J, Lee Y, Suzuki K, et al. Correction of a pathogenic gene mutation in human embryos. Nature. 2017;548(7668):413–9. https://doi.org/10.1038/nature23305
39. Cyranoski D, Ledford H. Genome-edited baby claim provokes international outcry. Nature. 2018;563(7733):607–8.
40. Kaiser J. A yellow light for embryo editing. Science. 2017;355(6326):675. https://doi.org/10.1126/science.355.6326.675-b
41. Carlson DF, Lancto CA, Zang B, Kim ES, Walton M, Oldeschulte D, et al. Production of hornless dairy cattle from genome-edited cell lines. Nat Biotechnol. 2016;34(5):479–81. https://doi.org/10.1038/nbt.3560
42. Panko B. Gene-edited cattle produce no horns. Science. 2016. https://doi.org/10.1126/science.aaf9982
43. Regalado A. Bill Gates doubles his bet on wiping out mosquitoes with gene editing. MIT Technology Review. 2016.
44. Inova Healthcare. Transforming healthcare with groundbreaking genomics research. http://www.inova.org/itmi/home. Accessed 4 Oct 2017.
45. Cyranoski D. China embraces precision medicine on a massive scale. Nature. 2016;529 (7584):9–10.
46. Jang J. 3D bioprinting and in vitro cardiovascular tissue modeling. Bioengineering. 2017;4(3): 71.
47. Aravamudhan S, Bellamkonda RV. Toward a convergence of regenerative medicine, rehabilitation, and neuroprosthetics. J Neurotrauma. 2011;28(11):2329–47.
48. Thakor NV. Neuroprosthetics: past, present and future. In: Jensen W, et al., editors. Replace, repair, restore, relieve–bridging clinical and engineering solutions in neurorehabilitation. Cham: Springer; 2014. p. 15–21.
49. Eyre HA, Forbes M, Raji C, Cork N, Durning S, Armstrong E, et al. Strengthening the role of convergence science in medicine. Convergent Sci Phys Oncol. 2015;1(2):026001.
50. Peplow M. Synthetic biology's first malaria drug meets market resistance. Nature. 2016;530 (7591):389–90. https://doi.org/10.1038/530390a
51. Biotechnology Industry Organization. Current uses of synthetic biology for renewable chemicals, pharmaceuticals, and biofuels. 2013. https://www.bio.org/sites/default/files/files/2013-03-03-Synthetic-Biology-Products.pdf. Accessed 9 Dec 2018.
52. Galanie S, Thodey K, Trenchard IJ, Filsinger Interrante M, Smolke CD. Complete biosynthesis of opioids in yeast. Science. 2015;349(6252), 1095–100. https://doi.org/10.1126/science.aac9373
53. iGEM. Colour generators. 2009. http://2009.igem.org/Team:Cambridge/Project/Pigments. Accessed 4 Oct 2017.

54. Molteni, M. Scientists upload a galoping horse GIF into bacteria with CRISPR. Wired; 2017.
55. Shipman SL, Nivala J, Macklis JD, Church GM. CRISPR-Cas encoding of a digital movie into the genomes of a population of living bacteria. Nature. 2017;547(7663), 345–9. https://doi.org/10.1038/nature23017
56. Regalado A. Scientists hack a computer using DNA. MIT Technol Rev. 2017. https://www.technologyreview.com/s/608596/scientistshack-a-computer-using-dna/. Accessed 9 Dec 2018
57. Abil, Z., Xiong, X., & Zhao, H. Synthetic biology for therapeutic applications. Mol Pharm. 2015;12(2):322–31. https://doi.org/10.1021/mp500392q
58. Lawrence J. Synthetic biology: microbes made to manufacture drugs. Pharm J. 2015;295 (7882). Online.
59. Perkel JM. Revolutionizating vaccine development with synthetic biology. Biocompare. 2015.
60. Rojahn SY. Synthetic biology could speed flu vaccine production. MIT Technol Rev. 2013. https://www.technologyreview.com/s/514661/synthetic-biology-could-speed-flu-vaccine-production/. Accessed 4 Oct 2017
61. Slomovic S, Pardee K, Collins JJ. Synthetic biology devices for in vitro and in vivo diagnostics. Proc Natl Acad Sci USA. 2015;112(47):14429–35. https://doi.org/10.1073/pnas.1508521112
62. Global Biodefense. The DoD and synthetic biology milestones. Global Biodefense; 2016.
63. Keasling JD. Synthetic biology for synthetic chemistry. ACS Chem Biol. 2008;3(1):64–76. https://doi.org/10.1021/cb7002434
64. Office of Technical Intelligence, Office of the ASD for R and E. Technology assessment: synthetic biology. In: U.S. Department of Defense, editor. 2015.
65. Strychalski E. Biological Control. 2016. https://www.darpa.mil/attachments/BiologicalControlProposersDay(Strychalski).pdf. Accessed 4 Oct 2017.
66. National Research Council. Giving full measure to countermeasures: addressing problems in the DoD program to develop medical countermeasures against biological warfare agents. Washington, DC: National Academy Press; 2004.
67. Department of Defense. Transformational Medical Technologies Initiative (TMT): OUSD (AT&L) FY2007. https://www.hsdl.org/?abstract&did=10926. Accessed 9 Dec 2018.
68. Reed JD. Department of Defense Biological Program. 2007.
69. National Research Council. The public health emergency countermeasure enterprise: innovative strategies to enhance products from discovery through approval. Washington, DC: National Academy Press; 2010.
70. National Research Council. Animal models for assessing countermeasures to bioterrorism agents. Washington, DC: National Academies Press; 2011.
71. US Army Medical Research and Material Command. Medical chemical biological defense research. http://mrmc.amedd.army.mil/index.cfm?pageid=medical_r_and_d.medchembio.overview. Accessed 5 Oct 2017.
72. Newmark, J. The integrated national biodefense portfolio initiative. 2009.
73. Global Biodefense. DoD medical countermeasure systems broad agency announcement. Global Biodefense; 2016.
74. Clark A. U.S. Department of Defense Expands Medical Countermeasure Capabilities. Chemical & Biological Defense News. 2016.
75. Department of Health and Human Services (Producer). BARDA guiding principles. 2013. https://www.phe.gov/about/barda/stratplan/Pages/barda-guiding-principles.aspx. Accessed 6 Oct 2017
76. Global Biodefense. DARPA pandemic prevention platform (P3) proposers day events. Global Biodefense. 2017. https://globalbiodefense.com/2017/02/07/darpa-pandemic-prevention-platform-p3-proposers-day-events/
77. Plitsch J, Evans T. DARPA seeks to establish new platforms for rapid development of medical countermeasures. Procurement Law and Policy Insights. 2017. https://www.insidegovernmentcontracts.com/2017/02/darpa-seeks-to-establish-new-platforms-for-developing-medical-countermeasures-within-sixty-days-of-identifying-a-biological-threat/

78. Dormitzer PR, Suphaphiphat P, Gibson DG., Wentworth DE, Stockwell TB, Algire MA, et al. Synthetic generation of influenza vaccine viruses for rapid response to pandemics. Sci Transl Med. 2013; 5(185):185ra168. https://doi.org/10.1126/scitranslmed.3006368
79. Becker MM, Graham RL, Donaldson EF, Rockx B, Sims AC, Sheahan T, et al. Synthetic recombinant bat SARS-like coronavirus is infectious in cultured cells and in mice. Proc Natl Acad Sci USA.2008;105(50):19944–9. https://doi.org/10.1073/pnas.0808116105
80. Yount B, Roberts RS, Lindesmith L, Baric RS. Rewiring the severe acute respiratory syndrome coronavirus (SARS-CoV) transcription circuit: engineering a recombination-resistant genome. Proc Natl Acad Sci USA. 2006;103(33):12546–51. https://doi.org/10.1073/pnas.0605438103
81. Noyce RS, Lederman S, Evans DH. Construction of an infectious horsepox virus vaccine from chemically synthesized DNA fragments. PLoS One. 2018;13(1):e0188453.
82. Sullivan NJ, Martin JE, Graham BS, Nabel GJ. Correlates of protective immunity for Ebola vaccines: implications for regulatory approval by the animal rule. Nat Rev Microbiol. 2009;7(5):393–400. https://doi.org/10.1038/nrmicro2129
83. Hartung T, Zurlo J. Alternative approaches for medical countermeasures to biological and chemical terrorism and warfare. ALTEX. 2012;29(3):251–60.
84. Food and Drug Administration. Extramural research: current MCM-related research projects. 2017. https://www.fda.gov/EmergencyPreparedness/Counterterrorism/MedicalCountermeasures/MCMRegulatoryScience/ucm391318.htm. Accessed 6 Oct 2017.
85. National Research Council. Advancing regulatory science for medical countermeasure development: workshop summary. Washington, DC: National Academies Press; 2011.
86. US National Institutes of Health. Request for Information (RFI): FDA Medical Countermeasures Initiative Regulatory Science Program. Washington, DC: U.S. National Institutes of Health; 2011. https://grants.nih.gov/grants/guide/notice-files/NOT-FD-11-001.html
87. Weiner T. Soviet defector warns of biological weapons. NY Times. 1998. http://www.nytimes.com/1998/02/25/world/soviet-defector-warns-of-biological-weapons.html
88. Takahashi H, Keim P, Kaufmann AF, Keys C, Smith KL, Taniguchi K, et al. Bacillus anthracis incident, Kameido, Tokyo, 1993. Emerg Infect Dis. 2004;10(1):117–20. https://doi.org/10.3201/eid1001.030238
89. International Society for Infectious Diseases. ProMED-Mail. 2017. http://www.promedmail.org/aboutus/. Accessed 6 Oct 2017.
90. Public Health Agency Canada. About GPHIN. 2017. https://gphin.canada.ca/cepr/aboutgphin-rmispenbref.jsp?language=en_CA. Accessed 6 Oct 2017.
91. World Health Organization. Epidemic intelligence – systematic event detection. 2017. http://www.who.int/csr/alertresponse/epidemicintelligence/en/. Accessed 6 Oct 2017.
92. European Commission. Crisis preparedness and response: medical intelligence in Europe. https://ec.europa.eu/health/preparedness_response/generic_preparedness/planning/medical_intelligence_en. Accessed 6 Oct 2017.
93. Collier NH. BioCaster. https://sites.google.com/site/nhcollier/projects/biocaster. Accessed 6 Oct 2017
94. Institute for Health Metrics and Evaluation. Biolding a new approach to research through high technology. 2017. http://www.healthdata.org/events/seminar/building-new-approach-research-through-high-technology. Accessed 6 Oct 2017.
95. HealthMap. HealthMap. 2017. http://www.healthmap.org/site/about. Accessed 6 Oct 2017.
96. Gilpin L. How an algorithm detected the Ebola outbreak a week early, and what it could do next. TechRepublic; 2014.
97. Barboza P, Vaillant L, Le Strat Y, Hartley DM, Nelson NP, Mawudeku A, et al. Factors influencing performance of internet-based biosurveillance systems used in epidemic intelligence for early detection of infectious diseases outbreaks. PLoS One. 2014;9(3):e90536. https://doi.org/10.1371/journal.pone.0090536

98. Kman NE, Bachmann DJ. Biosurveillance: a review and update. Adv Prev Med. 2012;301408. https://doi.org/10.1155/2012/301408
99. Foster V. ESSENCE – a DoD health indicatory surveillance system. 2004.
100. Gould DW, Walker D, Yoon PW. The evolution of BioSense: lessons learned and future directions. Public Health Rep. 2017;132(Suppl 1):7S–11S. https://doi.org/10.1177/0033354917706954
101. Centers for Disease Control and Prevention. National Syndromic Surveilllance Program: BioSense Platform. 2017. https://www.cdc.gov/nssp/biosense/index.html. Accessed 6 Oct 2017.
102. Centers for Disease Control and Prevention. Guidance for supplementary activities in support of the early warning infectious disease surveillance system. 2004. https://www.cdc.gov/phpr/documents/coopagreement-archive/fy2004/ewids-attachi.pdf.
103. Implementing Recommendations of the 9/11 Commission Act of 2007. 6 U.S. Code 195b, Pub. L. No. P.L. 110-53; 2007.
104. Department of Homeland Security. National Biosurveillance Integration Center. https://www.dhs.gov/national-biosurveillance-integration-center. Accessed 6 Oct 2017.
105. DHS Office of Health Affairs. National Biosurveillance Integration Center. https://www.dhs.gov/sites/default/files/publications/2015-Factsheet-NBIC_0.pdf
106. Defense Threat Reduction Agency. The biosurveillance ecosystem. http://www.dtra.mil/Portals/61/Documents/bsve-fact-sheet_draft_05-01-2014_pa-cleared-distro-statement.pdf
107. Pellerin C. DTRA scientists develop cloud-based biosurveillance ecosystem. DoD News. 2016. https://www.defense.gov/News/Article/Article/681832/dtra-scientists-develop-cloud-based-biosurveillance-ecosystem/
108. President Barack Obama. National strategy for biosurveillance. 2012. https://obamawhitehouse.archives.gov/sites/default/files/National_Strategy_for_Biosurveillance_July_2012.pdf
109. US National Institute of Allergy and Infectious Disease. NIAID biodefense research agenda for Category A agents. https://www.hsdl.org/?abstract&did=441285. Accessed 6 Oct 2017.
110. Federation of American Scientists. Biocontainment laboratories. 2007. https://fas.org/biosecurity/resource/research.htm. Accessed 6 Oct 2017.
111. US National Institute of Allergy and Infectious Diseases. Regional centers of excellence for biodefense and emerging infectious diseases. 2002. https://grants.nih.gov/grants/guide/rfa-files/RFA-AI-02-031.html
112. US National Institute of Allergy and Infectious Diseases. NIAID emerging infectious diseases/pathogens. 2016. https://www.niaid.nih.gov/research/emerging-infectious-diseases-pathogens. Accessed 6 Oct 2017.
113. US National Institute of Allergy and Infectious Diseases. Modeling immunity for biodefense. 2016. https://www.niaid.nih.gov/research/modeling-immunity-biodefense#overview. Accessed 6 Oct 2017.
114. Kim H, Bartsch MS, Renzi RF, He J, Van de Vreugde JL, Claudnic MR, Patel KD. Automated digital microfluidic sample preparation for next-generation DNA sequencing. J Lab Autom. 2011;16(6):405–14. https://doi.org/10.1016/j.jala.2011.07.001
115. Dutse SW, Yusof NA. Microfluidics-based lab-on-chip systems in DNA-based biosensing: an overview. Sensors (Basel). 2011;11(6):5754–68. https://doi.org/10.3390/s110605754
116. US National Institute of Allergy and Infectious Diseases. Genomic centers for infectious diseases. 2016. https://www.niaid.nih.gov/research/genomic-centers-infectious-diseases. Accessed 6 Oct 2017.
117. US National Institute of General Medical Sciences. Models of infectious disease agent study (MIDAS) overview. 2017. https://www.nigms.nih.gov/Research/specificareas/MIDAS/background/Pages/Overview.aspx. Accessed 6 Oct 2017.
118. Ferguson NM, Cummings DA, Fraser C, Cajka JC, Cooley PC, Burke DS. Strategies for mitigating an influenza pandemic. Nature. 2006;442(7101):448–52. https://doi.org/10.1038/nature04795

119. Halloran ME, Ferguson NM, Eubank S, Longini IM Jr, Cummings DA, Lewis B, et al. Modeling targeted layered containment of an influenza pandemic in the United States. Proc Natl Acad Sci USA. 2008;105(12):4639–44. https://doi.org/10.1073/pnas.0706849105
120. Zeitvogel K. Fogarty's RAPIDD program has catalyzed the field of infectious disease modeling. Global Health Matters. 2016. https://www.fic.nih.gov/news/globalhealthmatters/may-june-2016/pages/rapidd-infectious-disease-modeling.aspx
121. US National Science Foundation. Investigating the spread of infectious diseases with NSF, NIH, UK funding. 2011. https://www.nsf.gov/news/news_summ.jsp?cntn_id=121607. Accessed 6 Oct 2017.
122. US National Science Foundation. Ecology and evolution of infectious diseases. https://www.nsf.gov/news/special_reports/ecoinf/inf.jsp. Accessed 6 Oct 2017.
123. US National Science Foundation. Ecology and evolution of infectious disease. 2017. https://www.nsf.gov/funding/pgm_summ.jsp?pims_id=5269. Accessed 6 Oct 2017.
124. US Agency for International Development. Emerging pandemic threats. 2016. https://www.usaid.gov/news-information/fact-sheets/emerging-pandemic-threats-program. Accessed 6 Oct 2017.
125. Department of Defense. The department of defense chemical and biological defense: 2017 annual report to congress. 2017.
126. Homeland Security University Programs. Animal disease defense. 2017. https://www.hsuniversityprograms.org/centers/emeritus/zadd-animal-disease-defense/. Accessed 6 Oct 2017.
127. Homeland Security University Programs. Food defense. 2017. https://www.hsuniversityprograms.org/centers/emeritus/fpdi-food-defense/. Accessed 6 Oct 2017.
128. Homeland Security University Programs. Data and visual analysis. 2017. https://www.hsuniversityprograms.org/centers/emeritus/cvada-data-visual-analytics/. Accessed 6 Oct 2017.
129. Department of Homeland Security. National Biodefense Analysis and Countermeasure Center. https://www.dhs.gov/science-and-technology/national-biodefense-analysis-and-countermeasures-center. Accessed 6 Oct 2017.
130. Burans J. National Bioforensics Analysis Center. 2007.
131. US National Institute of Justice. Forensic applications of microbiomes. 2016. https://www.nij.gov/topics/forensics/evidence/Pages/microbiome.aspx#summary. Accessed 6 Oct 2017.
132. National Research Council. Science needs for microbial forensics: developing initial international research priorities. Washington, DC: National Academies Press; 2014.
133. US National Institute of Justice. Funding for forensic research and development, DNA analysis, capacity enhancement and other activities. 2016. https://www.nij.gov/topics/forensics/Pages/funding-history.aspx?tags=Forensics. Accessed 6 Oct 2017.
134. Ferrell R. Advancing fundamental research underlying forensic science. 2017.
135. US Center for Advanced Research in Forensic Science. U.S. Center for Advanced Research in Forensic Science. 2017. http://forensic-research.org/. Accessed 6 Oct 2017.
136. US Government. National Commission on Forensic Science. 2017. https://www.justice.gov/archives/ncfs. Accessed 6 Oct 2017.
137. CBS 60 Minutes (Producer). The pentagon's bionic arm. 2017. https://www.cbsnews.com/news/the-pentagons-bionic-arm/. Accessed 6 Oct 2017.
138. CBS (Producer). Imagining the future of restoration. 2012. https://www.cbsnews.com/videos/imagining-the-future-of-restoration/. Accessed 6 Oct 2017.
139. Clark, L. Mind-controlled artificial limbs fusing man and machine coming next year. Wired. 2012, November 28.
140. Miranda RA, Casebeer WD, Hein AM, Judy JW, Krotkov EP, Laabs TL, et al. DARPA-funded efforts in the development of novel brain-computer interface technologies. J Neurosci Methods. 2015;244:52–67. https://doi.org/10.1016/j.jneumeth.2014.07.019
141. Defense Advances Reserach Projects Agency. Revolutionizing prosthetics. https://www.darpa.mil/program/revolutionizing-prosthetics. Accessed 6 Oct 2017.

142. Brainline.org (Producer). DARPA: giving back life and movement with the revolutionizing prosthetics program. 2013. https://www.youtube.com/watch?v=95yLf4fzARk. Accessed 6 Oct 2017.
143. Johns Hopkins University Applied Physics Laboratory (Producer). Modular prosthetic limb. 2011. https://www.youtube.com/watch?v=DjzA9b9T3d8. Accessed 6 Oct 2017.
144. Ford TE, Colwell RR, Rose JB, Morse SS, Rogers DJ, Yates TL. Using satellite images of environmental changes to predict infectious disease outbreaks. Emerg Infect Dis. 2009; 15(9):1341–6. https://doi.org/10.3201/eid1509.081334
145. Midekisa A, Senay G, Henebry GM, Semuniguse P, Wimberly MC. Remote sensing-based time series models for malaria early warning in the highlands of Ethiopia. Malar J. 2012; 11:165. https://doi.org/10.1186/1475-2875-11-165
146. Bharti N, Tatem AJ, Ferrari MJ, Grais RF, Djibo A, Grenfell BT. Explaining seasonal fluctuations of measles in Niger using nighttime lights imagery. Science. 2011;334(6061): 1424–7. https://doi.org/10.1126/science.1210554
147. Human Rights Watch. Attacks on Ghouta: analysis of alleged use of chemical weapons in Syria. 2013.
148. US National Science and Technology Council. National biosurveillance science and technology roadmap. 2013. https://obamawhitehouse.archives.gov/sites/default/files/microsites/ostp/biosurveillance_roadmap_2013.pdf
149. Committee on Identifying the Needs of the Forensic Sciences Community. Strengthening foreign science in the United States: a path forward. 2009. https://www.ncjrs.gov/pdffiles1/nij/grants/228091.pdf
150. US National Science and Technology Council. Identifying science and technology opportunities for national preparedness. 2016. https://obamawhitehouse.archives.gov/sites/default/files/microsites/ostp/NSTC/npst_sandt_opportunities_report_final.pdf
151. US National Science and Technology Council. Implementation roadmap for the national critical infrastructure security and resilience research and development plan. 2016. https://obamawhitehouse.archives.gov/sites/default/files/microsites/ostp/NSTC/cisr_rd_implementation_roadmap_final.pdf
152. US National Science and Technology Council. Homeland biodefense science and technology capability review. 2016. https://obamawhitehouse.archives.gov/sites/default/files/microsites/ostp/NSTC/biodefense_st_report_final.pdf
153. President Donald Trump. National Biodefense Strategy. 2018. https://www.whitehouse.gov/wp-content/uploads/2018/09/National-BiodefenseStrategy.pdf. Accessed on 9 Dec 2018.
154. Georgetown University. Biohazardous threat agents & emerging infectious diseases. 2017. https://biohaz.georgetown.edu/biohaz-masters. Accessed 6 Oct 2017.
155. George Mason University. Master's in Biodefense. https://schar.gmu.edu/prospective-students/programs/masters-programs/biodefense-ms. Accessed 9 Dec 2018
156. CITI Program. Dual use research of concern (DURC). 2017. https://about.citiprogram.org/en/course/dual-use-research-of-concern-durc/. Accessed 6 Oct 2017.
157. American Biological Safety Association (Producer). Online education. 2017. https://absa.org/online-education/
158. International Federation of Biosafety Associations (Producer). Professional Certification. 2017. http://internationalbiosafety.org/index.php/147-professional-certification#. Accessed 6 Oct 2017.
159. Hoyle J. Next generation global health security network. http://ghsanextgen.wixsite.com/home. Accessed 6 Oct 2017.
160. Johns Hopkins University Center for Health Security. Emerging leaders in biosecurity fellowship. http://www.centerforhealthsecurity.org/our-work/emergingbioleaders/. Accessed 6 Oct 2017.
161. Palmer M. Synthetic biology leadership excellence accelerator program. https://www.synbioleap.org/. Accessed 6 Oct 2017.

Regulatory Pathways for Medical Products Used in Biodefense 14

Theresa Allio

14.1 Regulatory Framework for Medical Countermeasures (MCM)

Many of the medical countermeasure (MCM) products authorized for use in the United States (US) are for biodefense-related indications. A large number of these products target various strains of influenza A virus. Other indications include infections with *Bacillus anthracis*, *Yersinia pestis*, Zika virus, Ebola virus, human respiratory syncytial virus, and intoxication with shiga toxin. Many of these products are diagnostic medical tests to confirm exposure, whereas others are meant for treatment or prophylactic use [1–3].

Whenever an individual uses a medicinal product, some assurance is desired that the product is safe, reliable, and effective. Several major public health incidents throughout history have shown that determination of these parameters should not be solely left up to manufacturers [4]. These public health emergencies have led to the introduction of laws and regulations providing for the review of product-related data to ensure that labeling and promotional materials are truthful and promote safe use. MCMs are no exception to this scrutiny.

Most countries have established government agencies that are tasked with overseeing the review and approval of therapeutic products and medical devices, although each country has a slightly different organization. In the US, the Food and Drug Administration (FDA) has the responsibility for drug oversight for the entire nation. In Europe, the European Medicines Agency (EMA) is charged with centralized consideration of issues that affect the European Union as a whole, while individual country agencies consider issues specific to their local population. For example, the Agence National de Sécurité du Médicament et des Produits de Santé (ANSM) provides country-specific oversight in France, but the Bundesinstitut

T. Allio (✉)
Theresa Allio LLC, Lucinda, PA, USA

S. K. Singh, J. H. Kuhn (eds.), *Defense Against Biological Attacks*,
https://doi.org/10.1007/978-3-030-03053-7_14

für Arzneimittel und Medizinprodukte (BfArM) provides country specific oversight in Germany.

Other federal and local agencies may play a role in the distribution of MCMs. For example, in the US, the role of maintaining a national stockpile of MCMs and coordinating their distribution falls largely to the Centers for Disease Control and Prevention (CDC). Local state governments are also involved in emergency preparedness plans and may maintain their own stockpiles of certain products. However, the role of ensuring product safety and efficacy for the US population still remains with the FDA.

Although the FDA does not regulate the practice of medicine, it does influence it, as FDA decisions impact product availability. To be able to use a medicinal product on a wide scale basis in the US, it generally must be licensed, approved, or cleared for that use. Otherwise, the product would be considered "investigational" and would have certain restrictions associated with it. Other legal pathways for provision for use do exist [i.e., Emergency Use Authorizations (EUAs) and emergency dispensing orders], but they should not be thought of as long-term solutions to avoid seeking marketing authorization [5].

The provisions for investigational use, marketing authorization, and other types of authorization are contained within US law and regulation. The US Food Drug and Cosmetic Act (FD&C Act) and the Public Health Service Act (PHS Act) are the two primary sets of laws enacted by the US Congress for oversight of drugs, biologics, and medical devices. The US FDA is responsible for developing the regulations which provide distinct requirements to meet the FD&C Act and PHS Act requirements. These regulations are published in the Code of Federal Regulations. The FDA also issues guidance documents to share its current thinking on how regulation requirements may be fulfilled. However, one must remember that guidance documents are meant to provide help to an applicant but do not carry any legal force as laws and regulations do. This tiered system of law, regulation, and guidance are used by regulatory agencies in a number of countries and regions, including Canada and Europe.

Navigation of this legal framework for investigational use, marketing approval/licensure and other authorizations is essential for successful product development. These pathways involve formal applications to the FDA and an iterative process of data gathering and review to support advancement of a product through a development program and into the marketplace. One must remember that in the absence of a marketing license/approval or other FDA authorization it is illegal to market a product for an unapproved use or even to ship it across state lines. The sections below provide an overview of the various pathways available for making MCMs available to the public. Careful consideration must be put towards determining which pathway is most appropriate for a given scenario and product.

14.2 Mechanisms for Making MCMs Available to the Public

14.2.1 Investigational New Drug Applications and Investigational Device Exemptions (INDs and IDEs)

Investigational product applications are used for both unapproved products and unapproved uses of an approved product. In the US, these applications are known as Investigational New Drug applications (IND) for drug and biologic products, or Investigational Device Exemptions (IDEs) for medical devices. The regulations for INDs can be found at 21 CFR 312, whereas those for IDEs are located in 21 CFR 812. The FDA has published several guidance documents related to INDs and IDEs [6].

Filing an IND or IDE with the FDA is required to support clinical trials and to ship unapproved products across state lines. INDs and IDEs can sustain the entire duration of a product's development from initial clinical studies to pivotal Phase 3 clinical trials. In the interim, the applications are updated through information amendments. These applications include nonclinical and chemistry manufacturing and control (CMC) information and clinical protocols and data. The primary purpose of a development program, and thus the content of an IND or IDE, is to gather data to support the safety, efficacy, and assurance of ability to consistently manufacture a quality product with defined identity, strength, quality, purity, and potency or performance parameters. Many other countries, including Canada and those in Europe, handle things slightly differently, requiring applicants to submit a new clinical trial application (CTA) for each clinical study to be done, whereas the US system allows for all clinical protocols to be filed with a single application.

Generally speaking, an IND or IDE is needed whenever a clinical study is conducted. A preIND/preIDE dossier may be opened to support exchanges of information with the FDA prior to initiation of a clinical program, and thus far have been the mechanism for dialog related to Animal Rule programs. However, even an Animal Rule program will need an IND or IDE if it requires a clinical study during development, even if it is a simple pharmacokinetic study to support dose extrapolation. Investigational product applications may also include clinical protocols to support emergency use in the form of expanded access protocols as described in 21 CFR 312.310. However, expanded access protocols are not a means for making a product available on a large-scale basis, but only allow for emergency use in individual patients administered under a specific protocol. To make MCMs available on a larger scale, one must obtain a marketing license or be granted an EUA or emergency dispensing orders from the FDA.

14.2.2 Permission to Market for Specific Use (Approval, Licensing, and Clearance)

The specific term used to denote the legal ability to market a product is dependent on the product class (drug, biologic or device) and regulatory pathway used. For simplicity, these terms are used interchangeably throughout this chapter unless specifically indicated. Some entities may view product approval as a prerequisite to acquisition for preparedness stockpiles, although provisions introduced in the 2013 US Pandemic and All-Hazards Preparedness Reauthorization Act of 2013 removed this requirement for inclusion in the US national stockpile [7].

The US regulations for drug approvals are similar to those for biologics licensing, and so are grouped together in the discussion below. The US regulations for medical devices are unique enough that they merited their own separate section. Other countries draw similar lines of distinction between devices and drug/biologic products.

14.2.2.1 Drug Approvals and Biologic Licenses

Section 355 of the US FD&C Act and Section 351 of the PHS Act provide for the approval and licensing of drug and biologic products, respectively. These laws state that to be marketed, the product must first demonstrate substantial evidence of efficacy and safety for the proposed use. The regulations under 21 CFR 314 and 21 CFR 601 outline the specific pathways under which one could meet the substantial evidence requirement for drug and biologic products, respectively.

In the US, there are four main pathways for licensure of drugs and biologic products which are distinguishable from each other by the data necessary to support a marketing application. The traditional pathway relies on data from adequate and well-controlled clinical studies to demonstrate safety and efficacy of a product. As time went on, additional approval pathways were introduced to facilitate public access to products.

The second pathway is the abbreviated pathway. This pathway allows an applicant to submit an application that includes reference to data from another application or belonging to someone else, thus reducing the extent of data to be collected by the applicant. However, to use this pathway, an applicant would first have to secure legal right to reference said data before being able to submit such an application.

The third pathway is the accelerated approval pathway. This pathway was introduced in response to the HIV-1 crisis in the 1980s, and is specifically meant for products intended to treat serious or life-threatening illnesses. This pathway allows for conditional approval of a product, which is dependent on the applicant continuing to study safety and efficacy of the product in the post-marketing period. The marketing license granted under the accelerated approval pathway can be revoked if the applicant fails to fulfill the post-marketing commitments or if the new data do not continue to support an acceptable risk/benefit determination of the product.

The fourth pathway has been termed the FDA Animal Rule and is meant for use in situations when human efficacy studies are not ethical or feasible. The FDA Animal Rule came into effect in 2002, shortly after the September 11, 2001, terrorist attacks in the US.

Most recently, approvals under the Accelerated and Animal Rule pathways have predominated the drug and biologic MCM landscape. However, approvals under the Abbreviated pathway have also been seen, such as the 2003 approval of Radiogardase (Prussian Blue) to reduce the bioburden of CS-137 or thallium after known or suspected contamination [8].

One must remember that the FDA Animal Rule applies only to those new drug or biologic products for which (i) definitive human efficacy studies cannot be conducted because it would be unethical to deliberately expose healthy human volunteers to a lethal or permanently disabling toxic biological, chemical, radiological, or nuclear substance; and (ii) field trials to study the product's effectiveness after an accidental or hostile exposure have not been feasible (21 CFR 314.600 and 21 CFR 601.90). The FDA Animal Rule does not apply to products that can be approved using other licensing pathways (e.g., accelerated approval), it also does not apply to medical devices (see Sect. 14.2.2.2). Additional information on the data needed to support a marketing application under the Animal Rule are discussed further below.

On the international front, existence of pathways paralleling the FDA Animal Rule is mixed. EMA has a marketing authorization pathway known as Marketing Authorization under Exceptional Circumstances which can be used in situations when an applicant is unable to provide comprehensive data on the efficacy and safety under normal conditions of use (Article 14(8) of Regulation (EC) No 726/2004) [9]. This pathway may be used in three specific scenarios—(1) when the indication is encountered so rarely that comprehensive evidence cannot be reasonably expected, (2) in the present state of scientific knowledge, comprehensive information cannot be provided, or (3) in situations during which collection of such data would be "contrary to generally accepted principles of medical ethics to collect such information." Scenarios #1 and #3 have the most resonance with the FDA Animal Rule. In 2011, Canada added a similar set of regulations allowing for the approval of Extraordinary Use New Drugs (EUNDs) (Canadian law C.08.002.01(1)) [10].

The FDA has issued two guidance documents that provide insight into the Agency's expectations for applications seeking approval under the Animal Rule [11, 12]. Insight can also be gleaned from how prior applicants have successfully navigated this pathway. For other countries, one must consult the appropriate local and regional regulatory bodies for their requirements.

Under FDA's Animal Rule, requirements for chemistry, manufacturing, and controls (CMC) and toxicology data follow the same International Conference on Harmonisation (ICH) and FDA requirements as for products following traditional development pathways. The pathway distinguishes itself from others in the efficacy and clinical safety data packages.

Under FDA's Animal Rule, efficacy is established in animal studies instead of pivotal clinical trials used for other approval pathways. However, these are not simple studies to conduct. The animal studies must be adequate and well-controlled, and must satisfy the four regulatory criteria used to determine whether the animal data establish reasonable prediction of efficacy in humans (Table 14.1). Situations in which strict adherence to Good Laboratory Practices (GLPs) is not possible should be discussed with the FDA. If a product is approved under the Animal Rule, three post-marketing requirements would then be imposed (Table 14.2).

It should be clear that for Animal Rule approvals, a large majority of time is expended on developing the animal model itself—to demonstrate that it accurately reflects the human condition/disease of interest and that the product performs similarly in animals and humans. Although the FDA's drug development tool qualification pathway offers the potential to reduce some of the burden of this work on an applicant, these economies have not yet been realized as no animal models have been qualified to date. The drug development tool qualification process offers a means to share the knowledge gained from an initial investment of time and money for developing an

Table 14.1 Four criteria to determine whether animal data establish reasonable effectiveness of drug in humans (21 CFR 314.600 for drugs and 21 CFR 601.90 for biologics) [12]

1.	There is a reasonably well-understood pathophysiological mechanism of the toxicity of the substance and its prevention or substantial reduction by the product[a]
2.	The effect is demonstrated in more than one animal species expected to react with a response predictive for humans, unless the effect is demonstrated in a single animal species that represents a sufficiently well-characterized animal model for predicting the response in humans
3.	The animal study endpoint is clearly related to the desired benefit in humans, generally the enhancement of survival or prevention of major morbidity
4.	The data or information on the kinetics and pharmacodynamics of the product or other relevant data or information, in animals and humans, allows selection of an effective dose in humans

[a]Although not specifically stated in Criterion 1, human relevance must be established for both the pathophysiology in the animal model and mechanism of action by which the product elicits its efficacy

Table 14.2 Three requirements imposed if a product is approved under the Animal Rule (21 CFR 314.610(b)(1)-(3) for drugs and 21 CFR 601.91(b)(1)-(3) for biologics) [12]

1.	Post-marketing studies (e.g., field studies) to provide evaluation of safety and clinical benefit if circumstances arise in which a study would be feasible and ethical (i.e., in the event of an emergency arises and the drug is used). A plan or approach to conducting such a study must be included with the New Drug Application (NDA) or Biologics License Application (BLA)
2.	Restrictions to ensure safe use, if needed (e.g., restricting distribution to facilities or health care practitioners with special training, requiring specified types of follow-up, or imposing record keeping requirements)
3.	Information to be provided in the labeling to patient recipients that explains that for ethical or feasibility reasons, the drug's approval was based on efficacy studies conducted in animals alone. This drug labeling should also include all other relevant information required by FDA at the time of approval (e.g., directions for use, contraindications, a description of any reasonably foreseeable risks, adverse reactions, anticipated benefits, and drug interactions). This information must be provided before administration or dispensing, if possible

animal model to essentially provide a protocol or off-the-shelf model that could be used by others seeking approval for another product targeting the same indication (as long as no product incompatibilities exist). Qualified animal models would also provide an existing dataset to leverage off of, potentially reducing the information needed to extend the animal model to other uses (e.g., to study new products using alternative routes of administration). Although the pharmaceutical sector may appreciate the economies of scale offered through the use of a qualified animal model, the commercial incentives for developing a qualified animal model may lay more with contract research organizations and academic laboratories that provide nonclinical testing services. The provisions for animal model qualification were codified under Section 507 of the US FD&C Act by the Twenty-First Century Cures Act of 2016 (Public Law 114-255) [13]. Thus, animal model qualification is a distinct activity from product approval under the Animal Rule.

As of September 4, 2017, FDA has approved a total of 12 products under the Animal Rule (Table 14.3) [1, 14]. Eight of these applications were related to biologic

Table 14.3 Summary of products approved by FDA under the Animal Rule as of April 18, 2017 [1, 14]

Product	Active pharmaceutical ingredients	Date approved	Indication
Pyridostigmine bromide, USP	Pyridostigmine bromide	Feb 5, 2003	Prophylaxis against the lethal effects of soman nerve agent poisoning
Cyanokit	Hydroxocobalamin	Dec 15, 2006	Treatment of known or suspected cyanide poisoning
Levaquin	Levofloxacin	April 27, 2012	Treatment and prophylaxis of plague due to *Yersinia pestis* in adults and pediatric patients 6 months of age and older
	Raxibacumab	Dec 14, 2012	Treatment of adult and pediatric patients with inhalational anthrax due to *Bacillus anthracis* in combination with appropriate antibacterial drugs and for prophylaxis of inhalational anthrax when alternative therapies are not available or are not appropriate
Botulism antitoxin (BAT)	Botulism antitoxin heptavalent (A, B, C, D, E, F, G)-(equine)	Mar 22, 2013	Treatment of symptomatic botulism following documented or suspected exposure to botulinum neurotoxin serotypes A, B, C, D, E, F, or G in adults and pediatric patients
Cipro	Ciprofloxacin	Feb 2, 2015	Treatment and prophylaxis of plague due to *Y. pestis* in adults and pediatric patients from birth to 17 years of age
Anthrasil	Anthrax immune globulin (human)	Mar 25, 2015	Treatment of adult and pediatric patients with toxemia associated with inhalational anthrax
Neupogen	Filgrastim	Mar 30, 2015	To increase survival in patients acutely exposed to myelosuppressive doses of radiation (Hematopoietic Subsyndrome of Acute Radiation Syndrome)
Avelox	Moxifloxacin	May 8, 2015	Treatment and prophylaxis of plague due to *Y. pestis* in adult patients
Neulasta	Pegfilgrastim	Nov 13, 2015	To increase survival in patients acutely exposed to myelosuppressive doses of radiation (Hematopoietic Subsyndrome of Acute Radiation Syndrome)
BioThrax	Anthrax vaccine adsorbed	Nov 23, 2015	Post-exposure prophylaxis of disease following suspected or confirmed *B. anthracis* exposure, when administered in conjunction with recommended antibacterial drugs
Anthim	Obiltoxaximab	Mar 18, 2016	For adult and pediatric patients for the treatment of inhalational anthrax due to B. *anthracis* in combination with appropriate antibacterial drugs; for prophylaxis of inhalational anthrax due to *B. anthracis* when alternative therapies are not available or not appropriate

threats (infectious disease). Review of FDA press releases and other publically available information reveal that the applications supporting these approvals are largely in line with FDA guidance [1, 12]. A majority of programs involved pivotal GLP animal studies in animals of two species with some number of additional supporting studies [1]. Survival was the main endpoint measured in these studies, as it was believed to be the most predictive of human outcomes in these animal models. In addition to establishing efficacy, these studies also established conditions of use for the product (e.g., triggers to treat, dosing regimen) that were associated with efficacy outcomes. Human data were required to support dose extrapolation and safety and were gained from both de novo studies and experience with the product in other indications. In a few situations, limited human efficacy data were available from accidental human exposures or human challenge models. When de novo studies were done for a product, the human safety database (those receiving at least one dose) ranged from 90 to 500 subjects with the mode being ~300 [1]. Safety databases for products that have already been approved for another indication are much larger.

It must be remembered that an applicant's responsibilities continue into the post-marketing period. Three specific post-marketing requirements are written into the Animal Rule (Table 14.2). Only the need for elements to ensure safe use (ETASU) has been routinely waived by the FDA, as it has felt that the controls surrounding dispensing and use of the product for the approved use were sufficient to ensure patient safety (e.g., prescription required or administration limited to hospital setting or military personnel) [1]. This does not mean that ETASU may not be considered for future products. As for the remaining two post-marketing requirements, the patient labeling requirement is the most straightforward and easiest to comply with. The post-marketing study requirement is a little more complex.

The Animal Rule requires that an applicant conduct post-marketing studies (e.g., field studies) to provide information to support evaluation of safety and clinical benefit if circumstances arise in which a study would be feasible and ethical (i.e., in the case when an emergency arises and the drug is used). As a step towards enforcement of this requirement, applicants must include a plan or approach for this study in their licensing application. However, due to the confidential nature of licensing application content, it is not clear which study designs have been deemed acceptable by the FDA. In some cases, opportunistic pharmacokinetic sample analysis was determined to be an acceptable way to support dose extrapolation from animal studies [1]. However, pharmacokinetic sampling alone does not answer questions of efficacy and safety. A recent report from the Committee on Clinical Trials During the 2014–2015 Ebola virus disease (EVD) outbreak provides some insight on just how complicated the matter is [15].

At the time of this outbreak, there were no approved products available for treatment of EVD [15]. Several products for treatment, prevention, or diagnosis of EVD were in development and had not accumulated data sufficient to support a marketing application in any country. US and international government health organizations joined to determine how to approach studying potential therapeutic products in the midst of an epidemic. Surprisingly, the report seemed to largely support the conduct of a randomized, controlled clinical study as being the best study design to conduct during the course of an epidemic. Some based their recommendations on the fact that there were limited supplies of product and thought a randomized protocol would be the most ethical way to distribute the limited quantities. Others thought that outside of a clinical trial, patients may place too much emphasis on potential benefit without considering the risks associated with an unapproved therapy or vaccine. It is interesting that a controlled clinical study was still considered to be the model to pursue in situations during which suboptimal storage conditions may have affected product quality and medical staff inadequately trained to conduct of clinical trials could affect data quality. The negative view on use of historical controls was also intriguing, as although not ideal, the FDA has accepted historical controls to support recent approvals of products for both infectious diseases and rare diseases [16–18].

The obvious concern throughout the crisis was how to develop a reliable dataset upon which decisions could be based. The ad hoc National Academies of Sciences, Engineering and Medicine Ebola committee recognized that one of the things that greatly hindered a timely response to the epidemic was lack of agreement on what an adequate clinical design may look like. Some suggested solutions included developing a template library supporting a variety of protocol designs for various scenarios, which would facilitate study execution in a future crisis [15]. Perhaps the lessons learned and tools developed from the EVD crisis might help facilitate emergency response and inform the design of clinical studies aimed at fulfilling the post-marketing requirements of the FDA Animal Rule.

14.2.2.2 Medical Device Clearances and Approvals

The Sections 510 and 515 of the US FD&C Act establishes FDA's oversight of medical devices, while specific regulations for medical devices can be found in 21 CFR 800-1299. The FD&C Act defines medical devices as "an instrument, apparatus, implement, machine, contrivance, implant, in vitro reagent, or other similar or related article, including a component part or accessory which is:

(i) Recognized in the official National Formulary, or the United States Pharmacopoeia, or any supplement to them,
(ii) Intended for use in the diagnosis of disease or other conditions, or in the cure, mitigation, treatment or prevention of disease, in man or other animals, or
(iii) Intended to affect the structure or any function of the body of man or other animals, and which does not achieve its primary intended purposes through chemical action within or on the body of man or other animals and which is not dependent upon being metabolized for the achievement of any of its primary intended actions."

The US regulatory pathways for devices are product specific, with different tiers of requirements depending on product characteristics and risk associated with the proposed use. Device applications are divided into three classes. Class 1 devices are low risk and generally only require adherence to general class controls. Class

2 devices are considered intermediate risk and would require additional controls to provide reasonable assurance of the device's safety and effectiveness. Class 3 devices carry the highest risk and require more data to support marketing [19]. The applications to support marketing of a device are called 510(k) applications, of which there are three types—traditional, abbreviated or special circumstances, depending on the data available and the circumstances of use [20]. Notably, there is no Animal Rule equivalent in device regulations. Thus, MCM-related applications look much like those for non-MCM indications.

So far there have been several device MCMs that have been approved, including diagnostic tests for several strains of influenza A virus and brain injury, and devices meant to control severe bleeding. The applications filed to support these products represent all three application pathways (traditional, abbreviated, and special) [2]. Several more device MCMs have been authorized for use under FDA's authority to issue EUAs (see below).

14.2.2.3 Other Authorizations for Use (EUAs and Emergency Dispensing Orders)

Section 564 of the FD&C Act contains provisions to allow special use of drugs and other medical products during certain types of emergencies (see Table 14.4). Section 564 was added to the FD&C Act by the Project BioShield Act of 2004 [21], and was amended by PAHPRA 2013 [7]. The original Project BioShield law included provisions for EUAs. The 2013 PAHPRA amendments were introduced to increase accessibility to unapproved MCMs in times of emergency. The provisions contained within Section 564 should not be viewed as a replacement for seeking FDA approval for an MCM and should be pursued only in close consultation with the FDA and CDC [5]. Early inclusion of the CDC is important, as the agency will likely be intimately involved in the administration of any approved EUA (e.g., stockpiling, distribution). The CDC itself may even end up submitting the EUA application to the FDA.

It must be remembered that the provisions under Section 564 cannot be pursued until the Secretary of the US Health and Human Services (HHS) declares the existence of an emergency or threat that justifies authorization of emergency use [5]. Once the HHS Secretary declares the existence of an emergency, the responsibilities for administration fall to the FDA and the CDC. An emergency is considered any event that has the potential to affect national security or the health and security of US citizens living domestically or abroad. Thus, EUAs may be issued for events that are happening both inside and outside of US borders—such as during the 2013–2016 EVD epidemic. An EUA may be issued for both unapproved products and for unapproved uses of a previously approved product [2, 5].

The EUA process requires that a formal application be submitted to the FDA for review and approval. The application may be initiated by the CDC or another entity (e.g., pharmaceutical company or other government agency). The FDA then conducts a review of the (1) capability of the emergency or threat to cause a serious or life-threatening condition, (2) evidence of product effectiveness, (3) risk/benefit profile of the product, and (4) availability of alternative treatments/devices, which

Table 14.4 Summary of FD&C Act Section 564 as amended by the Pandemic and All-Hazards Preparedness Reauthorization Act of 2013 [7]

FD&C Section	Provision
564	**Emergency Use Authorization (EUA)** After the HHS Secretary has made a declaration of emergency or threat justifying emergency use, the FDA Commissioner may authorize the emergency use of an unapproved medical product or an unapproved use of an approved medical product for certain emergency circumstances
564A	**Provisions to facilitate preparedness and response activities *outside* of the EUA authority**
	• FDA may extend the expiration date of an eligible FDA-approved MCM stockpiled for use in a CBRN emergency and to establish appropriate conditions relating to such extensions, such as appropriate storage, sampling, and labeling
	• FDA may waive otherwise-applicable current good manufacturing practice (CGMP) requirements, such as storage or handling to accommodate emergency response needs
	• Allow for emergency dispensing of MCMs during an actual CBRN emergency event without requiring an individual prescription for each recipient of the MCM or all of the information otherwise required or by responders who may not otherwise be licensed to dispense, if permitted by state law in the state where such dispensing occurs or if in accordance with an order issued by FDA
	• Permit the CDC to create and issue "emergency use instructions" (EUI) concerning the FDA-approved conditions of use for eligible products
564B	**MCM Stockpiling** Permits government stakeholders to pre-position (e.g., stockpile, forward deploy) MCMs in anticipation of FDA approval, clearance, authorization of an investigational use, or the issuance of an EUA to support rapid deployment during an emergency
505-1(k)	**Authorities for REMS Waivers** FDA may waive Risk Evaluation and Mitigation Strategy (REMS) requirements for CBRN emergencies

may include consideration of limitations of use of other available therapies such as contraindications in certain populations. If the data are sufficient, the FDA will grant an EUA. The EUA may remain in effect for the duration of the HHS Secretary's emergency declaration, until the product is approved for that use, or until the FDA determines that the EUA should be revoked because the EUA no longer has an acceptable risk/benefit profile or if the applicant is no longer pursuing product development [5].

One should not regard the EUA as extra work, but rather a mechanism by which potential life-saving products can be made available to the public prior to approval without some of the constraints of an investigational application. For example, an EUA eliminates the need for administering a product under a specific protocol, although informed consent regarding the use of the product under an EUA should be acquired to the extent practicable [5]. Any information gathered to support an EUA would eventually be submitted in a marketing application, and can include data from animal efficacy studies. Finally, any data collected on the safety and efficacy of the clinical use of the product as administered under the EUA would also be information to include a marketing application. Thus, one could think of the EUA as one step on a product development pathway.

The FD&C Act as amended by [7] has several other mechanisms aimed at improving the use and increasing the availability of MCMs in the case of an emergency (see Table 14.4 for details). The activities under these provisions are primarily initiated by the FDA or CDC, in consultation with other government stakeholders. The FDA's authority to issue emergency dispensing orders provides another mechanism by which unapproved MCMs may be made available to the public. Other FDA authorities are aimed at easing constraints on supply and distribution that may occur during preparedness or response efforts, such as those allowing for extension of expiration date of a stockpiled drug, and waiver of current good manufacturing practices and/or Risk Evaluation and Mitigation Strategy requirements. The CDC's authority to issue "emergency use instructions" (EUI) are really meant to provide easy-to-follow information for an MCM that has been approved, cleared, licensed or authorized for use by the FDA, and cannot support an unapproved use on its own.

Again, the emergency dispensing orders are not to be considered the endpoint for an MCM development program. Emergency dispensing orders are meant to fill gaps in state and federal preparedness efforts and require tight coordination of all stakeholders. The FDA authorizes, but cannot direct or require, emergency dispensing of a product for an unapproved use. Emergency dispensing orders may also include logistical considerations such as decision trees, chains of command or special instructions for use to aid in application of the MCM. As with EUAs, emergency dispensing orders can be issued either at the time of an emergency or in preparation for a threat. In most cases, an emergency dispensing order remains in effect until FDA issues a subsequent order to revise or revoke it. The FDA may also specify the duration of an emergency dispensing order, with the opportunity for extension if needed [5].

14.3 Concluding Thoughts

It is clear that preparedness and response to CBRN emergencies involve a high degree of coordination between private and public sectors. Each participant has an important role to play-from the scientists and manufacturers who develop the data and product, to the government officials responsible for the approval/authorization and distribution of these products. We hope this chapter will help those interested in developing MCMs navigate what may seem like an overly rigid set of rules and see it for the dance of scientific and logistical considerations that it is.

References

1. Allio T. Product development under FDA's Animal Rule: understanding FDA's expectations and potential implications for traditional development programs. Ther Innov Regul Sci. 2016;50:660–70.
2. FDA. FDA medical countermeasures initiative (MCMi) program update fiscal year 2016. 2016. https://www.fda.gov/downloads/EmergencyPreparedness/Counterterrorism/MedicalCountermeasures/AboutMCMi/UCM546894.pdf. Accessed 04 Sept 2017.
3. Humeniuk R. Pediatric medical countermeasures: closing the gap. Paper presented at the 2015 Pediatric Acute Radiation Syndrome and Chemical Public Health Emergency Medical Countermeasures (MCM) Conference. National Institutes of Health, Rockville, Maryland, 10 June 2015.
4. FDA. Significant dates in U.S. Food and Drug Law History. 2014. https://www.fda.gov/AboutFDA/WhatWeDo/History/Milestones/ucm128305.htm. Accessed 04 Sept 2017.
5. FDA. Guidance for industry and other stakeholders "Emergency Use Authorization of Medical Products and Related Authorities". 2017. https://www.fda.gov/downloads/EmergencyPreparedness/Counterterrorism/MedicalCountermeasures/MCMLegalRegulatoryandPolicyFramework/UCM493627.pdf. Accessed 04 Sept 2017.
6. FDA Guidance Database. US FDA, Silver Spring. 2017. https://www.fda.gov/RegulatoryInformation/Guidances/default.htm. Accessed 04 Sept 2017.
7. PAHPRA. US Pandemic and All-Hazards Preparedness Reauthorization Act of 2013. Public Law No. 113-5; 2013.
8. FDA. Approval letter for radiogardase. 2003. https://www.accessdata.fda.gov/drugsatfda_docs/nda/2003/21-626_Radiogardase_Approv.pdf. Accessed 04 Sept 2017.
9. EMEA. Guideline on procedures for the granting of a marketing authorisation under exceptional circumstances. Pursuant to Article 14 (8) of Regulation (EC) NO 726/2004. 2005. http://www.ema.europa.eu/docs/en_GB/document_library/Regulatory_and_procedural_guideline/2009/10/WC500004883.pdf. Accessed 04 Sept 2017.
10. Health Canada. Guidance document submission and information requirements for extraordinary use new drugs (EUNDs). 2014. https://www.canada.ca/content/dam/hc-sc/migration/hc-sc/dhp-mps/alt_formats/pdf/brgtherap/applic-demande/guides/eund-dnue-eng.pdf. Accessed 04 Sept 2017.
11. FDA. FDA guidance for industry and FDA staff qualification process for drug development tools. 2014. https://www.fda.gov/downloads/Drugs/GuidanceComplianceRegulatoryInformation/Guidances/UCM230597.pdf. Accessed 04 Sept 2017.
12. FDA. FDA guidance for industry product development under the Animal Rule. 2015. https://www.fda.gov/downloads/Drugs/GuidanceComplianceRegulatoryInformation/Guidances/UCM399217.pdf. Accessed 04 Sept 2017.
13. 21st Century Cures Act of 2016. Public Law No. 114-255.
14. FDA. Animal Rule Information. 2017. https://www.fda.gov/EmergencyPreparedness/Counterterrorism/MedicalCountermeasures/MCMRegulatoryScience/ucm391604.htm. Accessed 04 Sept 2017.
15. National Academies of Sciences, Engineering, and Medicine. Integrating clinical research into epidemic response: the Ebola experience. Washington, DC: The National Academies Press; 2017. https://doi.org/10.17226/24739.
16. FDA. FDA News Release – FDA approves first treatment for a form of Batten disease. 2017. https://www.fda.gov/NewsEvents/Newsroom/PressAnnouncements/ucm555613.htm. Accessed 04 Sept 2017.
17. Mishra P, Murray J, Birnkrant D. Direct-acting antiviral drug approvals for treatment of chronic hepatitis C virus infection: scientific and regulatory approaches to clinical trial designs. Hepatology. 2015;62(4):1298–303.
18. Nace M. Breaking News: FDA approves Sarepta's Exondys 51 (Eteplirsen) as first therapy for Duchenne MD. Muscular Dystrophy News Today. 2016. https://musculardystrophynews.com/2016/09/19/fda-approves-sarepta-exondys-51-eteplirsen-dmd-first-therapy. Accessed 04 Sept 2017.

19. FDA. About FDA webpage – what does it mean for FDA to "classify" a medical device? 2015. https://www.fda.gov/aboutfda/transparency/basics/ucm194438.htm. Accessed 04 Sept 2017.
20. FDA. 510(k) Submission methods. 2017. https://www.fda.gov/medicaldevices/deviceregulationandguidance/howtomarketyourdevice/premarketsubmissions/premarketnotification510k/ucm134034.htm. Accessed 04 Sept 2017.
21. Project BioShield Act of 2004. Public Law No. 108-276.

Big Data and Biodefense: Prospects and Pitfalls

15

Kathleen M. Vogel

15.1 Introduction

Since the 1980s, with the rise and proliferation of genetic engineering techniques and technologies, national and international security concerns have grown regarding the misuse of biology by state and non-state actors [1–7]. Over this timeframe, and particularly within the past 20 years, the rise of new digital and mobile technologies in terms of sensors, geospatial technologies, advanced analytics, social media, and cell phones, have made the collection and analysis of a wide variety of information, to include different types of biological data, more readily accessible. These developments provide opportunities to think about biodefense preparedness in new and unique ways. Harnessing digital technologies more effectively to better identify and assess emerging biosecurity threats could lead to improved detection, response, and preparedness measures to eliminate or at least mitigate a bioweapons attack. This chapter will discuss the application of "big data" and "big data" analytics to biodefense problems in the specific domains of: (1) threat awareness; and (2) surveillance and detection. The chapter will begin by defining "big data" and biodefense, and then will examine the potential promises and pitfalls in biodefense applications and provide recommendations for the future of "big data" in addressing biodefense problems.

K. M. Vogel (✉)
School of Public Policy, University of Maryland at College Park, College Park, MD, USA
e-mail: kvogel12@umd.edu

S. K. Singh, J. H. Kuhn (eds.), *Defense Against Biological Attacks*,
https://doi.org/10.1007/978-3-030-03053-7_15

15.2 What Is "Big Data"?

Although conceptions of "big data" have been circulating since the 1990s, the term gained cache in technology and public circles in 2008 [8]. "Big data" is a term used to describe extremely large datasets[1] which can only be analyzed computationally, either individually or integrated with other datasets, to reveal previously unknown patterns, trends, and associations. What makes "big data" approaches unique is that they offer, "the capacity to collect and analyze data with an unprecedented breadth and depth and scale" ([10], p. 722). "Big data" can be collected through either active means (e.g., a user's posts on social media) or passive means (e.g., FitBit data).

George et al. [11] discuss that "big data" includes different categories of data: public data, private data, data exhaust, community data, and data generated through self-quantification. Public data are data typically held by governments, governmental organizations, and local communities and are publically released or available. Private data are data held by private firms, non-profit organizations, governments, and individuals that reflect private information that cannot be easily derived from public sources. Data exhaust refers to data that are passively collected for a different purpose, but can be recombined with other data sources to create new information (e.g., cell phone data or internet searching data). Community data is a distillation of unstructured data—especially text—that capture a variety of social trends, thoughts, and behaviors (e.g., consumer reviews on products). Self-quantification data are types of data that are revealed by the individual through quantifying personal actions and behaviors (e.g., FitBit data). As one can see there are a variety of "big data" types. "Big data" analytics involve the process of examining "big data" gathered over multiple time points (e.g., seconds, minutes, hours, days, months, years), using a variety of analytic techniques, developed by unique algorithms. Some examples include text analytics (mining data from different kinds of text based documents and materials), machine learning (the development of computer programs that can access data and use it to learn for themselves to generate new data), predictive analytics (the use of different statistical, modeling, and other quantitative approaches to analyze current data to make predictions about future), and natural language processing (using computer programs to understand human speech as it is written or spoken)—to name only a few.

[1]The MacKensey Global Institute expands this definition to note that: ""Big Data" refers to datasets whose size is beyond the ability of typical database software tools to capture, store, manage, and analyze. This definition is intentionally subjective and incorporates a moving definition of how big a dataset needs to be in order to be considered Big Data—i.e., we don't define Big Data in terms of being larger than a certain number of terabytes (thousands of gigabytes). We assume that, as technology advances over time, the size of datasets that qualify as Big Data will also increase. Also note that the definition can vary by sector, depending on what kinds of software tools are commonly available and what sizes of datasets are common in a particular industry. With those caveats, Big Data in many sectors today will range from a few dozen terabytes to multiple petabytes (thousands of terabytes)." ([9], p. 1).

Typically there are four characteristics associated with "big data" often described as the four Vs: Volume, Variety, Velocity, and Veracity. ***Volume*** refers to working with increasingly larger volumes of data (e.g., terabytes, petabytes, or larger). ***Variety*** refers to the ability to make sense of different sources and types of data, including structured forms (e.g., numbers, dates, and groups of words and numbers) and unstructured forms (e.g., text, video, audio, tweets) that can come from a wide variety sources, to include public and privately held sources. This data can consist of information that is solely digital in origin, as well as data that has been subsequently digitized. ***Velocity*** refers to the increasing frequency and speed of incoming data that needs to be analyzed, and the speed at which it can be analyzed. ***Veracity*** signifies the trustworthiness of the data. It is important to note at the outset that "big data" is often incomplete, imperfect, and error-prone, and how it is collected and stored in databases and repositories is not necessarily standardized. These shortcomings can make data access and sharing issues difficult—therefore, the use of "big data" has analytic challenges that are analogous, if not necessarily identical, to other types of data.

The benefits of using "big data"/"big data" analytics include the ability to gather, analyze, and compare large datasets, consisting of disparate sources, to identify new patterns and generate new knowledge about a problem. There is potentially a wealth of new information and knowledge that could be learned to improve biodefense using big datasets and technologies. There are, however, also some challenges to consider. Questions are often raised about the veracity of the data (including ways to thwart the collection of accurate data) and problems with the algorithms used by the analytics programs to create and process the data—which are issues that can be difficult for analysts to understand and assess. There are also potential information gaps in existing "big data" that cannot be filled simply by analytics programs. The problems and prospects of "big data" will be discussed further in more detail in subsequent sections of this chapter.

15.3 What Is Biodefense?

For the purposes of this chapter, I will define biodefense as the defensive measures taken in the United States to protect against an attack using biological weapons or biological agents. The overarching strategy that has set U.S. biodefense policy and programs since 2004 is Homeland Security Presidential Directive 10 (HSPD 10), *Biodefense for the 21st Century* [12].[2] This Presidential Directive described the "essential pillars" of the U.S. national biodefense program as: Threat Awareness, Prevention and Protection, Surveillance and Detection, and Response and Recovery. Threat Awareness involves gathering and analyzing information about potential adversaries and emerging developments in the life sciences and biotechnology that

[2]There have been some follow-on White House biodefense policy documents: *The National Strategy for Countering Biological Threats* [13]; *The National Strategy for Biosurveillance* [14].

can pose threats. Prevention and Protection involves developing a range of political tools (e.g., arms control, export control, international assistance) to impede adversaries from developing a bioweapons capability and launching an attack, as well as developing technologies and response measures to protect critical infrastructure from a potential attack. Surveillance and Detection is focused on creating early warning and detection technologies and systems to identify and provide attribution of a bioweapons attack. Response and Recovery deals with the various technologies, medical and non-medical countermeasures, policies and plans to enable quick and effective response and recovery in the event of a bioweapons attack. Each of these pillars is designed to work individually and collectively to support different kinds of defensive activities such as identification of emerging state and non-state actor threats, controlling access to dangerous pathogens, identification of suspicious disease outbreaks, protection of critical infrastructure used in biodefense work, and development of medical countermeasures in the event of an attack (See Fig. 15.1).

15.4 Intersections of "Big Data" and Biodefense

There have not been many articles or reports published specifically on the intersection of "big data" and biodefense. One publication, however, that first raised this connection was a 2014 report, *National and Transnational Security Implications of Big Data in the Life Sciences*, co-authored by the American Association for the Advancement of Science, the Biological Countermeasures Unit of the WMD Directorate of the Federal Bureau of Investigation, and the United Nationals Interregional Crime and Justice Research Institute [16]. The report assessed a range of national security risks and benefits of using "big data" in the life sciences, and examined possible solutions to address the risks. The report focused largely on possible risks associated with the use of "big data" analytics, such as the vulnerabilities of life science datasets and cyber infrastructures/cloud-based systems to cyber intrusion/cyber hacking, and the potential design and development of pathogens, toxins, or biologically active molecules as biological weapons from the integration and analysis of "big data" in the life sciences. The report also tended to focus its discussion on a broad brush stroke of "big data" and security issues in the life sciences, rather than a discussion of specific cases of beneficial applications or harms in biodefense. Therefore, this chapter will fill a gap in knowledge and focus on how one might develop and use "big data"/"big data" analytics for two pillars of U.S. biodefense: (1) Threat Awareness; and (2) Surveillance and Detection.

15.4.1 Threat Awareness and "Big Data"

According to HSPD 10, Threat Awareness involves obtaining timely, accurate, and relevant biological warfare related intelligence, as well as conducting assessments of current and future trends and patterns in the evolving bioweapons threat. These activities involve collecting and analyzing a wide range of information on potential

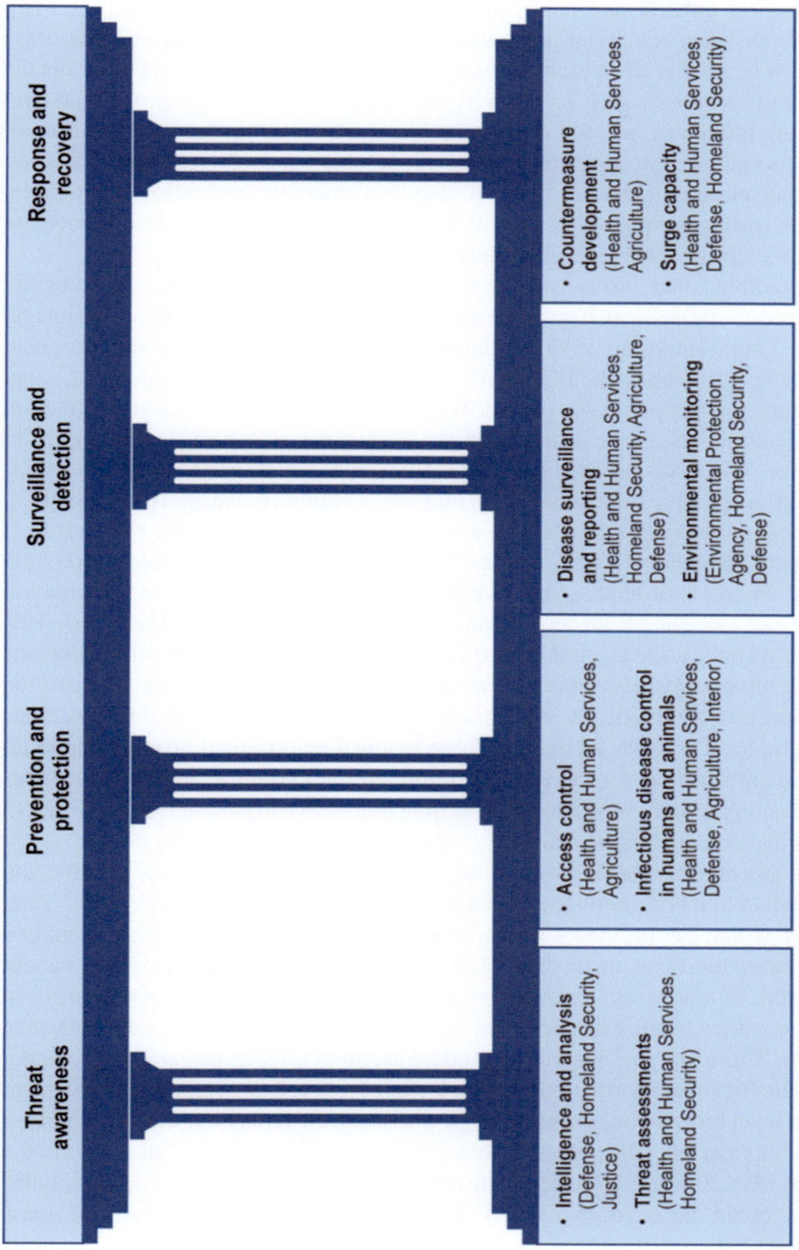

Source: GAO analysis of Homeland Security Presidential Directive 10. | GAO-16-547T

Fig. 15.1 Pillars of U.S. Biodefense ([15], p. 4)

adversaries and their potential interest in, and capability with, biology and biotechnologies, to "help position intelligence collectors ahead of the problem," as well as inform medical and non-medical countermeasure development [12]. "Big data" and "big data" analytics are well poised to assist with this effort. "Big data" approaches now offer even more ways to amass, count, organize, and store details relevant to biodefense, such as many types of biologically-relevant materials and related tangible resources used to develop biological weapons, as well as different sets of information (e.g., identities, composition, location, resources, capabilities) on potential adversaries. Different types of "big data" analytics technologies would be useful for sorting through a variety of classified and unclassified information on potential adversaries and emerging biotechnology threats.

For example, data fusion analytics, such as GIS technologies, can integrate heterogeneous datasets such as different types of sensor data, with video/image/text data. Laura Tateosian at North Carolina State University is developing a new GIS analytics prototype called GazeGIS that combines natural language processing, eye tracking, and geoparsing to provide users with additional contextual information as they read/process any text displayed on a computer screen [17]. When used in conjunction with eye-tracking hardware, the prototype displays real-time, gaze-contingent geographic data about particular city or country names. Alternatively, a user can hover over these words with a computer mouse to create the same effect if eye-tracking hardware is unavailable. The current prototype is able to provide overview maps, coordinates, and weather information, as well as information derived from Google search results; additional features are being developed. This system could help assist an intelligence analyst in processing a variety of intelligence reports on adversaries developing bioweapons capabilities—as the analyst reads the report, GazeGIS automatically identifies and creates a real-time map of the locations and geographical features identified. These mapped features can provide important contextual information for the text. In doing so, the GazeGIS prototype provides users the ability to have an enhanced real-time immersive experience about different geographical and contextual data while reading text and reports (Fig. 15.2).

There are other types of visual analytics platforms that can help identify and connect social relations between potential adversaries. One example is SAS Visual Analytics, which can provide a variety of data about and between social actors in a network based on social media data. The program can identify various features about the structure of a network, and who are key influencers or actors in the network to develop an understanding of micro and macro-level aspects of a social network (See Fig. 15.3). These types of analytics would be useful to intelligence analysts to better process intelligence information and understand the actors involved in a state or non-state level bioweapons program (e.g., who are critical players, who are involved in leadership/management roles, the broader community/network that supports the actors). A May 2017 NATO meeting discussed how various text and visual analytics programs could be used to identify threats stemming from weapons of mass destruction [18].

Nowviskie and Saathoff [19] showed how "big data" analytics can be used to aid in the investigation of insider threats involving bioterrorism. They examine the 2001

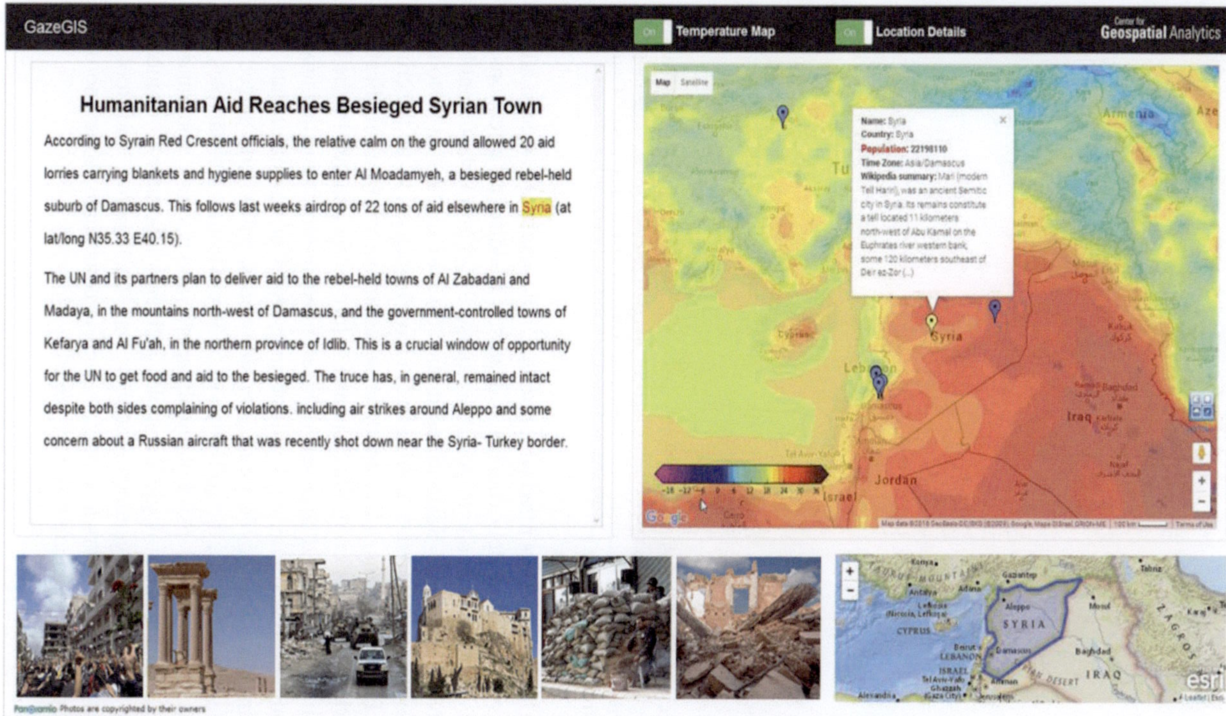

Fig. 15.2 GazeGIS Prototype: As analyst reads text on Syria, geographical and contextual information immediately pop up onto the computer screen. (Image and copyright approval obtained from Laura Tateosian.)

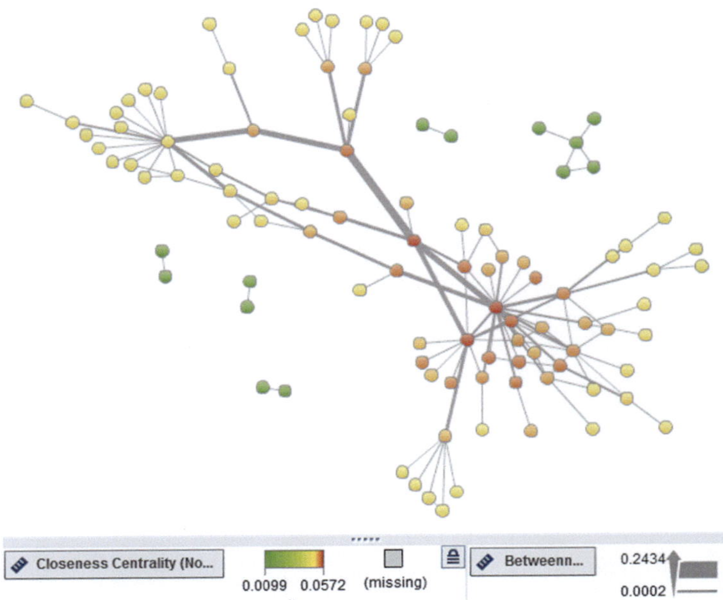

Fig. 15.3 Example of SAS Visual Analytics: Nodes and connectors show relationships between actors and the strength of the relationships in the social network(s). (Obtained from: http://blogs.sas.com/content/sascom/2014/02/19/exploring-social-networks-with-sas-visual-analytics/. Copyright approval obtained from Falko Schulz.)

Amerithrax case, in which letters filled with *Bacillus anthracis* spores were mailed in the U.S. postal system, ultimately causing 5 deaths, 17 additional infections, and risking health of 10,000 others due to possible exposure and prophylactic antibiotic therapy [20].[3] Nowviskie and Saathoff used this case to discuss how tools of computational text analysis and data mining could have been used to aid the criminal investigation on Dr. Bruce Ivins, a U.S. biodefense researcher who was implicated in the crime (and who ultimately committed suicide before being charged by the FBI). The FBI's investigation included large sets of heterogeneous structured and unstructured data including: 26,000 emails, 4 million megabytes of computer data from 10,000 witness interviews and 5750 grand jury subpoenas, 6000 items seized from Ivins' home and other locations, the collection of 5730 environmental samples from 60 site locations, medical and psychological reports about Ivins, and microbial forensic data generated from biological samples in Ivan's laboratory.[4] Combining and interpreting these heterogeneous datasets was challenging and took the U.S. government nearly 10 years to complete, and is considered to be one of the most complex investigations undertaken by the FBI.

[3] See: https://www.fbi.gov/history/famous-cases/amerithrax-or-anthrax-investigation

[4] Ibid.

Nowviskie and Saathoff argue that had "big data" analytics been available at that time, it could have streamlined the Amerithrax investigative process and made important connections and correlations between the disparate datasets in a more efficient manner. For example, Nowviskie and Saathoff describe how now there are suites of new "big data" analytic tools to assist the investigation that would involve sentiment analysis, authorship attribution, and pattern analysis at scale [21, 22]. Sentiment analysis is the study of opinions, sentiments, and emotions in text that can reveal affect or a psychological state. Although Ivins deleted many of his emails, it was later learned that copies of these messages were saved on his computer system. These messages could have been processed with new sentiment analysis tools to generate new understandings of Ivins mental state and behavior. New authorship attribution algorithms could have been applied to the *Bacillus anthracis* letter mailings and compared to Ivins' writing to see if stronger corroboration existed. Also, using data analytics to combine information on Ivins movements, financial transactions, and other data on his life could have revealed particular patterns in his life and work that went unnoticed by investigators at the time, or that only were connected much later in the investigation.

The applicability of these types of analytics to help intelligence assessments becomes clear when the flawed 2003 intelligence assessments on Iraq's bioweapons programs are examined. One could imagine a different outcome had "big data" analytic approaches been used to help with analysis of the Iraqi data. Although there are several reasons for the intelligence failure (see [23]) and collection on Iraqi data was difficult, there were critical datasets and data integration efforts that were not pursued by intelligence analysts. Richard Kerr, former Deputy Director of the CIA who was tasked to form a team to study the intelligence failures on Iraq's weapons of mass destruction (WMD) programs, found that there was little intelligence collection on open source social, cultural, political, and economic impacts on Iraq from the nearly 20 years of war and 10 years of sanctions, UN inspections, and international isolation, and no analyses of how the datasets from these impacts might affect the progress of a WMD program ([24], p. 48–49, 51). Kerr et al. note that although "softer" intelligence data on Iraqi societal issues, personalities, and elites was available in different databases, they were difficult for U.S. intelligence to access, process, and integrate with classified information. As a result, there was "no attempt to synthesize a broader understanding of Iraq out of the many detailed pieces. . ..the absence of such a contextual effort contributed to assessments that failed to recognize the significance of gaps in collection that may have been more evident when viewed from a larger perspective" ([24], p. 52). What resulted, then, was a narrow, and ultimately flawed, technical analysis of Iraq's WMD programs, based on thin classified data that was not corroborated and integrated with other information sources. At the time, if there might have been data mining and data fusion "big data" analytic tools available for intelligence, the results of that data processing might have produced analyses showing a more measured perspective on Iraq's WMD capabilities.

15.4.2 Cautions in Using "Big Data" for Threat Awareness

Despite the outlines potential benefits, one must also exert caution with relying exclusively on "big data" approaches. The social dimensions of science and technology that relate to measuring know-how/skills or technological diffusion are not readily captured by existing "big data" platforms and techniques. Acquisition of such data typically involves qualitative methodologies, such expert interviews, focus groups, laboratory observation (or other forms of ethnography). How "big data" approaches can be enabled to encourage these alternative, non-quantitative means of analysis for the life sciences and incorporate them with other pieces of information is an important point for discussion in considering the impact of "big data" on assessments. Lack of careful attention to both the tangible and intangible dimensions of the life sciences and biotechnology may result in erroneous intelligence and policy assessments.

To illustrate these challenges, I will discuss here a case study that I have investigated on advances in the life sciences: the 2002 artificial synthesis of a poliovirus by Professor Eckard Wimmer and his research group at the State University of New York [25]. This is a paradigmatic case to consider for security concerns about ongoing research and the availability of "big data" in the life sciences, particularly in the fields of synthetic genomics and synthetic biology. After the open scientific publication of this poliovirus experiment in the journal *Science*, the U.S. policy community raised concerns about whether terrorists might pick up the experimental methods used in this paper to re-create poliovirus, or other more deadly viruses. Using standard data gathering and analytic methods, it would appear that this synthesis experiment would be replicated by anyone with access to relevant materials and equipment, which are commercially available. However, closer ethnographic study of this case revealed important laboratory know-how that constituted the success of this synthesis experiment. Without this know-how, one would be unable to synthesize poliovirus using the experimental protocol regardless of how many pieces of DNA one might buy, what equipment one has, who one might buy these materials from, or the step-by-step procedure. What I have learned about this experiment is that in spite of a wealth of explicit information and readily available materials out in the public domain, critical portions of making this experiment work remain craft-like techniques that involve quite a bit of specialized, localized know-how that is difficult to transfer to other laboratories or contexts.

This is not to suggest that all synthetic genomics experiments will be this difficult to be used for harm. But, this analysis did reveal that there is a spectrum of skill difficulty involved in different kinds of biological work that needs to be considered in conducting threat assessments. The poliovirus case study illustrates the need to examine, at a detailed micro-level, the social context and relationships that comprise scientific and technological work and the complex practices and timescales that contextualize the replication and transfer of these techniques, even in cases when the science or technology in question seems at first approximation to be readily standardized, accessible, and transferable. As sociologists of science Nigel Gilbert and Michael Mulkay have long noted, the public and private accounts of science are very different; only the private discourse reflects the varied, contingent, and messy character that constitutes the social character of scientific work [26]. As a result, scientific publications, presentations, and other formal means of codified scientific knowledge that are readily accessible through many "big data" approaches do not tell the whole story. Therefore, caution should be used in applying "big data" analytics to published scientific experiments as these analytics may not capture important details crucial to accurate threat assessment.

Existing "big data" approaches may be able to collect a lot of explicit information about the life sciences and biotechnology, but are they able to collect "Whole Data" [27]? For example, not only is there a need to consider individual tacit knowledge involved in technical work, but also knowledge networks, organizational/leadership components and dynamics, as well as the larger socio-political-economic contexts at the micro and macro levels that can also shape technical work. Importance aspects of this data is also not available solely through quantitative means. Thus, a reliance only on "big data" approaches may obscure or make invisible important information that is difficult to measure in tangible forms. In addition, Nowviskie and Saathoff [19, p. 56] caution that "big data" analysis still, "depends on the subjective standing points of those who assemble datasets, design the processes by which they are analyzed, and interpret the results." They argue that "big data" analytics do not remove the need for experts to work with and analyze the data, but rather serve as an aid to investigations to provide additional information, details, and patterns that can help narrow the focus and attention of the work. Therefore, understanding the private, micro-level practices and macro-level context that underpin scientific and technological work, as well as relying on expertise to interrogate "big data" results, is critical to evaluating the nontrivial social aspects that can shape biosecurity threats.

15.4.3 Surveillance and Detection

According to HSPD 10, Surveillance and Detection is focused on creating early warning and detection technologies and systems to aid in determining the timing, location, magnitude, and attribution of a bioweapons attack. The ability to predict and track when and where a disease outbreak may occur and how a pathogen transmits can significantly improve response strategies at the local, national, and international levels. Understanding of pathogen transmission has generally required epidemiologic data, involving clinical and laboratory reports and on-the-ground investigations, to generate accurate forecasts and an understanding of transmission patterns that take into account the various biological, environmental, behavioral, and socio-cultural issues that that can dynamically change disease patterns.

Traditional public health surveillance systems remain primarily based on manually collected and coded data, which is slow to gather and expensive, and difficult to disseminate for analysis; also, data is often aggregated at the national or regional level [28]. Van Panhuis et al. [29] note that the acquisition and sharing of detailed epidemiological data from traditional public health systems can be particularly difficult and scarce in crisis/emergency situations or resource-limited environments;

also, information is often not readily released due to trust concerns or other socio-political-economic factors. In this context, alternative data sources, such as from social media and the Internet, serving as nontraditional surveillance systems can be useful to help understand disease dynamics in an outbreak.

Bansal et al. [28] discuss how there are different types of electronic data streams for disease surveillance that can be gathered through "big data" approaches: medical encounter data, participatory syndromic data, and non-health digital data. Medical encounter data include electronic records from healthcare facilities, medical insurance claims, hospital discharge records, and death certificates. Such data streams provide information on the disease in patients (including the specific diagnosis) and can be gathered and monitored at the individual level or aggregated geographically. Participatory syndromic data are crowd-sourced data in which volunteers self-report their symptoms. This data does not provide confirmation of disease attributed to specific pathogens but does provide individual-level health data in near real time. Finally, non-health digital data are data gathered from the use of Internet search engines, social media, or mobile phones. They involve data on health-related behavior, including location and travel patterns, which can provide additional details about potential disease transmission. A 2016 National Academies of Science workshop noted that other kinds of structured and unstructured non-health electronic data sources, such as data from school attendance records, veterinary clinics, pharmaceutics sales, global transportation patterns, and climate could also be combined with other types of individual and community-level medical information to yield a more holistic picture of health in a specific area that would enable early warning or alert systems of a disease outbreak ideally weeks or months in advance ([30], p. 5).

Chowell et al. [31] have illustrated the potential of using "big data" from various Internet sources to understand the 2013–2016 Ebola virus disease epidemic in Western Africa and the 2015 Middle East respiratory syndrome (MERS) outbreak in South Korea. They used Internet news reports, health bulletins, and online reports from various health authorities and respected news sources, coupled to epidemiological information from traditional public health systems, to gather and analyze near real-time "big data" on disease clusters of patients to identify MERS coronavirus transmission chains. "Big data" analysis indicated that all 150 MERS cases in the outbreak involved a single cluster in which all cases were linked to exposure in the healthcare setting—either via occupational or social/family exposure. In case of the Ebola virus disease outbreak, epidemiological data was scarce; that which was initially available was largely aggregated at the country level, with more localized data available later. With these data shortages and the need to quickly understand Ebola virus transmission characteristics, the researchers collected information on Ebola virus disease case clusters from Internet news reports published during the outbreak. They gathered data on suspected, probable, and confirmed cases of Ebola virus disease in the three most affected countries (Guinea, Liberia, and Sierra Leone). They focused on reports and news items available from the World Health Organization web site, and other available Ebola virus disease situational reports, as well as online authoritative media outlets.

The data revealed a lot of variation in transmission patterns across Western Africa—revealing over 100 distinct patient clusters contributing to disease transmission. Later on, these clusters were confirmed via traditional surveillance systems. From using a variety of on-line information Chowell et al. [31] found that the main exposure to Ebola virus occurred via family contacts, as well as funeral and hospital settings. Once proper infection control measures were in place, the number of new cases in these settings substantially decreased. This study illustrated how publically available real-time online reports and news items can contain useful and actionable information regarding exposure and transmission patterns during disease outbreaks. This approach of "digital epidemiology" [32] could prove particularly useful for forecasting and response in resource limited areas, in which detailed transmissions studies are limited, and in crisis situations of major disease outbreaks when traditional surveillance reports are delayed.

15.4.4 Cautions in Using "Big Data" for Surveillance and Detection

Along with the beneficial use of "big data" in disease surveillance, one must also consider some of the problems that can emerge with the use of "big data" platforms. Google Flu Trends (GFT) was a web-based service developed in 2008 that monitored and mined millions of Google users' health-related on-line searching combined with computer modeling, and that coupled this data to laboratory-confirmed influenza cases and influenza tracking information from the Centers from Disease Control and Prevention (CDC) to analyze whether influenza A virus was present in a given population. GFT was similar to weather-forecasting methods that use real-time observational and environmental data to continually update and correct weather predictions. By aggregating Google search queries, GFT attempted to make accurate predictions about influenza activity in 29 countries to help localities and countries reduce the impact of seasonal and pandemic influenza (e.g., increasing vaccinations and supplies of antiviral drugs, issuing information about infection control measures, advocating school or other public closings) [33]. The original Google paper about GFT stated that its predictions were 97% accurate compared with CDC data, and that GFT was able to predict regional influenza outbreaks up to 10 days before they were reported by the CDC [34].

Subsequent independent reports, however, asserted that GFT predictions have sometimes been very inaccurate—e.g., in 2009 when GFT completely missed a non-seasonal influenza outbreak of H1N1 influenza, over the interval of 2011–2013 when GFT consistently overestimated influenza A virus prevalence, over one interval in the 2012–2013 influenza season when GFT predicted twice as many doctors' visits as the CDC recorded, and when GFT missed the peak and severity of the 2013 influenza season [33, 35–37]. Lazer et al. [36] investigated these failures and found there was a persistent pattern of GFT performing well for 2–3 years, and then failing significantly; overall, GFT appeared to perform worse over time. These failures were attributed to several reasons. First, there were inherent problems with Google's GFT algorithm, which was quite vulnerable to overfitting. For example, the algorithm was fitting unrelated terms like "high school basketball" to influenza or to different health conditions. With millions of search terms being fit to the CDC's data, many of which were unlikely to be related to actual influenza cases, one can imagine how this led to problems in the accuracy of influenza predictions. Similar to the GFT errors, in one other well-known example of "big data" gone wrong, Leinweber [38] demonstrated that data mining techniques could show a strong but spurious correlation between the changes in the S&P 500 stock index and butter production in Bangladesh. Therefore, analysts who do not possess expert domain knowledge, may be misled by correlations from "big data" analysis.

Google also did not take into account changes in users' search behaviors over time [35]. Also, Lazer et al. [36] pointed out how the GFT algorithm was part of a larger business operation at Google, in which engineers made continual changes to Google's algorithm to improve its commercial service and advertising revenue and to increase its customers. As a result, the company was constantly testing and creating tweaks in the algorithm to improve searching. These modifications, however, could then skew the functioning of GFT, unintentionally making some search terms more prevalent and thereby hindering the overall accuracy of GFT. Such background adjustments are not a problem only for GFT but can also plague other social media datasets such as Twitter or Facebook, which are also constantly being changed and updated.

Lazer et al. [36] wrote that one of the key problems of GFT was related to "big data hubris," which is the, "implicit assumption that "big data" are a substitute for, rather than a supplement to, traditional data collection and analysis ([36], p. 1203). Allain-Jacques Valleron, an epidemiologist at the Pierre and Marie Curie University in Paris and founder of France's Sentinelles monitoring network, concurs with Lazer et al.: "It is hard to think today that one can provide disease surveillance without existing systems. . .The new systems depend too much on old existing ones to be able to live without them" [33]. This is not to say that the GFT algorithm could not be improved to rectify all or many of its original problems, but the GFT failure also raises cautions about relying too heavily on "big data" approaches alone. Scott Dowell, deputy director for surveillance and epidemiology at the Bill and Melinda Gates Foundation, advocates that "big data" approaches are, "most useful when combined with old-fashioned 'shoe leather' epidemiology—researchers walking door-to-door collecting primary data to understand the spread of an outbreak" ([30], p. 8).

15.4.5 Additional Concerns to Consider Regarding the Use of "Big Data"/"Big Data" Analytics for Biodefense

Beyond the Google Flu Trends mishap, there are additional challenges that researchers and policymakers need to understand regarding the use of "big data" and "big data" analytics for biodefense. Crawford and Finn [39] provide a useful overview of the limitations of what they call "crisis data"—the use of social and mobile data to understand disasters and other crisis events—of which a bioterrorism/

biological weapons attack would be relevant. Crawford and Finne discussed how social media data can provide important tactical information about a crisis and suggest potential response measures for that moment. They caution, however, that social media datasets represent a specific time point in an event, usually represented by a spike in Twitter messages or the use of particular hashtags. This focus can lead to a narrow conceptualization of the disaster, which leaves out an important understanding of the origin of the crisis or an understanding of the aftermath in which more of the impact and longer-term implications of the disaster is realized.

Also Crawford and Finne pointed out that "big data" researchers often work in a very different context and location than those involved in the response and recovery in the midst of the disaster; therefore, the researchers can only mine a small fraction of the entire experience and effect of the disaster, leading to a partial and incomplete record of the event. Also Crawford and Finne noted that social media data is not always representative of a local population. For example, Twitter tends to be used by younger, wealthy, and more urban adult populations, meaning that infants, older, rural, and more marginalized populations may not be represented by Twitter posts ([28], p. S377). Also, most social media data lacks demographic information, such as age and sex, which is usually important in epidemiological studies. All of these factors could be a significant shortcoming for response planning in a bioweapons attack given that infants/children, older, rural, and more marginalized/developing country populations might be excluded from social media datasets and those populations are typically more vulnerable and could be significantly harmed by a bioweapons attack.

Social media platforms can also be blind to important animal or zoonotic disease outbreaks ([28], p. S378). This blindness could lead to missing important patterns of disease outbreaks or recognizing patterns when none actually exist. Therefore, it is important to be aware of the properties and limitations of social media data or other kinds of "big data," what other kinds of datasets are needed, and taking all of these factors into account for designing appropriate response measures (see [40]).

Crawford and Finne also pointed out the problem of bots in Internet and social media platforms that can hamper accurate interpretation of events. Bots are software applications that perform an automated task; they are often used to perform repetitive tasks such as accessing websites to gather their content for search engine indexing (e.g., web crawlers), or other types of information gathering, organizing, and displaying. There are also malicious bots that come with viruses. Bots are designed to simulate human responses and activity and are a significant presence in many online spaces. Crawford and Finne demonstrated that a large number of tweets in datasets are derived from bots. De Micheli and Stroppa [41] conducted a research study of Twitter and found 20 million "fake" accounts—approximately 9% of Twitter's active users. As Crawford and Finne wrote, "Those bots are friending and retweeting other bots, producing a complex bot culture which is an emerging phenomenon unto itself." ([39], p. 496). Bots create significant problems for interpretation and filtering of social media data—is the tweet coming from a human or from a bot?

A similar problem is the emergence of "fake news," i.e. manipulated media posted by individual and group internet users to spread disinformation and/or propaganda to achieve a particular social, cultural, political, or economic goal

[42]. Also, users can also shape/disguise their online identities and create false information [16, 43, p. 40–41]. This misinformation could be used to hide covert bioweapons activities or redirect public health attention in other directions. One can imagine that in the event of a possible bioweapons attack, a key question will be whether social media information represents real data, manipulated data, or artificial bot-generated data. The problem of fake information is not a new development—there have long been efforts to create fake stories about bioweapons. In their book on the Soviet biological weapons program, Leitenberg et al. [44] reveal a disturbing US covert deception and disinformation campaign that operated from the late 1960s through the early 1970s. This program was directed at misleading the Soviet Union regarding the objectives of the US offensive bioweapons program and to suggest that a covert US bioweapons program existed even after the United States had unilaterally renounced its biological weapons program in 1969, leading the Soviets to invest time and money on a response. This deception program was led by the Federal Bureau of Investigation with assistance of the US Army, and used a double agent to convey false information to the Soviets. The authors found that the Central Intelligence Agency, years later, debated whether this campaign might have contributed to the Soviet decision to violate the Biological and Toxin Weapons Convention and move its weapons activities underground, or at least provided political cover to justify the decision to increase its biological weapons program. Such disinformation campaigns arguable are easier to conduct in the internet age. Therefore, it is important to remember that "big data" approaches do not remove the problem of disinformation, but, in many cases, may make it more problematic.

15.5 Conclusion

This chapter has illustrated how the use of "big data" can enhance biodefense in the areas of threat awareness and surveillance and detection. This chapter has also highlighted some of the pitfalls of relying on "big data" approaches without considering the limitations of such data, and how "big data" approaches can obscure or ignore other important information for analysis. Although there are many public pronouncements that "big data" analytics will soon replace human assessments, a more realistic and useful understanding would be to see "big data" approaches as one of several tools that be brought to bear on biodefense efforts. Rather than replace traditional efforts, Lazer et al. [36] have argued that we will likely see more "hybrid" systems that integrate "big data" with traditional data analysis. In doing so, the real analytic innovation will be using data from all sources, to include traditional public health data, coupled to social science, environmental, contextual, and ethnographic data, and otherwise, "small data", with "big data" [27, 36, 45–50]. In this way, the biases and short-comings of each method can be used to balance each other to arrive at a more accurate picture of a disease outbreak or bioweapons attack. Lazar et al. argued that the real data revolution, then, will be, "using data from all traditional and new sources, and providing a deeper, clearer understanding of our world" [47]. This kind of data analysis will also require new kinds of multi-disciplinary analytic teams, who possess relevant subject matter expertise and understand long term trends and patterns [51], to be created and work together on these problems. To date, multi-method, multi-platform research studies are rare [52, 53], particularly in biodefense, but the payoff would be significant.

Uncertainties related to "big data," as with so many aspects of science, call for an approach of humility [54], in which we (e.g., scientists, academics intelligence practitioners, policymakers) recognize the limitations of science and when it is necessary to look beyond science for answers to complex, "wicked" problems. As Jasanoff noted, "[t]hese technologies compel us to reflect on the sources of ambiguity, indeterminacy and complexity" [54]. This does not mean that we need to be reject using "big data" or "big data" approaches to solve biodefense problems, but that we need to be very wary of over-relying on these techniques and technologies and losing sight of other complementary ways to generate knowledge.

References

1. Achenbach J, Sun LH. Scientists synthesize smallpox cousin in ominous breakthrough. The Washington Post. 2017. https://www.washingtonpost.com/news/speaking-of-science/wp/2017/07/07/scientists-synthesize-smallpox-cousin-in-ominous-breakthrough/?utm_term=.95d61eea13c8.
2. Central Intelligence Agency. The darker bioweapons future. 2003. https://www.hsdl.org/?abstract&did=442021.
3. Hilts PJ. Biological weapons reweighed. The Washington Post. 1986.
4. Makunda G, Oye KA, Mohr SC. What rough beast: synthetic biology and the future of biosecurity. Polit Life Sci. 2009;28(2):2–26.
5. National Research Council. Biotechnology research in an age of terrorism. Washington, DC: The National Academies Press; 2004. https://doi.org/10.17226/10827.
6. The Commission on the Intelligence Capabilities of the United States Regarding Weapons of Mass Destruction. Report to the President. 2005. https://fas.org/irp/offdocs/wmdcomm.html.
7. U.S. Department of Defense. Advances in biotechnology and genetic engineering: implications for the development of new biological warfare agents. 1996. www.acq.osd.mil/cp/docs/reports/biotech96.pdf.
8. Lohr S. How big data became so big. The New York Times. 2012. http://www.nytimes.com/2012/08/12/business/how-big-data-became-so-big-unboxed.html.
9. Manyika J, Chui M, Brown B, Bughin J, Dobbs R, Roxburgh C, Hung Byers A. Big data: the next frontier for innovation, competition, and productivity. The McKinsey Global Institute. 2011. http://www.mckinsey.com/business-functions/digital-mckinsey/our-insights/big-data-the-next-frontier-for-innovation.
10. Lazer D, Pentland AS, Adamic L, Aral S, Laszlo Babasi A, Brewer D, Christakis N, Contractor N, Fowler J, Gutmann M, Jebara T, King G, Macy M, Roy D, Van Alstyne M. Life in the network: the coming age of computational social science. Science. 2009;323 (5915):721–3.
11. George G, Haas MR, Pentland A. Big data and management. Acad Manag J. 2014;57(2):321–6.
12. The White House. Homeland Security Presidential Directive 10 (HSPD 10): biodefense for the 21st century. 2004. https://fas.org/irp/offdocs/nspd/hspd-10.html.
13. The White House. The national strategy for countering biological threats. 2009. https://obamawhitehouse.archives.gov/the-press-office/president-obama-releases-national-strategy-countering-biological-threats.

14. The White House. The national strategy for biosurveillance. 2012. https://obamawhitehouse.archives.gov/the-press-office/2012/07/31/national-strategy-biosurveillance.
15. U.S. General Accountability Office. Biodefense: The Nation Faces Multiple Challenges in Building and Maintaining Biodefense and Biosurveillance, Statement of Chris Currie, Director, Homeland Security and Justice, Testimony Before the Committee on Homeland Security and Governmental Affairs U.S. Senate GAO-16-547T. 2016. http://www.gao.gov/assets/680/676548.pdf.
16. American Association for the Advancement of Science. Federal Bureau of Investigation, and the United Nations Interregional Crime and Justice Research Institute. National and transnational security: implications of big data in the life sciences. 2014. https://www.aaas.org/report/national-and-transnational-security-implications-big-data-life-sciences.
17. Tateosian L, Glatz M, Shukunobe M, Chopra P. GazeGIS: a gaze-based reading and dynamic geographic information system. In: Burch M, Chuang L, Fisher B, Schmidt A, Weiskopf D, editors. Eye tracking and visualization. ETVIS 2015. Mathematics and visualization. Berlin: Springer; 2017. p. 129–47.
18. NATO. Distributed data analytics for combating weapons of mass destruction. STO meeting proceedings, MP-IST-131. 2017.
19. Nowviskie B, Saathoff GB. Interpretation and insider threat: rereading the anthrax mailings of 2001 through a "big data" lens. In: Akhgar B, Saathoff GB, Arabnia H, Hill R, Saniforth A, Bayerl P, editors. Application of big data for national security. 1st ed. Amsterdam: Elsevier; 2015. p. 55–67.
20. Murch RS. Amerithrax: the investigation of bioterrorism using *Bacillus anthracis* sproes in mailed letters. In: Katz R, Zilinskas RA, editors. Encyclopedia of bioterrorism. New York: Wiley; 2011. p. 25–30.
21. Moretti F. Graphs, maps, trees: abstract models for a literary history. London: Verso Books; 2005.
22. Moretti F. Distant reading. London: Verso Books; 2013.
23. Vogel KM. Phantom menace or looming danger: a new framework for assessing bioweapons threats. Baltimore, MD: Johns Hopkins University Press; 2013.
24. Kerr R, Wolfe T, Donegan R, Pappas A. Collection and analysis on Iraq: issues for the US intelligence community. Stud Intell. 2005;49(3):47–54.
25. Vogel KM. Framing biosecurity: an alternative to the biotech revolution model? Sci Public Policy. 2008;35(1):45–54.
26. Gilbert NG, Mulkay M. Opening Pandora's box: a sociological analysis of scientists' discourse. New York, NY: Cambridge University Press; 1984.
27. Boyd D, Crawford K. Critical questions for big data: provocations for a cultural, technological, and scholarly phenomenon. Inf Commun Soc. 2012;15(5):662–79.
28. Bansal S, Chowell G, Simonsen L, Vespignani A, Viboud C. Big data for infectious disease surveillance and modeling. J Infect Dis. 2016;214(Suppl 4):S375–9. https://doi.org/10.1093/infdis/jiw400.
29. van Panhuis WG, Paul P, Emerson C, Grefenstette J, Wilder R, Herbst AJ, Heymann D, Burke DS. BMC Public Health. 2014;14:1144.
30. National Academies of Sciences, Engineering, and Medicine. Big data and analytics for infectious disease research, operations, and policy: proceedings of a workshop. Washington, DC: The National Academies Press; 2016. https://doi.org/10.17226/23654.
31. Chowell G, Cleaton JM, Viboud C. Elucidating transmission patterns from Internet reports: Ebola and Middle East respiratory syndrome as case studies. J Infect Dis. 2016;214(Suppl 4): S421–6. https://doi.org/10.1093/infdis/jiw356.
32. Salathé M, Bengtsson L, Bodnar TJ, Brewer DD, Brownstein JS, Buckee C, et al. Digital epidemiology. PLoS Comput Biol. 2012;8(7):e1002616. https://doi.org/10.1371/journal.pcbi.1002616.
33. Butler D. When Google got flu wrong. Nature. 2013;494(7436):155–6. http://www.nature.com/news/when-google-got-flu-wrong-1.12413

34. Ginsberg J, Mohebbi MH, Patel RS, Brammer L, Smolinski MS, Brilliant L. Detecting influenza epidemics using search engine query data. Nature. 2009;457:1012–4. https://doi.org/10.1038/nature07634.
35. Cook S, Conrad C, Fowlkes AL, Mohebbi MH. Assessing Google Flu Trends performance in the United States during the 2009 Influenza Virus A (H1N1) pandemic. PLoS One. 2011;6: e23610.
36. Lazer D, Kennedy R, King G, Vespignani A. The parable of Google Flu: traps in big data analysis. Science. 2014;343(6176):1203–5.
37. Olson DR, Konty KJ, Paladini M, Viboud C, Simonsen L. Reassessing Google Flu Trends data for detection of seasonal and pandemic influenza: a comparative epidemiological study at three geographic scales. PLoS Comput Biol. 2013;9(10):e1003256. https://doi.org/10.1371/journal.pcbi.1003256.
38. Leinweber D. Stupid data miner tricks: overfitting the S&P 500. J Invest. 2007;16(1):15–22. https://doi.org/10.3905/joi.2007.681820.
39. Crawford K, Finn M. The limits of crisis data: analytical and ethical challenges of using social and mobile data to understand disasters. GeoJournal. 2015;80:491–502. https://doi.org/10.1007/s10708-014-9597-z.
40. Dixon D. Analysis tool or research methodology? Is there an epistemology for patterns? In: Berry D, editor. Understanding digital humanities. London: Palgrave Macmillan; 2012.
41. De Micheli C, Stroppa A. Twitter and the underground market. 11th Nexa lunch seminar, Turin, Italy. 2013. http://nexa.polito.it/nexacenterfiles/lunch-11-de_michelistroppa.pdf.
42. Marwic A, Lewis, R. Media manipulation and disinformation online. 2017. https://datasociety.net/output/media-manipulation-and-disinfo-online/k.
43. Mazur E. Collecting data from social networking web sites and blogs. In: Gosling SD, Johnson JA, editors. Advanced methods for conducting online behavioral research. Washington, DC: American Psychological Association; 2010. p. 77–90.
44. Leitenberg M, Zilinskas RA, Kuhn JH. The Soviet biological weapons program: a history. Cambridge, MA: Harvard University Press; 2012.
45. Belk RW. Qualitative research in advertising. J Advert. 2017;46(1):36–47.
46. Manovich L. Trending: the promises and the challenges of big social data. In: Gold MK, editor. Debates in the digital humanities. Minneapolis, MN: University of Minnesota Press; 2012. p. 460–75.
47. Murthy D. Digital ethnography. An examination of the use of new technologies for social research. Sociology. 2008;42(5):837–55. https://doi.org/10.1177/0038038508094565.
48. Orgad S. How can researchers make sense of the issues involved in collecting and interpreting online and offline data? In: Markham AN, Baym NK, editors. Internet inquiry. Conversations about method. Los Angeles, CA: Sage; 2009. p. 33–53.
49. Snijders C, Matzati U, Reips U-D. "Big data": big gaps of knowledge in the field of Internet science. Int J Internet Sci. 2012;7(1):1–5.
50. Tufekci Z. Big questions for social media big data: representativeness, validity and other methodological pitfalls. In: ICWSM '14: Proceedings of the International AAAI Conference on Weblogs and Social Media; 2014.
51. Arbesman S. Stop hyping big data and start paying attention to long data. Wired. 2013. https://www.wired.com/2013/01/forget-big-data-think-long-data/.
52. Adar E, Weld DS, Bershad BN, Gribble SS. Why we search: Visualizing and predicting user behavior. In; Proceedings of the 16th International Conference on World Wide Web, 161–70. WWW '07. New York, NY, USA: ACM; 2007.
53. Kairam SR, Morris MR, Teevan J, Liebling D, Dumais S. Towards Supporting Search over Trending Events with Social Media. In: Seventh International AAAI Conference on Weblogs and Social Media; 2013.
54. Jasanoff S. Technologies of humility. Nature. 2007;450:33.

Big Data and Artificial Intelligence for Biodefense: A Genomic-Based Approach for Averting Technological Surprise

16

Willy A. Valdivia-Granda

16.1 Introduction

Global security faces several dynamic challenges ranging from persistent regional low-intensity conflict to climate change and infectious disease outbreaks, dual-use technologies, the emergence of new centers of power, the aggregate consequences of state failure, and the rise of non-state actors. Despite military and international law enforcement deterrence efforts to stop the development and use of biological weapons, some countries and groups are actively creating a "gray zone" for hybrid warfare which includes the use of microbes [1, 2]. By distorting military, economic, and criminal activities, hybrid warfare makes deterrence risky and costly. Proliferators can supply non-state actors engaged in sectarian or ethnic extremism with biological agents to cause harm while avoiding international condemnation and sanctions [3–7]. This increasingly polarized and complex environment is more unpredictable, and in some instances more threatening, since biological weapons can be used not only against combatants or armed personnel but also civilians, animals, and plants without regard of the global health and economic consequences [8, 9].

The proliferation of access to new capabilities in biotechnology, including computational and synthetic biology, from countries, groups, and individuals, increases the threat of misuse. As stated in the Quadrennial Defense Review [10] and by the Defense Science Board [11], new biotechnology breakthroughs could make dangerous bioagents fast-moving and widely available to state and non-state actors more easily than a decade ago. This assessment is also shared by the Director of National Intelligence not only because the unique security and strategic challenges posed by biotechnology but also because their low signature can outpace regulation, control, verification, and intelligence collection. Through denial and

W. A. Valdivia-Granda (✉)
Orion Integrated Biosciences, Inc., Manhattan, KS, USA
e-mail: willy.valdivia@orionbio.com

S. K. Singh, J. H. Kuhn (eds.), *Defense Against Biological Attacks*,
https://doi.org/10.1007/978-3-030-03053-7_16

deception, proliferators can attempt to conceal information and deceive intelligence and policymaking communities about their intentions and capabilities. As the pace of technological development accelerates, skill requirements and expenses decrease, which can result in individuals achieving asymmetric capabilities that render ineffective the countermeasures developed by a nation [12, 13]. The above situation arises from the very large array of pathogens and toxins which exist in nature and that can be modified to enhance their virulence or host range with computational and synthetic biology techniques.

The collection of information for intelligence has a significant impact in shaping the policies to counter the impact of biological weapons. Analysts who underestimate the power of denial and deception increase their vulnerably to overconfidence and knowledge bias. For example, despite the evidence, the Soviets assumed that the public announcement of the abolishment of the US biowarfare program meant that the program had merely gone "clandestine," perhaps because of some "technological breakthrough" [14]. This led to a sharp increase in the activity of the Soviet biowarfare program and espionage in the West [15, 16], the opposite and unexpected reaction of what the USA assumed would happen. Although aware of the situation, the US intelligence and academic analysts failed to determine the scale and extent of the Soviet bioweapons program [14] and the complex cycles of dual-use technology transfer to other countries. This occurred while both the USA and Russia were defining the Biological Weapons Convention treaty limits on research on specific biological agents.

There is a need for techniques averting capability and intention related to bioweapon development, biowarfare, and bioterrorism. However, proliferation assessments have been plagued by deficient estimates [14, 17]. Due to its complexity, it could take years for the intelligence community to fully understand the biotechnological landscape and many years for the research community to develop countermeasures against advanced biothreats [7, 18]. This chapter highlights aspects related with development of a global biothreat awareness enterprise beyond biosurveillance: Big Data associated with the acquisition of genomic and other disparate information and the deployment of artificial intelligence techniques to human-machine teaming to detect and avoid technological surprise.

16.2 Big Data and Genomic Information for Biodefense

Information is a key determinant of technological superiority and a potent deterrent against proliferators of biological weapons. Both genomics and biosurveillance generate significant amounts of data; however, few tools can use this information to generate strategic and tactical assessment for making risk-based biodefense decisions. More than 1.2 million strains of known dangerous pathogens circulate not only naturally but with more than 80,000 researchers in at least 20,000 government, military, academic, and commercial facilities worldwide performing passage through in vitro culture and genetic manipulation. This has generated much discussion about the potential accidental or intentional release of a biological agent.

Similarly, billions of news feeds from social media and 10,000 newspapers from 170 countries represent a unique stream of information. Despite their scale, both genomic information and news feeds present several reliability and analytical challenges for integration and use in biodefense.

Next-generation DNA sequencing technology resulted in the exponential accumulation of microbial genomic information. Coupled with new analysis tools, new insights about determinants of pandemic potential, host range, and virulence were generated. Genomic sequencing projects have grown from interrogating few genomes to thousands of genomes which are leading to massive all-by-all comparisons. Despite the progress on data decentralization, biological repositories with relevance for biodefense remain isolated silos of information, where their heterogeneity and inaccuracy remain the biggest hurdle [19, 20]. Approximately 60% of genomic records associated with microorganisms of relevance for biodefense lack source-of-origin information, 18% have incorrect assignment to serotype or species, and 36% lack descriptive information for other types of analytics (e.g., animal or in vitro passage) [8]. Despite the critical relevance of annotations in analytics, reliability and impact of attribution, and bioforensic analytics, limited progress has been made in correcting, disambiguating, and verifying genomic databases. Furthermore, genomic information from a Big Data perspective lacks the operational coordination among the varied systems hosting the various databases, the developers of the analytical packages, and the specific needs of intelligence analysts to relay key aspects of the data to decision makers using this information. Obviating time-dependent sampling techniques and quality issues of metadata and the legacy of disparate genomic datasets might result in inaccurate analysis leading to overconfidence, cognitive bias, and blinded consensus [21, 22].

Timely situational awareness is a cornerstone of biodefense and involves collecting, aggregating, and processing data from both official and informal sources (e.g., social media). While Internet-based communicable disease biosurveillance was conceived in the mid-1990s [23], only after the 2001 anthrax attacks, this approach expanded significantly. However, available biosurveillance systems show that their accuracy and timeliness in the reconstruction of outbreak transmission chains are affected by source moderation, language, regions of occurrence, and the use of text processing tools to discard noise [24–26]. Even when mining the same information, biosurveillance tools showed limited overlap and inaccurately reported the emergence and scale of Ebola, Zika, yellow fever, and cholera outbreaks [23, 27–31]. Due to the trade-off between accuracy and timeliness [23], biosurveillance tools still have limitations in biothreat awareness, and increasingly their future and operational relevenace is been questioned. Although some tools use natural language processing, automated web scrapping, and deep learning, it is still unclear which variables are key to infectious disease outbreak forecasting (Fig. 16.1).

The impact of a biodefense enterprise depends not only on how the data is collected, ingested, disambiguated, and processed but also on how it is relayed to different operational users with different technological backgrounds. The accurate and timely assessment of the technological condition, capability, and intention of state and non-state players can facilitate the deployment of adequate

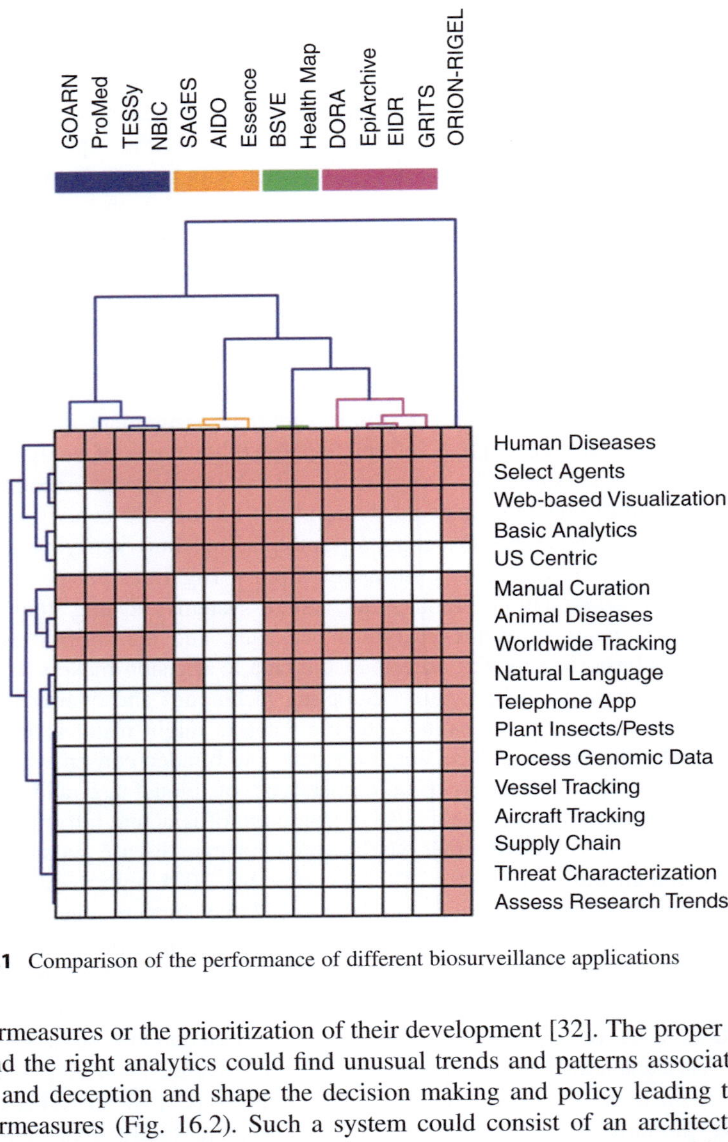

Fig. 16.1 Comparison of the performance of different biosurveillance applications

countermeasures or the prioritization of their development [32]. The proper integration and the right analytics could find unusual trends and patterns associated with denial and deception and shape the decision making and policy leading to better countermeasures (Fig. 16.2). Such a system could consist of an architecture that ingests data, generated by sensors and devices detecting pathogens and integrated with other signal and measurement intelligence (SIGMINT), i.e., open source (OSINT), geospatial (GEOINT), and human (HUMINT) intelligence [8]. However, pathogen detection signals at the molecular level are not constant. Since detection technologies, such as PCR, antibody-based assays, and next-generation DNA sequencing, vary in sensitivity and specificity, there is a wide range of errors found in the generation of data and its interpretation [33, 34]. Addressing the above issue requires standardized metadata formats that facilitate the integration of

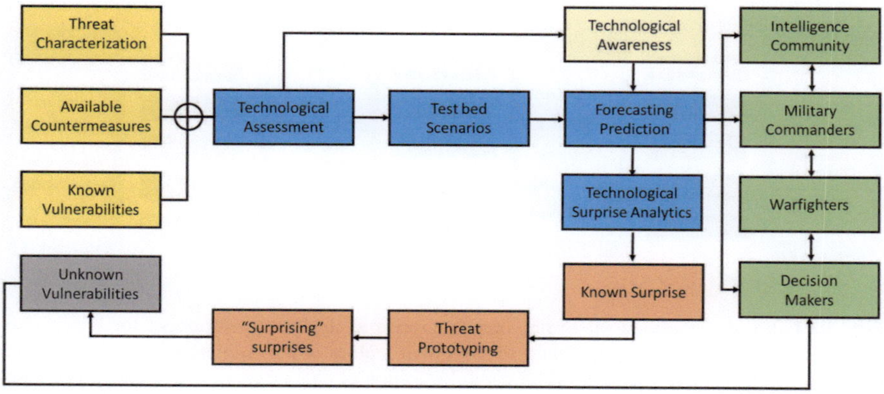

Fig. 16.2 Biodefense enterprise general components. This architecture requires an "ensemble analysis" that includes dependent or independent analytical executions. The idea is that the result is an ensemble score from the results of different models, no matter how each of these models is derived. The efficacy of the pipeline is dependent of how the system gathers, stores, disseminates, and digests the body of information that exists outside traditional intelligence channels

multidimensional, single-point, continuous, semi-structured, unstructured, qualitative, and quantitative data sources into a biodefense enterprise [20].

Systematic attempts are being made to anticipate and understand the potential direction, characteristics, and effects of technological change, innovation, adoption, and use. As emerging technologies quickly advance, their life cycle tends to be reduced, making it difficult to differentiate "innovation" from "improvement." Forecasting of violent civil wars and nonviolent democratic reversals with 80% accuracy requires models with few but key variables [35]. While this could be same for averting biological threats, analysts tend to create "lists of lists" of potential dual-use biological technologies. Different threat scenarios are not sufficient in the context of the vulnerabilities due to the cognitive biases of those involved.

Due to the large number of microbial species that can affect global health and the economy, the scientific community developed strategies to rank a list of microorganisms and toxins of biodefense relevance. This select agent list is derived from literature analysis, disease and death rates, dissemination, public perception, and modeling of the potential impact on military and civilian populations, livestock, and plants [36, 37]. However, the select agent approach does not include alternative agents or scenarios [18]. Considering that there are at least more than 1500 distinct types of infectious pathogen species, risk-based prioritization seems a daunting task considering the limited number of models and scenarios present. Furthermore, pathogen ranking does not (and cannot) cover emerging unknown microbes and toxins that can be weaponized and used in different types of attacks to disrupt the health of humans, animals or plants. To understand the risk of these threats, it is necessary to use multiple types of information and analytics to guide quantitative risk-based investments for biodefense (Fig. 16.3).

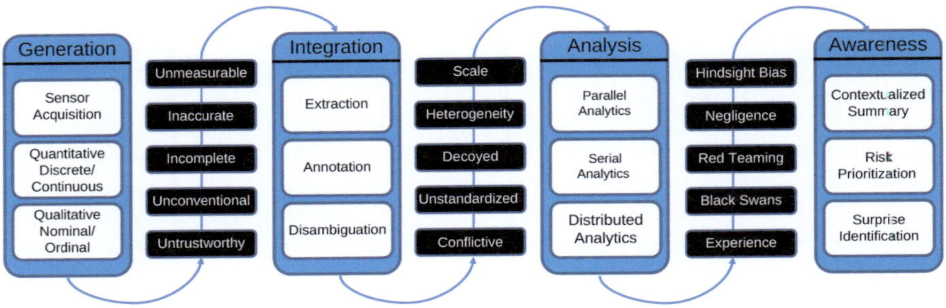

Fig. 16.3 Process overview of technological surprise analysis and awareness for biodefense. Information storage and decentralized exchange will require efficient interoperable architectures capable of supporting Big Data needs of careful selection of data categories. The data preparation stage of predictive modeling is often viewed as the most time-consuming, with experts describing the time expended in this stage ranging from 50% to 90% of task time. Blue boxes represent required logical steps and opportunities for data processing. Black boxes represent potential challenges. To be useful in Big Data, biodefense stakeholders must agree on the general procedures to support data exchange, including informed consent, privacy protections, security and confidentiality breaches or misuses, and appropriate categorization. Tools, including noSQL, Hbase, MongoDB, Hive, and PIG, are likely to continue to grow, but for interoperability will demand customized development

16.3 Artificial Intelligence for Biodefense

Defined in 1956 by John McCarthy as "the science and engineering of making intelligent machines," artificial intelligence (AI) consists of algorithms that perform data analysis tasks through recursive self-improvement [38]. Thanks to the unprecedented computer capacity and the emergence of quantum computing, AI machines can perform in parallel, distributed, or centralized operations that outperform humans [19, 39–44]. While originally developed with training sets for identifying and reclassifying patterns, performing regression, or generating forecasts, a second generation of ready-to-use AI models can solve problems for which no prior solution exists, or no prior steps or partial steps have been taken [45]. AI can improve the understanding of the biothreat landscape at the tactical and strategic levels. At the tactical level, AI algorithms can provide scenarios to support engagement decisions of low-signature forces operating independently in far forward and contested environments. At the same time, AI can reduce judgment biases during the risk assessment of biological threats and perform dynamic and static targeting of transportation vessels that can affect the agricultural trade, the food defense, and the supply chain. At the strategic level, AI predictive analytics can identify emerging technologies as early as possible providing contextualized support for intelligence analysts, military, and policy makers. While the chapter does not seek to review AI algorithms in detail, some examples of how these techniques can be used in biodefense are provided.

16.3.1 Analysis of Genomic Information

Due to their small signature, it is difficult to track biological materials of concern as easily as it is for fissile nuclear and radiological material. Nonetheless, precision in the detection and characterization of biothreats not only can reduce their morbidity and mortality but enhance deterrence. However, thanks to next-generation sequencing, an enormous amount of genomic data encoded by millions of different microbial strains is been generated. AI processing this information can provide unprecedented insights to identify "unknown surprises" and "surprising surprises" (e.g., previously unknown organisms that can be considered select agents and toxins). This requires the identification of specific motif fingerprints or genomic signatures associated with a known pathogen's biological characteristics that discriminate biothreats from "background noise" of the genomes from other microbial species. AI can use the available information to generate of all unknown combinatorial variants to determine the mutational landscape of an organism or concern.

16.3.2 Anomaly Detection Using Metagenomic Information

Spatial information is fundamental for the epidemiology of infectious diseases. As Big Data becomes more prevalent, public health officials could integrate genomic metadata with highly resolved spatial scales measuring diversity climatic and environmental parameters. At the same time, methods for the analysis of social networks, technology diffusion, and implementation offer news possibilities for the analysis of emerging biothreats [46]. Integrating this information while correcting the errors and ambiguities in these sources could provide new insights on technological development. Genomic information and its associated metadata are two disambiguating parameters with limited use to understand the technical and functional structure of organizations engaged in biodefense-relevant research. Both a decentralized or centralized biosurveillance and forecasting analytics could generate stakeholder-relevant descriptors. However, it will be necessary to fund these standardization and correction efforts to gain a big picture of a particular bio-event with a contextualized operational framework [47, 48].

16.3.3 Human-Machine Teaming for Technological Forecasting

Peering into the future is not easy. Technological surprise often involves the unanticipated and dynamic fusion of ideas from disparate subject areas that can individually be quantified and ranked within a tempo-spatial context. However, emerging technologies often arise from activities at the individual or very small group level. Horizon scanning using AI can be useful to prioritize specific countermeasures and identify unknown vulnerabilities against natural, genetically modified, and combinatorial bioweapons. This requires time-dependent "scores" associated with the level, capability, and intent of a proliferator [49] placed into a technological component [8]. Assessing trends is arguably dependent on understanding specific laboratory capabilities and shifts in personnel over time that are not necessarily associated with publications. The high degree of uncertainty in the assumptions regarding emerging technologies could affect the predictive value of these models.

AI scoring could be used to indicate the relative preference among alternatives or include the relationship between various factors that might be mutually dependent. Such approach can enhance human-machine teaming and increase the ability of an analyst in identifying misleading scenarios. AI also can increase the odds of the detection of "knowledge spillover" for short- and long-term forecasting [50]. Both a decentralized and centralized biodefense forecasting architecture should deal with the correction of ambiguous historic metadata and add stakeholder-relevant descriptors. However, it will be necessary to fund standardization and correction efforts to gain a big picture of a particular bio-event with a contextualized operational framework [47, 48]. This approach challenges the conventional wisdom of technological and organizational dynamics through the identification of hidden commonalities between genomic databases and OSINT to provide new strategic and tactical assessments. Situational awareness during human-machine teaming must include the ability to interpret data to create relevant, tailored visualization. These tools should provide the capability to visualize and manipulate data from many sources to create operational scenarios suited to the specific biothreat.

16.4 Conclusion

Global security faces several dynamic challenges. Twenty-first-century security threats include known and unknown zoonotic, food-borne, and emerging infectious diseases that can affect humans, animals, and plants. Precision in the detection and characterization of these agents is key not only to reduce their morbidity and mortality but enhance deterrence of proliferators. Addressing this challenge requires a biodefense enterprise that accurately assesses the risk and likelihood of different biothreats resulting from the fusion of emerging technologies. Genomic and geospatial data are two very important sources of information because their volume, dynamics, complexity, and accuracy offer an unparalleled opportunity for mapping disease outbreaks, attribution and to perform technical assessments relevant for biodefense. A *biodefense enterprise* using DNA information, Big Data, and artificial intelligence techniques can allow organizations and individuals to take full advantage of a large collection of disparate, unstructured qualitative and quantitative information. This approach could become useful for assessing potential vulnerabilities in our strategies, postures, and tactics, but more importantly avert denial and deception of proliferators. Analysts, military commanders, and decision makers can gain deeper understanding of potential adversary options and behavior and determine their intentions and capabilities at the strategic and tactical level.

Acknowledgments I am grateful to Drs. Juergen Richt and Romelito Lapitan for the insightful discussions and suggestions that improved this manuscript. The research associated with this book chapter was partially funded by the Department of Homeland Security Center of Excellence for Emerging and Zoonotic Animal Disease (CEEZAD), the Kansas Bioscience Authority, the Defense Threat Reduction Agency (DTRA) Biological Engagement Program, and the Department of Homeland Customs and Border Protection Office Ag/Bio-Terror Countermeasures (ABTC).

References

1. Anastasio M. Seven defense priorities for the new administration. 2017.
2. Martellini M. Cyber and chemical, biological, radiological, nuclear, explosives challenges. New York, NY: Springer; 2017.
3. Gonzalez JP, Souris M, Valdivia-Granda W. Global spread of hemorrhagic fever viruses: predicting pandemics. Methods Mol Biol. 2018;1604:3–31.
4. United States. Government Accountability Office. High-containment laboratories national strategy for oversight is needed: report to congressional requesters. Washington, DC: U.S. Govt. Accountability Office; 2009.
5. Hottes AK, Rusek B, Sharples FE, National Academy of Sciences (U.S.), Committee on International Security and Arms Control, National Academy of Sciences (U.S.), Committee on Anticipating Biosecurity Challenges of the Global Expansion of High-Containment Biological Laboratories. Biosecurity challenges of the global expansion of high-containment biological laboratories summary of a workshop. Washington, DC: National Academies Press; 2012. p. 204.
6. Laszlo AH, Derrington IM, Ross BC, Brinkerhoff H, Adey A, Nova IC, Craig JM, Langford KW, Samson JM, Daza R, et al. Decoding long nanopore sequencing reads of natural DNA. Nat Biotechnol. 2014;32(8):829–33.
7. Valdivia-Granda WA. Biosurveillance enterprise for operational awareness, a genomic-based approach for tracking pathogen virulence. Virulence. 2013;4(8):745–51.
8. Valdivia-Granda WA. Bioinformatics for biodefense: challenges and opportunities. Biosecur Bioterror. 2010;8(1):69–77.
9. Cropley D. The dark side of creativity. New York: Cambridge University Press; 2010.
10. United States, Congress, House, Committee on Armed Services Committee on Naval Affairs Committee on National Security Committee on Military A. The 2014 Quadrennial Defense Review: Committee on Armed Services, House of Representatives, One Hundred Thirteenth Congress, second session, hearing held April 3, 2014. Washington: U.S. Government Printing Office; 2014. Washington, DC: For sale by the Superintendent of Documents, U.S. Government Printing Office; 2014
11. Seven defense priorities for the new administration: report of the defense science board. Washington, DC: Defense Science Board; 2016.
12. Carlson R. Biodefense net assessment: causes and consequences of bioeconomic proliferation; 2012.
13. Leehy AG, Wildstein JJ, Schiffer M, United States. Department of Defense, Office of the Secretary of Defense. Military and security developments in China. Hauppauge, NY: Nova Science Publishers; 2012.
14. Vogel KM: Phantom menace or looming danger?: A new framework for assessing bioweapons threats. Baltimore: Johns Hopkins University Press; 2013.
15. Kouzminov A. Biological espionage: special operations of the Soviet and Russian foreign intelligence services in the west. New Delhi: Manas Publications; 2006.
16. Domaradskij IV, Orent LW. Achievements of the soviet biological weapons programme and implications for the future. Rev Sci Tech. 2006;25(1):153–61.

17. United States. Defense Science Board, United States. Office of the Under Secretary of Defense for Acquisition Technology and Logistics. Report of the defense science board 2008 summer study on capability surprise. Washington, DC: Office of the Under Secretary of Defense for Acquisition, Technology, and Logistics; 2009.
18. Colf LA. Preparing for nontraditional biothreats. Health Secur. 2016;14(1):7–12.
19. Fan W, Bifet A. Mining big data: current status, and forecast to the future. SIGKDD Explor Newsl. 2013;14(2):1–5.
20. Aggarwal CC. Outlier ensembles: position paper. SIGKDD Explor Newsl. 2013;14(2):49–58.
21. Fan J, Liu H. Statistical analysis of big data on pharmacogenomics. Adv Drug Deliv Rev. 2013; 65(7):987–1000.
22. Cohen R, Ruths D. Classifying political orientation on Twitter: it's not easy! Montreal, QC: McGill University; 2013.
23. Pollett S, Althouse BM, Forshey B, Rutherford GW, Jarman RG. Internet-based biosurveillance methods for vector-borne diseases: are they novel public health tools or just novelties? PLoS Negl Trop Dis. 2017;11(11):e0005871.
24. Bahk CY, Scales DA, Mekaru SR, Brownstein JS, Freifeld CC. Comparing timeliness, content, and disease severity of formal and informal source outbreak reporting. BMC Infect Dis. 2015; 15:135.
25. Chowell G, Cleaton JM, Viboud C. Elucidating transmission patterns from internet reports: Ebola and Middle East respiratory syndrome as case studies. J Infect Dis. 2016;214(suppl_4): S421–6.
26. Cleaton JM, Viboud C, Simonsen L, Hurtado AM, Chowell G. Characterizing Ebola transmission patterns based on internet news reports. Clin Infect Dis. 2016;62(1):24–31.
27. Butler D. When Google got flu wrong. Nature. 2013;494(7436):155–6.
28. Lyon A, Nunn M, Grossel G, Burgman M. Comparison of web-based biosecurity intelligence systems: BioCaster, EpiSPIDER and HealthMap. Transbound Emerg Dis. 2012;59(3):223–32.
29. Barboza P, Vaillant L, Le Strat Y, Hartley DM, Nelson NP, Mawudeku A, Madoff LC, Linge JP, Collier N, Brownstein JS, et al. Factors influencing performance of internet-based biosurveillance systems used in epidemic intelligence for early detection of infectious diseases outbreaks. PLoS One. 2014;9(3):e90536.
30. Nsoesie EO, Brownstein JS, Ramakrishnan N, Marathe MV. A systematic review of studies on forecasting the dynamics of influenza outbreaks. Influenza Other Respir Viruses. 2013;8(3): 309–16.
31. Smolinski MS, Crawley AW, Olsen JM. Finding outbreaks faster. Health Secur. 2017;15(2): 215–20.
32. Patel D, Olson S, Institute of Medicine (U.S.), Planning Committee on Information-Sharing Models and Guidelines for Collaboration: Applications to an Integrated One Health Biosurveillance Strategy, Institute of Medicine (U.S.), Board on Health Sciences Policy. Information sharing and collaboration: applications to integrated biosurveillance: workshop summary. Washington, DC: National Academies Press; 2012.
33. McElroy K, Thomas T, Luciani F. Deep sequencing of evolving pathogen populations: applications, errors, and bioinformatic solutions. Microb Inform Exp. 2014;4(1):1.
34. Velusamy V, Arshak K, Korostynska O, Oliwa K, Adley C. An overview of foodborne pathogen detection: in the perspective of biosensors. Biotechnol Adv. 2010;28(2):232–54.
35. Goldstone JA, Bates RH, Epstein DL, Gurr TR, Lustik MB, Marshall MG, Ulfelder J, Woodward M. A global model for forecasting political instability. Am J Polit Sci. 2010; 54(1):190–208.
36. National Research Council (U.S.), Committee on Scientific Milestones for the Development of a Gene Sequence-Based Classification System for the Oversight of Select Agents, National Research Council (U.S.), Board on Life Sciences. Sequence-based classification of select agents: a brighter line. Washington, DC: National Academies Press; 2010.
37. Rotz LD, Khan AS, Lillibridge SR, Ostroff SM, Hughes JM. Public health assessment of potential biological terrorism agents. Emerg Infect Dis. 2002;8(2):225–30.

38. Ciliberto C, Herbster M, Ialongo AD, Pontil M, Rocchetto A, Severini S, Wossnig L. Quantum machine learning: a classical perspective. Proc Math Phys Eng Sci. 2018;474(2209):20170551.
39. Dunjko V, Briegel HJ. Machine learning & artificial intelligence in the quantum domain: a review of recent progress. Rep Prog Phys. 2018;81(7):074001.
40. Kanter JM, Veeramachaneni K. Deep feature synthesis: towards automating data science endeavors. In: 2015 IEEE International Conference on Data Science and Advanced Analytics (DSAA); 2015. p. 1–10.
41. Williams AM, Liu Y, Regner KR, Jotterand F, Liu P, Liang M. Artificial intelligence, physiological genomics, and precision medicine. Physiol Genomics. 2018;50(4):237–43.
42. Murdoch TB, Detsky AS. The inevitable application of big data to health care. JAMA. 2013; 309(13):1351–2.
43. National Research Council (U.S.). Committee for Science and Technology Challenges to U.S. National Security Interests. Report of a workshop on big data. Washington, DC: National Academies Press; 2012.
44. Gill KS. Uncommon voices of AI. AI & Soc. 2017;32(4):475–82.
45. Swearingen T, Drevo W, Cyphers B, Cuesta-Infante A, Ross A, Veeramachaneni K. ATM: a distributed, collaborative, scalable system for automated machine learning. In: IEEE International Conference on Big Data (Big Data), 11–14 December 2017; 2017. p. 151–62.
46. Fan L, Wu W, Lu Z, Xu W, Du D-Z. Influence diffusion, community detection, and link prediction in social network analysis. In: Sorokin A, Pardalos PM, editors. Dynamics of information systems: algorithmic approaches, vol. 51. New York: Springer; 2013. p. 305–25.
47. Garrity GM, Field D, Kyrpides N, Hirschman L, Sansone SA, Angiuoli S, Cole JR, Glockner FO, Kolker E, Kowalchuk G, et al. Toward a standards-compliant genomic and metagenomic publication record. Omics. 2008;12(2):157–60.
48. Ladner JT, Beitzel B, Chain PS, Davenport MG, Donaldson EF, Frieman M, Kugelman JR, Kuhn JH, O'Rear J, Sabeti PC, et al. Standards for sequencing viral genomes in the era of high-throughput sequencing. MBio. 2014;5(3):e01360–14.
49. Ziegler BE: Methods for bibliometric analysis of research: renewable energy case study. Massachusetts Institute of Technology; Cambridge, MA 2009.
50. Sun Y, Han J. Mining heterogeneous information networks: a structural analysis approach. SIGKDD Explor Newsl. 2013;14(2):20–8.

MIX
Papier aus verantwortungsvollen Quellen
Paper from responsible sources
FSC® C105338
FSC
www.fsc.org

If you have any concerns about our products,
you can contact us on
ProductSafety@springernature.com

In case Publisher is established outside the EU,
the EU authorized representative is:
Springer Nature Customer Service Center GmbH
Europaplatz 3, 69115 Heidelberg, Germany

Printed by Libri Plureos GmbH
in Hamburg, Germany